Sustainability Science and Engineering
Defining principles

Sustainability Science and Engineering

Series Editors: M.A. Abraham and R. Sheldon

Sustainability Science and Engineering
Defining principles

Edited by

M.A. Abraham

Department of Chemical and Environmental Engineering, The University of Toledo, Toledo, OH, USA

ELSEVIER

Amsterdam – Boston – Heidelberg – London – New York – Oxford – Paris
San Diego – San Francisco – Singapore – Sydney – Tokyo

ELSEVIER B.V.	ELSEVIER Inc.	ELSEVIER Ltd	ELSEVIER Ltd
Radarweg 29	525 B Street	The Boulevard	84 Theobalds Road
P.O. Box 211, 1000 AE	Suite 1900, San Diego	Langford Lane, Kidlington,	London WC1X 8RR
Amsterdam, The Netherlands	CA 92101-4495, USA	Oxford OX5 1GB, UK	UK

First edition 2006

British Library Cataloguing in Publication Data
A catalogue record is available from the British Library.

ISBN-13: 978-0-444-51712-8
ISBN-10: 0-444-51712-X
ISSN: 1871-2711

∞ The paper used in this publication meets the requirements of ANSI/NISO Z39.48-1992 (Permanence of Paper).
Printed in The Netherlands.

Preface for Series

In the grand scheme of history the twentieth century will probably be remembered as the age of unbridled consumption of the planet's natural resources and devastating pollution of the environment. Attention was first drawn to the negative side effects of industrial and economic growth on our natural environment in the 1960s and 1970s with the publication of books such as *Silent Spring* by Rachel Carson and *The Closing Circle* by Barry Commoner. Nonetheless, it took three decades for the environmental movement to gather sufficient momentum to have a serious industrial and societal impact. In the final decade of the century a new paradigm began to emerge, based on the concepts of Green Chemistry and Sustainable Development. In hindsight, a turning point was the publication, in 1987, of the report *Our Common Future* by the World Commission on Environment and Development. It was recognized in this report that industrial and societal development must be sustainable over time. Sustainable development was defined as 'development that meets the needs of the present without compromising the ability of future generations to meet their own needs'. A decade later this concept was endorsed and further elaborated in the report, *Our Common Journey* by the Board on Sustainable Development of the US National Research Council.

'Sustainable Development' has subsequently become a catch phrase of the new millennium and many corporations are keen to show that their operations are 'sustainable'. Indeed, one could say that it is industry's answer to the environmental challenge.

Thomas Graedel has defined the two central tenets of sustainability as: (i) using natural resources at rates that do not unacceptably deplete supplies over the long term and (ii) generating and dissipating residues at rates no higher than can be assimilated readily by the natural environment. The core concept is the analogy between processes in the biosphere and in the technosphere. In the words of Barry Commoner, "in nature no organic substance is synthesized unless there is provision for its degradation; recycling is enforced".

For example, the use of fossil resources—oil coal and natural gas—as sources of energy and chemical feedstocks is clearly unsustainable even over a relatively short time span of the next 50 years. In the coming decades it needs to be supplanted by the use of agriculture-based renewable raw materials. A switch to processing of 'renewables' in biorefineries is also desirable for other reasons, such as biocompatibility, biodegradability and lower toxicity compared to oil-derived feedstocks and products. As noted above, production of chemicals from renewable raw materials tends to leave a smaller environmental footprint.

The time is clearly ripe for a series of books addressing the underlying science and engineering of sustainable development. To this end a series with the general title, 'Sustainability: Science and Engineering' has been commissioned. The first volume in the series is devoted to defining the principles of sustainable engineering and illustrating how these principles can be incorporated into the design of sustainable products and processes. Forthcoming volumes will deal with other topics of relevance to sustainable development, such as renewable raw materials and biorefineries, life cycle analysis, water usage and management and industrial ecology. Hopefully, this will provide a further stimulus for the development and implementation of sustainable technologies to the benefit of future generations of inhabitants of 'Spaceship Earth'

Roger A. Sheldon,
Delft, November 2005

Preface

The works collected in this book represent the results of the ECI-sponsored conference, "Green Engineering: Defining the Principles." Conceived as the first interdisciplinary conference to discuss the role of green engineering as an over-arching effort within all of engineering, this four-day workshop brought together a diverse group of engineers and scientists from industry, academia, government, and non-governmental organizations, with backgrounds in chemical, mechanical, civil, and electrical engineering, chemistry, social sciences, and business.

The development of the conference was led by a large cross-disciplinary organizing committee. Chaired by myself and Dr. William Sanders III, Director of the Office of Pollution Prevention and Toxics of the Environmental Protection Agency, the organizing committee worked for over 1 year to bring together this diverse group of contributors. Members of the organizing committee included David Allen, Department of Chemical Engineering, University of Texas — Austin; Sharon Austin, OPPT/US EPA; John Carberry, DuPont Experimental Station; Tim Gutowski, M.I.T. Advanced Composites Lab; Barbara Karn, US EPA, ORD, NCER; Barry Marten, Siemens Automotive Corp.; Nhan Nguyen, OPPT/US EPA; Walter Olson, Department of Mechanical Engineering, University of Toledo; Ron Williams, General Motors. I am indebted to the organizing committee for their participation in this effort. Without their help and support, the conference could never have been completed, and this book would not have been possible. Through a series of monthly conference calls, and substantial behind-the-scenes work, the members of the organizing committee worked tirelessly to bring about the conference.

The current book represents the final result from this conference. The principles outlined in the Introduction chapter, and expanded in Part II, describe the collective wisdom of conference attendees. These principles represent a consensus view of the principles of green engineering as determined during these deliberations, and should not be construed as the individual opinions of any of the participants or sponsors. A great deal of appreciation is provided to all conference participants, both for their efforts during the conference and their continuing work thereafter to complete the development of the principles.

Finally, my appreciation to those who have contributed to this book effort. In addition to the authors of the chapters contained herein, many additional people worked diligently to bring the book to fruition. The conference organizing committee promoted the development of the book, and several members of the committee, particularly Walt Olson and Nhan Nguyen suggested individuals who could provide valuable contributions. Each chapter was read by at least one evaluator who offered comments and suggestions. Publication of this book by Elsevier, and the support of their publishing staff, is also gratefully acknowledged.

The current book is intended to be the first book in a new series of books that further elaborate on "Sustainable Science and Engineering," to be published by Elsevier, and to be coordinated by Roger Sheldon (Delft University, Netherlands) and I. Additional books on Life Cycle Analysis, Biorefineries, and Water Use, are currently being developed or contemplated. The expanding discipline and growth of this area is due in large part to the contributions of the conference participants, who were at the forefront of this endeavor.

Conference participants

The conference was attended by a wide range of participants, from a range of backgrounds including industry, academia, government, and NGOs. All of the following conference attendees contributed to the principles that form the basis of this book, and are acknowledged for their participation in the discussion and development of these principles. The affiliations noted are those from the time of the conference, and may have changed in the ensuing period.

Martin A. Abraham	University of Toledo
Paul T. Anastas	White House Office of Science and Tech
T.K. Bandyopadhyay	Institute for Steel Development and Growth
Jane Bare	US EPA
Brian Blakey	GE Global Research
Leirad Carrasco-Martinez	University of Texas at El Paso
Rebecca Chamberlin	Los Alamos National Laboratory
Alex Chase	Baxter Healthcare Corporation
David J.C. Constable	SmithKline Beecham
Heather M. Cothron	UT-Battelle/Oak Ridge National Laboratory
Robert M. Counce	University of Tennessee
Nadia N. Craig	University of South Carolina
Theo Dillaha	VPI&SU
Dionysios D. Dionysiou	University of Cincinnati
Milorad P. Dudukovic	Washington University
Delcie Durham	National Science Foundation
Ahmed Elsawy	Tennessee Technological University

Steve Forbes	UTEP
Charles Freiman	ECI Site Staff
Michael H. Gregg	VPI&SU
Arnulf Grubler	IIASA
Tim Gutowski	M.I.T. Advanced Composites
Henry Hatch	NAI
Lauren G. Heine	Zero Waste Alliance
Robert P. Hesketh	Rowan University
Karen High	Oklahoma State University
Kathryn Hollar	Rowan University
Sukhvinder Kandola	University of Leicester
Barbara Karn	US EPA, NCER
Sibel Koyluoglu	Ford Motor Co.
Donald Liou	UNC Charlotte
Heath Lloyd	University of South Carolina
Barry Marten	Siemens Automotive Corp
Victor Martinez	University of Texas at El Paso
Peter Melhus	Bay Area Alliance for Sustainable Communities
Tom Merkle	University of Toledo
Arup Kumar Misra	Assam Engineering College
Kenneth L. Mulholland	Kenneth Mulholland & Associates
Nhan Nguyen	U.S. EPA, OPPT
Walter Olson	University of Toledo
Walter H. Peters	University of South Carolina
Gerhard Piringer	Tulane University
Mark Pitterle	Virgina Tech
Dmytro Pylypenko	California State University, Chico
Ferdinand Quella	Siemens AG
Anu Ramaswami	University of Colorado
Brindaban Ranu	Indian Association for the Cultivation of Science
Stephen Ritter	Chemical & Engineering News
Joseph Rogers	American Institute of Chemical Engineers
James A. Russell	University of South Carolina
Endalkachew Sahle-Demessie	National Risk Mgmt. Research Lab, US EPA
William Sanders	U.S. Environmental Protection Agency
Konrad Saur	Five Winds International
David Shonnard	Michigan Technological University
Michael Silsbee	Pennsylvania State University
Raymond Smith	U.S. EPA
Laura Steinberg	Tulane University
Mayadevi Suseeladevi	National Chemical Laboratory
Siret Talve	EcolabS Ltd.
Harry Van Den Akker	Technische University, Delft
Jorge A. Vanegas	Georgia Tech

P. Aarne Vesilind — Bucknell University
Nele Zechel — Brandenburg Technical University of Cottbus
Julie B. Zimmerman — University of Michigan

Acknowledgments

The conference attendees were fortunate to receive substantial financial support for travel and registration. Funding was received from the National Science Foundation (Grant number DMI-0303838), the Environmental Protection Agency (Grant number X8 83076001-0), the Department of Energy (Los Alamos National Laboratory), and the American Chemical Society's Green Chemistry Institute. The American Institute of Chemical Engineers, the Society of Automotive Engineers, and the American Society of Mechanical Engineers were technical cosponsors for this conference. The support of all sponsors, both technical and financial, is gratefully acknowledged and appreciated.

Martin Abraham

List of Contributors

P. Aarne Vesilind, *R. L. Rooke Professor of Engineering, Department of Civil and Environmental Engineering, Bucknell University, Lewisburg, PA 17837, USA*

M.A. Abraham, *Department of Chemical and Environmental Engineering, The University of Toledo, Toledo, OH, USA*

P.T. Anastas, *Green Chemistry Institute, American Chemical Society, 1155 Sixteenth Street, NW, Washington, DC 20036, USA*

B. Bras, *George W. Woodruff School of Mechanical Engineering, Georgia Institute of Technology, Atlanta, GA 30332-0405, USA*

N.N. Craig, *Department of Mechanical Engineering, University of South Carolina, 300 Main Street, Columbia, SC 29208, USA*

D.J.C. Constable, *GlaxoSmithKline, Corporate Environment, Health and Safety 2200 Renaissance Blvd. Suite 105, King of Prussia, PA 19406, USA*

H.M. Cothron, *Engineering and Infrastructure Business Unit, Science Applications International Corporation, 151 Lafayette Dr., Oak Ridge, TN 37831*

B.C. Coull, *School of the Environment, University of South Carolina, 901 Sumter Street, Room 702G, Columbia, SC 29208, USA*

R.M. Counce, *Chemical Engineering Department, University of Tennessee, 419 Dougherty Hall, Knoxville, TN 37996-2200, USA*

V.L. Cunningham, *GlaxoSmithKline, Corporate Environment, Health and Safety 2200 Renaissance Blvd. Suite 105, King of Prussia, PA 19406, USA*

A.D. Curzons, *Southdownview Way, Worthing BN14 8NQ, UK*

Concepción Jiménez-González, *Five Moore Dr., Research Triangle Park, NC 27709, USA*

D.D. Dionysiou, *Department of Civil and Environmental Engineering, 765 Baldwin Hall, University of Cincinnati, Cincinnati, OH 45221-0071, USA*

M.P. Dudukovic, *Chemical Reaction Engineering Laboratory (CREL), Washington University, St. Louis, MO, USA*

K. Geiser, *Lowell Center for Sustainable Production, University of Massachusetts Lowell*

M.H. Gregg, *Engineering Education, Virginia Tech, Blacksburg, VA 24061, USA*

S.A. Hamill, *Founder, Sustainable Labs, 5440 SW Buddington St., Portland, OR 97219, USA*

R.E. Hannah, *GlaxoSmithKline, Corporate Environment, Health and Safety 2200 Renaissance Blvd. Suite 105, King of Prussia, PA 19406, USA*

L. Heine, *Director of Applied Science, Green Blue Institute, 600 E. Water St., Charlottesville, VA 22902, USA*

L.G. Heine, *Green Blue Institute (GreenBlue), 600 E. Water St. Suite C Charlottesville, VA 22902, USA*

J.R. Hendry, *Assistant Professor, Department of Management, Bucknell University, Lewisburg, PA 17835, USA*

R.P. Hesketh, *Chemical Engineering, Rowan University, 201 Mullica Hill Rd., Glassboro, NJ 08028, USA*

G.A. Keoleian, *Center for Sustainable Systems, School of Natural Resources and Environment, University of Michigan, 440 Church St., Ann Arbor, MI 48109-1041, USA*

S.L. Landes, *Ford Motor Co., Dearborn, MI, USA*

P. Melhus, *Department of Urban and Regional Planning, San José State University, One Washington Square, San Jose, CA 95192, USA*

J.J. Michalek, *Design Decisions Laboratory, Department of Mechanical Engineering, Carnegie Mellon University, Pittsburg, PA, USA*

L. Moens, *National Bioenergy Center, National Renewable Energy Laboratory (NREL), 1617 Cole Boulevard, Golden, CO 80401, USA*

W.R. Morrow, *Environmental and Sustainable Technologies Laboratory (EAST), Department of Mechanical Engineering, The University of Michigan at Ann Arbor, Ann Arbor, MI, USA*

S.A. Morton III, *Chemical Engineering Department, Lafayette College, 266 Acopian Engineering Center, Easton, PA 18042, USA*

K.L. Mulholland, *Kenneth Mulholland & Associates, Inc., 27 Harlech Drive, Wilmington, DE 19807, USA*

W.H. Peters, *Department of Mechanical Engineering, University of South Carolina, 300 Main Street, Columbia, SC 29208, USA*

W. Olson, *Department of Mechanical, Industrial and Manufacturing Engineering, The University of Toledo, Toledo, OH 43606, USA*

P.A. Ramachandran, *Chemical Reaction Engineering Laboratory (CREL), Washington University, St. Louis, MO, USA*

A. Ramaswami, *Urban Sustainable Infrastructure Engineering Project (USIEP), Department of Civil Engineering, University of Colorado at Denver & Health Sciences Center, Denver CO 80217, USA*

J.A. Russell, *Department of Mechanical Engineering, University of South Carolina, 300 Main Street, Columbia, SC 29208, USA*

M. Sibel Bulay Koyluoglu, *Ford Motor Co., Dearborn, MI, USA*

S.J. Skerlos, *Environmental and Sustainable Technologies Laboratory (EAST), Department of Mechanical Engineering, The University of Michigan at Ann Arbor, Ann Arbor, MI, USA*

D.V. Spitzley, *Center for Sustainable Systems, School of Natural Resources and Environment, University of Michigan, 440 Church St., Ann Arbor, MI 48109-1041, USA*

C. Stewart Slater, *Chemical Engineering, Rowan University, 201 Mullica Hill Rd., Glassboro, NJ 08028, USA*

C. Tunca, *Chemical Reaction Engineering Laboratory (CREL), Washington University, St. Louis, MO, USA*

J.A. Vanegas, *Department of Architecture, College of Engineering, Texas A&M University, 3137 TAMU, College Station, TX 77843-3137, USA*

M.L. Willard, *AXIS Performance Advisors, Inc., 2515 NE 17th Ave Portland, OR 97212, USA*

B.J. Yates, *Department of Civil and Environmental Engineering, 765 Baldwin Hall, University of Cincinnati, Cincinnati, OH 45221-0071, USA*

J.B. Zimmerman, *National Center for Environmental Research, Office of Research Development, United States Environmental Protection Agency, 1200 Pennsylvania Avenue, NW (8722F), Washington, DC 20460, USA*

Contents

PART I:
THE PRINCIPLES

Sustainability Science and Engineering: Defining principles
Martin A. Abraham (Editor)
© 2006 Elsevier B.V. All rights reserved
DOI 10.1016/S1871-2711(05)01001-9

Chapter 1

Principles of Sustainable Engineering

Martin A. Abraham

*Department of Chemical and Environmental Engineering, The University of Toledo,
Toledo, OH, USA*

An engineer is a person whose job is to design or build machines, engines or electrical equipment, roads, railways or bridges, household products, agricultural materials, and other goods that society desires, using scientific principles. The engineer develops processes using a collection of raw materials and energy and converts them to the desired form. The manufacture of products that society desires is accompanied by the production of wastes, some of which cannot be avoided. Historically, the engineer has been able to optimize the process to be most efficient based on the maximum profit that can be gained by the conversion process, taking materials of little or no value and transforming them into materials of greater value.

Within the past roughly 30 years, since the mid-1970s, the engineer has also been called upon to transform materials into higher value products, while simultaneously meeting a set of environmental regulations that limited the amount of pollutants that could be emitted. Since, the restrictions were placed on the effluents from the process, substantial effort was made to capture the effluent and convert it to an innocuous material prior to the release to the environment. However, little attention was given to the effect of the product in the environment and the inherent risk of the manufacturing process.

This started to change in the early 1990s, with the advent of pollution prevention concepts, which made clear that the best opportunity to protect the environment was to stop pollution at its source. In the Pollution Prevention Act of 1990, the United States Congress established the United States' governmental policy on pollution and the environment [1]. It established the hierarchy shown in Fig. 1, making clear that source reduction was the highest form of pollution prevention (P2). As generally defined, P2 includes increased efficiency in the use

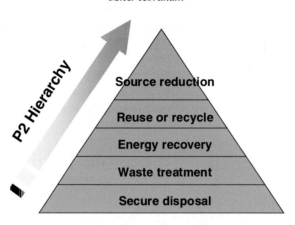

Fig. 1. The pollution prevention hierarchy.

of raw materials, energy, water, and other resources and the protection of resources through conservation. In 1997, the publication of Anastas and Warner's Twelve Principles of Green Chemistry [2], placed the concept of pollution prevention into a context through which scientists and engineers could seek ways to minimize the potential for environmental harm in the manufacture of chemical products.

Green engineering [3] is a discipline that spans all of the traditional engineering specialties. The chemical engineer can design a chemical process to minimize the use of organic solvents, to incorporate the principles of green chemistry, or to decrease energy consumption. A mechanical engineer may evaluate the life cycle of a manufactured product and evaluate the environmental impacts of manufacturing relative to those of use by the consumer. Civil engineers, long the architect of waste treatment processes, are now evaluating new techniques to reuse wastes and to remediate abandoned brownfields. Even electrical engineers have a role to play, as they investigate new battery storage devices that are more efficient or better technologies for conversion of solar energy to electricity.

These examples illustrate how various engineers have attacked the issues of green engineering from a discipline-specific viewpoint. However, the principles of green engineering transcend disciplinary boundaries. While life cycle analysis may be considered the domain of the mechanical engineer, many chemical engineers are using similar techniques in their evaluation of chemical processes. Similarly, civil and mechanical engineers are working together to develop new transportation systems, and chemical and electrical engineers are mutually involved in fuel cell research leading to decreased reliance on petrochemicals.

Sustainability, or sustainable development, is frequently used as a business driver throughout the world. Sustainable development is generally defined according to the original definition of the Brundtland Commission (1989): providing for human needs without compromising the ability of future generations to meet their needs. From a business perspective, the concepts of sustainability are often described as the triple bottom line [4],

- economic viability: the business aspects of a project,
- social concerns: human health and social welfare,
- natural or ecological issues: environmental stewardship.

The triple bottom line recognizes that a business decision must be based on more than the ability of the business to make a profit. It must recognize the environmental and social implications of that decision and factor those impacts into the overall decision-making process.

The introduction of the ISO 14000 standards in 1998 gave businesses an opportunity to review their practices within the concept of sustainable development, and provided the impetus to evaluate their environmental performance. The ISO 14000 standard calls for an environmental management system that includes continuous improvement, managerial commitment, planning, implementation, measurement, and evaluation [5]. As a result of changes in the environmental management structures, forward-thinking business concerns shifted their environmental programs from response to government regulation to proactively seeking ways to improve environmental performance. In many cases, those corporations that instituted environmental management systems, such as those suggested within ISO 14000, found a positive impact on their corporate profit [6].

Over the years, the concepts of Green Chemistry, Green Engineering, and Sustainability have merged together to provide an assembly of terms that are often used interchangeably to describe similar ideas. Figure 2 provides a conceptual framework that allows one to consider each of the elements in an appropriate context. As depicted in this diagram, there is substantial overlap among the various concepts, yet each has unique contributions to make in the area of environmental stewardship. For example, only the green chemist has the skills to develop new chemical reactions that will lead to new materials and new techniques for producing chemicals for today's society. At the crossroads of green chemistry and biology lies the toxicologist who can evaluate the potential harm that a new material might have on human health. The green engineer understands the concepts of life cycle, and can use these to identify the overall environmental impact of a product or process, from "cradle to grave". However, neither green chemists nor green engineers normally place their work within the context of societal and social impacts, the realm of the sustainability specialist (perhaps, a sustainabilist). Thus, we see that it would be inappropriate to identify one of these concepts as all encompassing and the others as subsets of

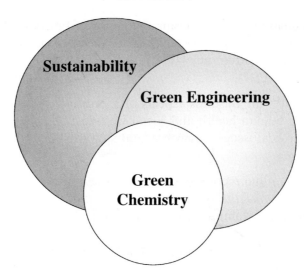

Fig. 2. A schematic representation of the varying roles of Green Chemistry, Green Engineering, and Sustainability.

the first; rather, these terms represent different perspectives on the issues, challenges, and solutions of environmental stewardship.

It is the ability to incorporate social implications into the business decision that distinguishes sustainable engineering from green engineering. Recently, Abraham [7] provided a brief discussion of the different contexts in which the elements of the triple bottom line can be considered. "While green chemistry addresses issues of natural capital, and green engineering addresses both natural capital and economic viability, sustainability also addresses the human condition and implores the individual to improve the quality of life for all inhabitants". Sustainability requires one to consider the non-technical aspects of technology, the social implications of their work, and the constraints presented by economics, society, and the environment. Sustainable engineering seeks solutions that are beyond those of green engineering, by looking outside the scope of the process or product being developed and considering the system as a part of the global ecosystem, which includes all of humanity. The sustainable engineer will be asked to design processes that do not specifically maximize *profit*, but rather maximize *benefits*, defined based on all of the elements of the triple bottom line: environment, human health, economics, or a combination thereof [8].

In order to accomplish the sustainable design goals, the sustainable engineer will first require technical depth, and then require consideration of the sustainability issues that envelop the technical questions. To provide a framework for their design, one can also describe a set of principles, or design concepts, that the sustainable engineer can use to evaluate processes and products and look for solutions that meet the goals of the triple bottom line. The "Principles of

Sustainable Engineering", developed from the ECI-sponsored conference "Green Engineering: Defining the Principles" provides such a set of guidelines. The principles were a consensus agreement by the diverse group of engineers that attended the meeting, and the discussion that occurred in the months following that effort. We distinguish the Principles of Sustainable Engineering from those of Green Engineering [3] and Green Chemistry [2] based on the scope of the debate that they inform.

The SanDestin conference: developing the principles

During the summer of 2002, several attendees of a regional Green Engineering conference began a conversation about the need to develop a set of defining principles, resulting in the proposal for a workshop-oriented conference to include all engineering disciplines. While numerous definitions of green engineering exist, and several principles statements for sustainability have been produced, no one was aware of a concise statement of principles for green engineering. These principles were collected into common thematic areas to identify overarching principles for discussion during the conference, and to serve as a basis for the development of green engineering principles.

Approximately 65 engineers and scientists convened at the Sandestin Resort for the first conference on "Green Engineering: Defining the Principles", during the week of May 19, 2003. Over 4 days, a series of oral and poster presentations were used to stimulate discussion of green engineering principles, while attendees met during breakout sessions and over lunch to discuss specific elements of green engineering principles. By the end of the conference, a set of principles that summarized the discussions from all of the breakout sessions was compiled and agreed to by the attendees.

After the meeting, the draft principles were forwarded to all conference participants for further review and evaluation. Responses were received from many participants, some with minor editorial suggestions and others with large changes in the principles documents. There was also substantial discussion among the participants and the organizing committee as to whether these principles were final, or whether additional work was required. Recognizing that the principles that were developed at the conference represented the collective wisdom of the attendees based on the discussion during the meeting, the principles were maintained as originally agreed, with only minor editorial changes.

While the original principles were maintained, a separate discussion ensued around how these principles should be titled. Although the purpose of the conference was to define green engineering principles, Anastas and Zimmerman released a set of green engineering principles approximately 1 month prior to the conference [3]. In addition, there was a group of participants that believed that the principles went beyond the realm of green engineering, to consider issues of social justice. As a result, it was agreed that the principles would be labeled as sustainable engineering principles, as published below.

The principles:

Sustainable Engineering transforms existing engineering disciplines and practices to those that promote sustainability. Sustainable Engineering incorporates development and implementation of technologically and economically viable products, processes, and systems that promote human welfare, while protecting human health and elevating the protection of the biosphere as a criterion in engineering solutions.

To fully implement sustainable engineering solutions, engineers use the following principles:

1. Engineer processes and products holistically, use systems analysis, and integrate environmental impact assessment tools.
2. Conserve and improve natural ecosystems while protecting human health and well-being.
3. Use life cycle thinking in all engineering activities.
4. Ensure that all material and energy inputs and outputs are as inherently safe and benign as possible.
5. Minimize depletion of natural resources.
6. Strive to prevent waste.
7. Develop and apply engineering solutions, while being cognizant of local geography, aspirations and cultures.
8. Create engineering solutions beyond current or dominant technologies; improve, innovate and invent (technologies) to achieve sustainability.
9. Actively engage communities and stakeholders in development of engineering solutions.

There is a duty to inform society of the practice of sustainable engineering.

The Principles of Sustainable Engineering provide a paradigm in which engineers can design products and services to meet societal needs with minimal impact on the global ecosystem. The principles cannot be taken as independent elements, but rather, should be considered as a philosophy for the development of a sustainable society. The principles are not prescriptive. They do not provide engineers with a definitive methodology for deriving a sustainable design. Rather, they provide engineers with overarching concepts that can be used along with traditional design principles to develop new products and services to be applied for the growth and development of human society, while simultaneously minimizing the impact of these designs on the global ecosystem.

Goals and structure of the book

This book is intended to be a comprehensive volume that places the issues of sustainability into a context appropriate for engineers. It is based on the principles developed at the SanDestin conference, which are intended to be incorporated into the design of products and processes. The authors were asked to utilize examples that illustrate the opportunities presented by these principles, and traditional engineering calculations that demonstrate the methods by which these principles can be incorporated into engineering design. The goal is to create a framework in which engineers can do engineering in a sustainable fashion.

The book is divided into three sections:

- overview and general material,
- expanded discussion on the principles, and
- case studies and examples that illustrate the principles.

All of the participants in the conference were invited to contribute to the book. Additional authors were sought based on recommendations of conference attendees to provide material that was deemed necessary for completeness but was not provided by the conference attendees.

The first section places the principles in the context of the conference in which they were developed, and presents some additional overview material. The second section provides an elaboration on each of the principles developed at the conference, during which the authors use the principle as a guide, utilize examples to illustrate the principles, and provide some reference for the principle in the context of current literature. In the third section, we provide applications of these principles as they are used in both academia and industry. Again, authors were encouraged to use examples and demonstrate through computation the commercial advantage obtained through the application of the principles.

Concluding comments

The principles presented within this book should not be considered as a final set of materials for a fixed body of knowledge. Rather, the world of sustainable engineering is continuously evolving and developing into a valuable decision-making tool. The development of metrics and the expansion of methods are creating new opportunities for the sustainable engineer. Likewise, the changing world economy and the continuing fluctuation in global climate create new opportunities and needs to implement the methods, tools, and principles in engineering analysis and design.

In a similar vein, the examples included within the book should not be considered as a comprehensive discussion of all the ways in which the principles can be applied. These examples only describe the state of the art at the current time within the applications in which they were developed. Other excellent examples exist throughout the literature, and new examples are continually being developed. As the field continues to expand, these principles will be applied in new areas to create further examples of sustainable engineering — engineering to meet the needs of the triple bottom line.

References

[1] D.T. Allen, K.S. Rosselot, Pollution Prevention for Chemical Processes, Wiley, New York, 1997 434pp.
[2] P.T. Anastas, J.C. Warner, Green Chemistry: Theory and Practice, Oxford University Press, Oxford, 1998.
[3] P.T. Anastas, J. Zimmerman, Environ. Sci. Technol. 37 (2003) 94A–101A.
[4] A. Barrera-Roldan, A. Saldivar-Valdes, Proposal and Application of a Sustainable Development Index, Ecol. Indicators 2 (2002) 251–256.
[5] K. Hersey, A Close Look at ISO 14000, Professional Safety 43 (7) (1998) 26–29.
[6] V.N. Bhat, Does It Pay to be Green?, Int. J. Environ. Studies: Sections A & B, 56 (4) (1999) 497–508.
[7] M.A. Abraham, Environmental Progress, December, 2004.
[8] C. Hendrickson, M. Conway-Schempf, L. Lave, F. McMichael, Introduction to Green Design, Green Design Initiative, Carnegie-Mellon University, Pittsburgh PA, 1999.

Sustainability Science and Engineering: Defining principles
Martin A. Abraham (Editor)
© 2006 Published by Elsevier B.V.
DOI 10.1016/S1871-2711(05)01002-0

Chapter 2

The Twelve Principles of Green Engineering as a Foundation for Sustainability

P.T. Anastas[a], J.B. Zimmerman[b,c]

[a]Green Chemistry Institute, American Chemical Society, 1155 Sixteenth Street, NW, Washington, DC 20036, USA
[b]Department of Civil Engineering, University of Virginia, Thornton Hall D219, 351 McCormick Road, PO Box 400742, Charlottesville, VA 22904-4742
[c]National Center for Environmental Research, Office of Research Development, United States Environmental Protection Agency, 1200 Pennsylvania Avenue, NW (8722F), Washington, DC 20460, USA

In recent years, there has been increasing attention focused on the need to address with the challenges of sustainability. Sustainability as defined by the Brundtland Commission calls for meeting the needs of the current generation while preserving the ability of future generations to meet their needs. There are those who argue that our current path is unsustainable and those who contend that these concerns are unfounded. In either case, changes can be made through innovations in science and technology to mutually benefit the environment, the economy, and the global society. Given this scenario there are four possible outcomes (Table 1). If change is needed and does not take place, the result will be an unsustainable future (Quadrant 1). If the current path is unsustainable and the necessary changes are made to mitigate the potential consequences, sustainability will be realized (Quadrant 2). However, if change was not required and no action was taken, sustainability may be realized without the ancillary benefits (Quadrant 3). Finally, the shifts made toward sustainability when no change is required have the potential to realize many ancillary near-term benefits (Quadrant 4). In other words, by engaging science and technology for sustainability, the possible outcomes include ancillary benefits and a sustainable future without sacrifices in economic growth and quality of life.

Table 1
The four possible outcomes of engaging in science and technology for sustainability

Quadrant 1 Change needed – Did not happen UNSUSTAINABLE	*Quadrant 2* Change needed – Happened SUSTAINABLE

Quadrant 3
Change not needed – Did not happen
SUSTAINABLE; NO ANCILLARY
  BENEFITS REALIZED
	Quadrant 4
	Change not needed – Happened
	SUSTAINABLE; REALIZED
	  ANCILLARY BENEFITS

Most popular constructs of sustainability are in agreement that there are three major aspects – environmental, economic and societal. Inherently, if an action is not advancing each of the three "pillars" of sustainability it could not be viewed as advancing sustainability overall. Therefore, the difficult questions are involved with the short-term versus long-term consequences of actions, regional versus global, and known consequences versus unforeseeable consequences.

The question then becomes how one makes the goals of sustainability become a reality in practice. While various disciplines of sociology, economics, behavioral sciences, etc. will be important, it is difficult to envision a scenario where science and technology do not play a fundamental and essential role in moving toward sustainability.

1. What is green engineering?

Green Engineering is the design, discovery, and implementation of engineering solutions for sustainability. The approach is scalable applying across molecular, product, process, and system. This approach is as broad as the disciplines of engineering themselves and includes molecular and bio-engineering, civil, electrical, mechanical, environmental, and systems engineering.

The Principles of Green Engineering [1] (Table 2) provide a framework for understanding and represent a reflection of those engineering techniques that are being used to become more sustainable. While there are significant, creative, and important examples of engineering solutions that are being developed, they are neither comprehensive nor systematic. The 12 Principles should be thought of not as rules, laws or inviolable standards. Instead they are a set of guidelines for thinking in terms of sustainable design criteria that, if followed, can lead to useful advances for a wide range of engineering problems.

The Principles will be a set of parameters in a complex system where there will be synergies in which progress toward achieving the goal of one principle will augment progress toward several other principles. In other cases, there may

Table 2
The Twelve Principles of Green Engineering [1]

Principle 1 – Designers need to strive to ensure that all material and energy inputs and outputs are as inherently non-hazardous as possible.

Principle 2 – It is better to prevent waste than to treat or clean up waste after it is formed.

Principle 3 – Separation and purification operations should be a component of the design framework.

Principle 4 – System components should be designed to maximize mass, energy and temporal efficiency.

Principle 5 – System components should be output pulled rather than input pushed through the use of energy and materials.

Principle 6 – Embedded entropy and complexity must be viewed as an investment when making design choices on recycle, reuse or beneficial disposition.

Principle 7 – Targeted durability, not immortality, should be a design goal.

Principle 8 – Design for unnecessary capacity or capability should be considered a design flaw. This includes engineering "one size fits all" solutions.

Principle 9 – Multi-component products should strive for material unification to promote disassembly and value retention – (minimize material diversity).

Principle 10 – Design of processes and systems must include integration of interconnectivity with available energy and materials flows.

Principle 11 – Performance metrics include designing for performance in commercial "after-life".

Principle 12 – Design should be based on renewable and readily available inputs throughout the life-cycle.

be trade-offs between the application of two principles. Those trade-offs can only be resolved by the specific choices and values of the practitioners within the context of their specific situation or within their society. In the end, all sustainability is local and the framework of the Twelve Principles of Green Engineering is a tool to aid in consciously and transparently addressing those design choices relevant to fundamental sustainability.

1.1. Important caveats

In order to address the issues of sustainability, Green Engineering needs to approach the fundamentals of design in a manner that takes into account:

• Life-cycle considerations
• Multi-scale applications: e.g. products, processes and systems

Without such an approach, a designer risks the inadvertent consequence of doing the wrong things but doing them very well. In other words, one could effectively apply Principles of Green Engineering on a single life-cycle stage e.g. manufacturing, yet the overall consequence for human health, the environment and the broader elements of sustainability would still be degraded. Life-cycle considerations cannot therefore be ignored.

In addition, the designer of engineering solutions must keep in mind the broader array of scales that can be addressed. If applied to a narrowly defined target, the Twelve Principles of Green Engineering could have assisted in making a more sustainable vacuum tube for electronics without the broader perspective that allowed for the development of the transistor. As leapfrog technologies become increasingly important in addressing fundamental sustainability challenges such as energy, food, water, resource depletion, etc. the broad perspective of multi-scale application of the Twelve Principles is essential.

2. Principle by principle analysis

It is useful and necessary to discuss and exemplify each of the Twelve Principles of Green Engineering individually to illustrate and explain what they mean and how they can and are being applied. However, there is a danger in this approach. The reader must note that there is not currently the portfolio of examples of engineering designs that are accomplishing all Principles simultaneously, even at an incremental level. Therefore, examples used for the illustration of a particular principle should not necessarily be considered an adequate representation of how best to achieve the goals of the other principles.

Principle 1. Designers need to strive to ensure that all material and energy inputs and outputs are as inherently non-hazardous as possible.

There are two ways to reduce risk (1) reduce hazard and (2) reduce exposure. With risk being a product of hazard and exposure two things become clear. As your hazard approaches infinity you must reduce your risk to approach zero by reducing exposure. Conversely, as you are able to make your hazard approach zero, you can allow your exposure to approach infinity.

Why is this important?: Our approaches to controlling the consequences of hazardous chemicals in the environment have largely been to try and control exposure. The engineering that has resulted in trying to lower the probability of exposure to a wide range of hazards including toxics, reactives, flammables and explosives has often been brilliant and Herculean. The bad part is that it is also expensive. A recent report from the White House Office of Management and Budget [2] places the cost of compliance with environmental regulations of the command and control approach to be approximately 2% of GDP. Perhaps more importantly, even the most brilliant engineering bandages can and do fail.

One of the most well known instances of an accident attributed to the malfunction of engineering systems is the accidental release of 2.5 million curies of radioactive gas from Three Mile Island. This failure was initiated due to either a mechanical or electrical failure and prevented the steam generators from removing heat. There is general consensus that the accident on Three Mile Island was further exacerbated by incorrect decisions made because the operators were overwhelmed with irrelevant, misleading, or incorrect information [3]. Another often cited example of engineering failure and human error leading to catastrophic results is the Challenger Space Shuttle accident in 1987. The Challenger explosion was caused by a fuel leak emanating from the solid rocket booster field joint. Engineers investigated and identified that the possibility of this type of joint failure prior to the Challenger launch could be hazardous to a successful launch [4]. Engineers informed managers of this problem; however, managers did not believe it posed a significant risk to the success of the launch. These examples demonstrate that engineering systems can fail and that these failures can be further compounded by human error. This argues for the need to engineer molecules, products, processes, and systems such that safety is not dependent on controls or designs that can fail or be sabotaged, either intentionally or accidentally.

It is merely a question of when and how often failure occurs since failure rate for exposure controls are simply a probability function. As chemistry and engineering progressed through the 19th and 20th centuries and industry used larger quantities of increasingly hazardous substances, the cost of containment increased as the need to make these containment technologies approach perfection also increased. This fundamental dilemma is not an equation that will change. What needs to change is the approach.

The alternative approach is necessarily making your materials and chemicals as innocuous as possible. As you decrease the intrinsic hazard of the substances that are being used, you decrease the pressure on exposure controls. The ultimate case of completely benign materials is that they need no exposure controls at all. In all due respect to the medieval physician Paracelsus who first said, "Everything is toxic, it simply is a matter of the dose." often paraphrased as "The dose makes the poison," we may never reach perfection of completely benign substances. However, the advances that can and are being made in moving toward inherently benign chemicals is significant and dramatic.

The ability to ensure that the materials and energy sources are as inherently benign as possible stems from our advances in a molecular level understanding of hazard. As we have in recent years begun to more fully understand how modification of molecular structure can directly modify the ability of a chemical or chemical class to manifest hazard, we have enabled designers to make this a performance criterion.

This fundamental approach to addressing sustainability through the use of inherently benign substances is important because it changes the equation for reducing risk from the circumstantial to the intrinsic. Instead of changing the circumstances or condition of use of a hazardous substance such as handling, transport, personal protective gear, smokestack scrubbers, etc. a reduction in the intrinsic hazard of a substance allows even the worst practices, the worst circumstances, to have fundamentally reduced consequences. If the material is released directly into a drinking water source but cannot cause harm and does not breakdown into hazardous substances, the consequences are minimal. If a material is used wastefully but is inherently non-depleting and completely renewable, the consequences of this behavior are dramatically reduced. A few of the promising examples of inherently benign substances in the marketplace include bioplastics, ethyl lactate solvents and non-chlorine water disinfection.

Bioplastics may be used to make products ranging from clothes to eating utensils to car parts and are biodegradable in a wide range of biologically active conditions. If widely used, bioplastics may reduce plastics in the waste stream by up to 80% and have the potential to reduce carbon dioxide in the atmosphere through removal during the growing [5]. If all plastics were made from bio-based polylactic acid, oil consumption used in the manufacturing process could decrease by 90–145 million barrels per year [5]. Perhaps the best example of a successful bioplastic is NatureWorks, based on polylactic acid, where lactic acid, a common food additive, is extracted from a renewable resource such as corn. The products produced from NatureWorks are compostable and can be recycled back to monomer and into polymers [6].

Millions of pounds of toxic industrial solvents may be able to be replaced by environmental friendly solvents made with ethyl lactate, an ester of lactic acid. Unlike other solvents, which may possess human health concerns, damage the ozone layer or pollute groundwater, ethyl lactate is considered benign. Lactate ester solvents have numerous attractive environmental advantages including being biodegradable, easy to recycle, non-corrosive, non-carcinogenic and non-ozone depleting [7]. As further evidence of the benign characteristics of this chemical, ester lactate is approved by the US Food and Drug Administration for use in food products. Most importantly, ethyl lactate solvents are cost and performance competitive with traditional solvents in a wide range of applications [7].

The disinfection of drinking water and municipal wastewater provides critical public health protection. Disinfection destroys bacteria and viruses, helping to protect ecosystems and prevent the spread of waterborne disease. The most commonly used disinfectant for both drinking water and wastewater treatment is chlorine. Although chlorination has been a very successful water disinfection technique, chlorine in drinking water has been linked to a number of potential concerns. Since mid-1970s, formation of trihalomethanes, possible human carcinogens, as a result of chlorination of natural waters has been documented [8–10]. A number of commercially available alternative processes, such as membrane processes, are able to remove bacteria, viruses and protozoa as well as a range of chemical contaminants. Alternatives to chemical disinfection, such as UV irradiation, are also being used for disinfection of drinking water. Such 'non-conventional' processes and disinfection methods can in principle be used to replace, or at least greatly reduce, the use of chemical disinfection of drinking water eliminating the potential for human health and environmental impacts associated with chlorination. These strategies also significantly reduce the vulnerability associated with large quantities of chlorine gas that is intended for water disinfection but may represent a significant potential for release, either accidental or intentional and malicious.

Principle 2. It is better to prevent waste than to treat or clean up waste after it is formed.

While nearly everything in our folklore informs us that waste is something to be avoided, (e.g. "waste not want not") and that prevention is an essential strategy (e.g. "an ounce of prevention is worth a pound of cure") our chemical enterprise has evolved where these adages are not always evident or applied. In certain sectors of the chemical and chemicals related industries such as pharmaceuticals, and electronics the waste profiles are fairly dramatic.

In an analysis of the semiconductor manufacturing process, Ayres, Willams and Heller, [11] concluded that for every 2 g computer chip that was made, 1.7 kg of waste was generated not including the water and material used in energy generation to make the chip. In an assessment by Sheldon over several industry sectors, he introduced the concept of the e-factor [12]. Sheldon estimated that a pharmaceutical manufacturing process may generate 100 kg of waste per kilograms of product. This analysis is limited to within a process, or "gate-to-gate", and would certainly increase through a more inclusive lifecycle analysis of material and energy investment and expenditure.

When a material input results in waste, a manufacturer or user has to "pay" for the substance three times. The first payment is when it is purchased as a virgin material. The second payment is the costs associated with separating the non-product from the desired product. The third payment is to dispose of the material as a waste in a manner that is responsible and in compliance with laws, regulations, and policies.

Waste is a man-made concept. Waste simply refers to a material or energy flow for which a useful, that is commercially viable for industry, purpose has not yet been identified. One could view waste as merely a temporary state in which the creativity of our designers have not yet realized the solution. Industrial ecology concepts highlight the brilliance in nature of eliminating the concept of waste. Within industrial ecology there is the recognition that when one organism or natural system generates a waste product, a substance not of use to the organism or system, another organism has evolved, that uses that waste as a feedstock or some other useful purpose. It has taken nature billions of years to evolve to the balanced cycles we have now and we have the advantage of using nature as model for our own engineered systems.

Our current mechanisms of dealing with waste call for:

- Large scale disposal often to land and water with many of our most toxic substances requiring landfill liners to keep the hazardous substances from contaminating the soil and the groundwater.
- These sites often require monitors to be installed that provide continuous information on the status of the landfill.
- In due course as a function of probability there are leaks detected in which contamination to the soil or groundwater has taken place.
- Remediation either through on site clean-up or removing megaton quantities of soil to be stored in containers or burned then takes place.
- In the case where storage is the option, the material once again has continuous monitoring installed.

A system that creates the quantities and nature of waste as we do today can best be summed up in two words, design flaw. From the molecular level, the physical chemical property of hazard is a design flaw. From a process operations perspective, a process that does not fully utilize the material that it takes in and generates waste is a design flaw. From a systems perspective, the non-integration of our processes such that there is not the conservation of materials and energy flows from one process to another that maintains value and utility is a design flaw. Principle 2 as well as others in the framework allow the engineer/ designer to move toward elimination of these flaws through elegant engineering design.

Principle 3. Separation and purification operations should be a component of the design framework.

Virtually, every current manufacturing process will require a separation and purification step. It may be as complex as extracting trace quantities of pharmaceutically active chemicals from a rare plant or as straightforward as cleaning off waste photoresist from a silicon chip. In every case, however, this separation and purifications step requires time, material, and energy that is otherwise not intrinsically necessary. Product separation and purification is one of the single largest consumers of energy and material in many manufacturing processes. The complexity and energy consumption of a separation operation depends significantly on the degree of dispersion of the components being separated. If substances are mixed on the atomic level and the concentration is low, energy consumption increases making the separation process increasingly energetic and/or materially intensive. Whenever product separations and purification can be minimized through appropriate upfront design, there can be significant savings in material, energy, and time.

For recycle and reuse strategies to be cost effective and practical, the ability to separate materials quickly and accurately is essential. Specific examples of designs that may hinder separation are:

• glass container with twist off metal cap that leaves ring of metal;
• metal coatings on plastic;
• multi-material systems that are permanently bonded such as automobile dash boards composed of plastic, metal and wood,
• irreversible fasteners such as rivets, chemical bonds and welds.

Some elegant design solutions have illustrated how through applying a Green Engineering perspective products, processes and systems can be designed for self-separation or ease of separation.

• Systems where the catalyst separates from the product at the end of a reaction so that it can be reused and does not create impurities in the product [13].
• Products that have self-cleaning surfaces by mimicking the water-repellent leaves of many plants. This is often referred as the "lotus effect." Recent advances in nanotechnology enables the application of extremely fine micro-structured finishes to materials to imitate the surface of lotus leaves. Thus far, the lotus effect has been explored for house paints, tiles, glass panes and plastic sheets [14].
• Using fasteners such as screws or heat-activated reversible fasteners that allow for the mechanical separation of materials without damage or contamination.

Principle 4. System components should be designed to maximize mass, energy and temporal efficiency.

While efficiencies of all types have always been a component of good design, the understanding of what can be accomplished continues to evolve. Historically chemical engineering may have asked how to optimize a batch reaction for a particular step in a manufacturing process for example. The products of that batch may be isolated and used in the next batch reaction in the future or the products may be transported to another facility or customer for further transformation. In our major companies today, we have chemical manufacturing sites, the size of entire cities in themselves in places like Ludwigshafen, Midland and the Chamberworks. Now there continues to be an evolution that moves our understanding of efficiency to continuous process and to process intensification. Here there are envisioned and realized transformations taking place at or near the individual molecule level where "mixing" is accomplished through micro-fluidic channels and nano-chambers. In these systems, multiple cascading reactions in self-assembling systems are now conceivable for a wide array of products.

Perhaps the most visible and successful embracing of this principle through the years has been in electronics products. Computers the size of large rooms a few decades ago now fit into the palm of a hand or less. With Moore's Law (Fig. 1) outlining efficiency growth in productivity holding true for many years now, there are certainly lessons to be learned from the electronic sector.

However, it is important to remember the overarching caveat that we need to take a lifecycle perspective. While the performance of the electronic products accomplish the goal of mass, energy and time efficiency – it is clear that the manufacturing processes does not – as referred to earlier with regard to the 1.7 kg chip. In addition, the systems perspective would also illustrate that the same increases and product efficiencies are driving the consumption of new units (computers, monitors, etc.) when little more than a new CPU is required making the overall system material and energy inefficient.

Principle 5. System components should be output pulled rather than input pushed through the use of energy and materials. (Le Chatlier's Principle).

A principle by a great French chemist is Le Chatlier as stated as:

> "If a system in equilibrium is subjected to a stress the equilibrium will shift in the direction which tends to relieve that stress."

In other words, when applied to chemical systems, if reactants are added to the system, the "balance" will be lost temporarily. The balance can be recovered

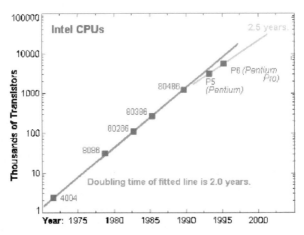

Fig. 1. Graph demonstrating Moore's Law. Taken from http://www.physics.udel.edu/wwwusers/Watson/scen103/intel-new.gif. [15].

when the system forms more product to bring back an equilibrium condition. In this way, the additional reactants are converted to form additional products, restoring equilibrium. This becomes an exceptionally useful and powerful mechanism in trying to drive a reaction to generate the optimal amount of product. In the case of the reaction shown below in equilibrium, one could try to shift the equilibrium toward products by adding a greater quantity of reactants such as A or B.

$$A + B \Rightarrow C + D$$

This approach is effective in achieving the desired outcome and is commonly done in making chemicals. However, this approach frequently has the undesired consequence of using more of a reactant (e.g. A), than is needed to make the product and therefore it ends up as waste.

$$\underset{(1)\uparrow}{A} + \underset{(2)\rightarrow}{B} \Rightarrow C + D$$

An alternative approach for exploiting LeChatlier's principle is to shift the equilibrium by pulling the reaction rather than driving it. In this case product would be removed from the system as soon as it is formed.

$$A + B \Rightarrow \underset{(2)\rightarrow}{C} + \underset{(1)\downarrow}{D}$$

In this case, one would have succeeded in generating increased product without consuming unnecessary material and generating unusable waste. This theory is being put into practice in such places as biological fermentation processes to make lactic acid (Fig. 2). In traditional batch reactors for fermentation, the products build-up and inhibit the ongoing reaction. Continuous fermentation is a process where the products are continually removed from the process preventing the product inhibition and pulling the reaction to completion.

While LeChartlier's Principle has been borne out in chemical and biochemical systems, it is posited that the beauty of this principle is also a useful design tool generally for a wide range of other processes and systems. The concept of Just-in-Time (JIT) manufacturing can be understood as an example of this principle. In JIT manufacturing, production is based on responding to real-time demand wherever feasible. By doing this, it is possible to greatly eliminate waste due to overproduction and lowers warehousing costs. In this system, supplies are closely monitored and quickly altered to meet changing demands smaller and more accurate resupply deliveries are made just when they are needed. This manufacturing approach also reduces the resource wastes associated with manufacturing and stocking products that will become obsolete on the shelf. This obsolescence may be the result of rapid technical advances but it is more likely the result of shifts in societal desires for particular aesthetic qualities such as color, size, and weight as well as "upgrading" to the newest gadget. A prime

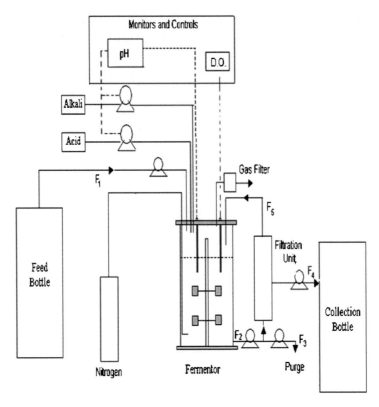

Fig. 2. Schematic diagram of continuous lactic acid fermentation by applying of Le Chatlier's Principle F1: feed rate (the dilution rate was maintained by maintaining the feed rate); F2: rate at which broth was taken out of the fermentor; F3: rate of purging (the sample from the purge were regularly analyzed for biomass and metabolites concentrations inside the fermentor); F4: rate at which permeate was taken out from the filtration unit; F5: rate at which biomass was recycled back to the fermentor. Taken from Narayanan et al. [16].

example of this is the rapid introduction of new cellular phone models. In the US, cell phone use has surged from 340,000 subscribers in 1985 to over 175 million today. The average life span of a cell phone is 18 months. It is estimated that 100 million cell phones will become obsolete and discarded each year in the US by 2005 and over 500 million cell phones will be stockpiled in US homes [17].

This approach can be used not only on the molecular scale and the manufacturing scale but can be understood in everyday terms as well. Elevators generally come to a floor when a person pressing the call button summons them. Having elevators constantly stopping on every floor opening and closing doors day and night would accomplish a goal of allowing people to get on and off but no one would argue that it is the most effective way to accomplish the goal. In light of this, one now sees that escalators are being redesigned such that they begin moving when a person approaches the escalator rather than being in constant motion.

Wherever there is a system in equilibrium, whether it is material equilibrium, energy equilibrium, gravitational equilibrium, etc. a design choice is available to take advantage of LeChatlier's Principle to minimize the resource inputs necessary to generate the desired outcome.

Principle 6. Embedded entropy and complexity must be viewed as an investment when making design choices on recycle, reuse or beneficial disposition.

As there is a growing realization, especially in the sustainable design community, that materials need to be designed for recycle, reuse and ultimately beneficial disposition we can no longer view all of the products we make as being equal and warranting the same treatment. In other words, while the public may believe that recycling is an environmentally friendly practice, it often has not been the most strategically wise option for a variety of reasons. There are numerous stories of municipal or campus recycling programs where the recycled materials wind up in a landfill.

The fact is that a brown paper bag, newspaper or corrugated box does not have the same level of material complexity as a computer or a television. The amount of complexity built into a product whether at the macro, micro, or molecular scale is usually a function of resource expenditures. Based on what levels of energy, time, and material has been invested, one can make generalizations about the proper way to handle the material.

- High complexity, low entropy – reuse
- Lower complexity – value-conserving recycling where possible or beneficial disposition.

Natural systems can also be recognized as having complexity. While the energy, material and time invested may belong to nature, it does not generally change the equation. Many biological feedstocks for chemicals and materials have high functionality and complex stereochemistry. Losing this complexity for lower value use such as "claiming its energy value" by burning it as fuel, may not be the wisest or optimal choice available. In addition, moving to biobased feedstocks lessens dependence on petroleum-based raw materials and uses renewable, non-depleting, feedstocks that prevents emissions of carbon dioxide into the atmosphere.

There are many examples of demonstrated success in using biobased, renewable feedstocks for chemicals, fuels, and materials. One of the more unusual innovations to date is the development of computer chips made from chicken feathers and plant oils. This new generation of computer microchips aims to replace silicon-based chips given that the resistance of electrical signals in the hollow tubes of chicken feathers is less than that of silicon. Preliminary performance tests demonstrate that electrical signals move twice as quickly through the chicken-feather chip as through a conventional silicon chip [18]. While it is unclear if this technology will prove to be commercially successful, it demonstrates the effectiveness of utilizing biobased, renewable resources, in this case a waste product, to make high-tech, high-value products with little resource or energy input by leveraging the existing natural characteristics.

Principle 7. Targeted durability, not immortality, should be a design goal.

There is always an important element of product design that strives to make products as robust as needed to withstand conditions that may be encountered. This has often resulted in unadvisable over-design where products persist and accumulate in an undesirable way. Products that last well beyond their useful commercial life often result in environmental problems ranging from solid waste to persistence and bioaccumulation. Several strategies to augment this include ease of repair and maintenance. However, it is important to consider the life cycle implications of these life-extending strategies as well. In other words, the designer should strive to balance targeted lifetime with durability and robustness under anticipated operating conditions.

There are numerous examples of products that have been designed for a short-term purpose but with a long-term lifetime. An example of products designed with a significantly longer lifetime than their intended purpose would suggest are single-use disposable diapers. These diapers typically consist of several materials, including non-biodegradable polymers, and represent the single largest non-recyclable fraction of municipal solid waste [19]. Similarly, packaging represents roughly one-third of municipal waste in the US [20]. However, packaging plays a vital role in our society protecting the integrity, cleanliness, and freshness of goods, allowing for long distance shipping while reducing waste and spoilage. Given the benefits of packaging in today's society, this suggests the need to design packaging that meets these needs and then at the end of its useful life does not pose lasting environmental problems.

Recent studies suggest that the quantity of packaging is no longer growing faster than the output of goods in the US. However, the slowdown is due to reduction in the weight of individual packages, not a reduction in the number of goods that are packaged [20]. While this would suggest that packaging would continue to play a significant role in the economy of goods, innovative packaging solutions are being explored with success. For instance, NatureWorks [21] developed a bio-based packaging material that is derived from polylactic acid and degrades completely, leaving only carbon dioxide, water and residues compatible with soil. Another example of biobased packing that has been successfully marketed is biodegradable, starch-based pellets, and foams used as protective packing material. These products, such as Eco-Fill, consist of food-grade inputs and can be readily dissolved in domestic/industrial water systems at the product's end of life [22]. These have been shown to be commercially competitive with traditional polystyrene packing for many applications including being used as sorbents in disposable diapers [23]. These materials demonstrate that designing for durability does not necessarily lead to a product that is persistent beyond its useful lifetime.

Principle 8. Design for unnecessary capacity or capability should be considered a design flaw. This includes engineering "one size fits all" solutions.

While product agility and product flexibility can be desirable, the cost in terms of materials and energy for unusable capacity and capability can be high. There is also a tendency to design for the worst-case scenario such that the same product or process can be utilized regardless of spatial or temporal scenarios. That is, the product, process, or system is designed for the maximum imaginable conditions, regardless of the fact that those conditions may be highly unlikely or even unrealistic in the vast majority of situations that the product, process, or system will operate. When safety is an issue, this approach can at least be rationalized.

However, overdesign of this type exists in many consumer products in the US given the vast differences of conditions within the country. For example, laundry detergent formulations are designed to perform anywhere in the US regardless of hard water conditions. As such, the formulations must be designed to work in the most extreme hard water conditions found in the Midwestern US (see Fig. 3) even in a formulation being sold in the Northeast. While it may be difficult to design regional laundry detergent formulations and the logistics of distribution may be more complex, this approach leads to significantly higher use, waste, and disposal of chemicals to remove hard water in the wastewater management system across the nation. Of even greater significance is that the builders used to remove water hardness in common laundry detergents were phosphates.

Phosphates have high nutrient value and can lead to eutrophication in water bodies when added in high enough quantities. Until recently, laundry detergents were formulated with enough phosphates to ensure performance in the most severe hard water conditions leading to serious strain on health and diversity of the Nation's water bodies. Accordingly, about half the states in the US, and much of Europe, have limited or banned the use of phosphates in laundry detergents [24]. However, in most areas of the US, phosphates are still allowed in laundry detergents for institutional or commercial laundries and phosphates are still allowed in dishwasher detergents throughout the country [24]. By overdesigning these products to meet performance criteria under the harshest of conditions, a significant amount of resources, including material, energy, and money, is expended for capacity that is unnecessary in the majority of scenarios.

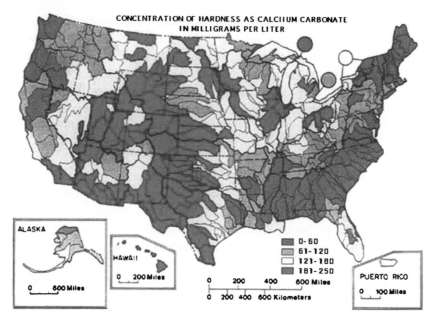

Fig. 3. Mean hardness as calcium carbonate in milligrams per liter of NASQAN stations as last recorded in 1975. Taken from http://water.usgs.gov/owq/map1.jpeg. [25].

This unnecessary capability has impacts throughout the lifecycle whether in consuming more energy in production, transporting additional weight, or disposing of unused materials. By designing to meet local rather than universal conditions, it is likely that performance goals can be met at reduced investments of resources since the product, process or system can be optimized, specifically, to those conditions.

Principle 9. Multi-component products should strive for material unification to promote disassembly and value retention (minimize material diversity).

Multiple materials are often used to obtain the desired properties of the final product such as durability and aesthetics. However, the design of products with minimum components and minimum number of materials, including reducing or eliminating the use of adhesives, coatings, and sealants, has impacts throughout the life cycle. For example, the utilization of multiple materials in a single design requires additional materials for the joining or fastening of all of the individual components. During the use phase, multiple material components can lead to issues of leaching and contamination. Finally, at end of life, a variety of materials can lead to difficulty in separation and recycling.

In the production phase, products with multiple materials often require adhesives to join the materials, sealants to maintain the integrity of a joint, or coating one material with another for a desired surface finish. Conventional adhesives, sealants, and coatings pose significant environment impacts since they are based organic solvents and release volatile organic compounds during the application and use phase. By enhancing the integration and unification of materials, the need to use auxiliary substances is significantly minimized, thereby reducing the overall environmental impact of the product.

By not integrating the desired properties into a single material and relying on additional substances, there is the potential for these secondary materials to leach from the product.

Within individual plastics there are various chemical additives, including thermal stabilizers, plasticizers, dyes, and flame-retardants. These chemicals can leach from the plastic both during the use phase and at end of life. One class of this type of chemical is brominated flame-retardants that are commonly used in electronic products as a means for reducing flammability. Brominated flame-retardants are released to the environment increasing human exposure through landfill leachate and incineration where they are concentrated in the food chain. Various scientific observations indicate that polybrominated diphenylethers (PBDE) might act as endocrine disruptors [26] and further studies have demonstrated that the levels of PBDEs in human breast milk are doubling every 5 years [26]. As a means to address these issues, the desired functionality that would be accomplished through the additive is being integrated into the polymer backbone [27]. Tailoring polymer properties can have a positive environmental effect in cases where leaching of additives may be an issue and ease of recycling is important.

Material diversity also becomes an issue when considering end-of-useful-life decisions, which determines the ease of disassembly for reuse and recycle. Options for final disposition are increased through up-front designs that minimize material diversity yet accomplish the needed functions. Selected automobile designers are reducing the number of plastics by developing different forms of polymers to have new material characteristics that improve ease of disassembly and recyclability. This technology is currently applied to the design of multilayer components, such as door and instrument panels. For example, components can be produced using a single material, such as metallocene polyolefins, that are engineered to have the various and necessary design properties. Through the use of this monomaterial design strategy, the need to disassemble the door or instrument panel for recovery and recycling is minimized (Fig. 4) [28].

Principle 10. Design of processes and systems must include integration and interconnectivity with available energy and material flows.

It would be unwise to view the design of any product, process or system with the assumption that it was operating within a vacuum. In other words, there needs to be a conscious recognition by the engineer that the design is being

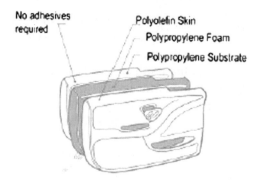

Fig. 4. Schematic diagram of a door panel made entirely out of metallocene polyolefins Taken from McAuley [28].

implemented within a context of material and energy flows. By utilizing readily available material and energy and integrating them into the process or system, the green engineer can increase overall system efficiency, lessen the costs and environmental impact of the material and energy. This ensures that these resource inputs are being used at their highest value and do not end up as waste products. In addition, through design for integration and interconnectivity, there is also the ability to minimize the need for long range transport of materials and energy which otherwise would result in environmental and economic costs and potential exposures.

There are numerous examples of integration of material and energy flows within facilities such as chemical plants where heat exchangers are routinely implemented and waste is recovered and recycled in process as a feedstock. However, there is perhaps no better current example of large scale cross-process design for integration of material and energy flows than that of Kolunborg, Denmark. This city represents the manifestation of industrial ecology whereby whole industries and commercial applications are interconnected such that the local materials that may be considered a waste of one process becomes a value-added feedstock for a nearby process. The same is true with the energy flows that are shared beneficially between various industrial and residential sectors of the community (Fig. 5).

The example of Principle 10 illustrated on the municipal level shows the scalability of Green Engineering design from small products and processes to much larger and complex systems.

Principle 11. Performance metrics include designing for performance in commercial "after-life".

If the manufacture of a product is benign to human health and the environment and the product itself is equally innocuous, there remains a need to ensure that once the commercial purpose is concluded that this product does not have

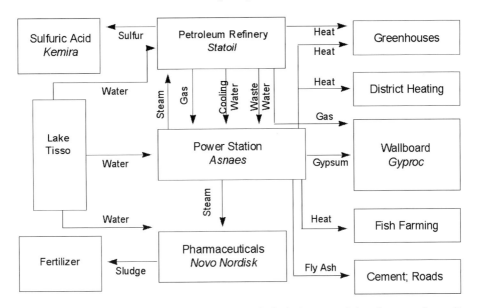

Fig. 5. Schematic diagram of Kalundborg, Denmark depicting material and energy flows. From Wernick and Ausubel [29].

negative environmental impacts at the end of its useful life. These less than desirable consequences can include contributions to solid waste, leaching of hazards to the environment, persistence and bioaccumulation. One classic illustration would be the case of chlorofluorocarbons (CFCs) as refrigerants and propellants. While these substances were recognized as being safer from the perspective of accident potential and widely understood as non-toxic, the problems that we now understand related to their ability to deplete stratospheric ozone were manifested in the life-cycle stage subsequent to commercial after-life. It is for reasons analogous to the CFCs case that green engineering recognizes the need for upfront considerations of recycle, reuse, and beneficial disposition.

Designing for commercial "after-life" can include strategies such as modular, upgradeable platforms. In these cases, a design does not have to be replaced or retired simply because a component is now obsolete. By allowing for a product or process to be upgraded through the component substitution, the useful life can be extended and future performance standards can be met. By prolonging the useful life and meeting or exceeding performance demands, the replacement rate of the entire product, process or system can be significantly reduced and the environmental impact minimized to that of a single component rather than the whole. This requires upfront design considerations since the potential performance and development of upgraded technology is uncertain. Designing such that these improvements can be easily and readily implemented suggests that there are environmental benefits throughout the life cycle.

Principle 12. Design should be based on renewable and readily available inputs throughout the life-cycle.

The nature of the origin of the materials and energy inputs can be a major influence on the sustainability of products, processes, and systems. Whether a substance or energy source is renewable or depleting can have far-reaching effects. Every unit of finite substance used in a consumptive manner incrementally moves the supply of that substance toward depletion. Certainly, from a definitional standpoint, this is not sustainable. In addition, because virgin substances require repetitive extractive processes, using depleting resources causes ongoing environmental damage. Biological materials are often cited as renewables. However, if a waste product from a process can be recovered and used as an alternative feedstock or recyclable input while retaining its value, this would certainly be considered renewable from a sustainability standpoint.

Similarly, whether material and energy inputs are available and can be assimilated locally can impact the environmental, economic, and social impacts of a design. By utilizing local materials in the design of products, processes, and systems, the designs can be more easily maintained and repaired as well as having a greater likelihood of long-term cultural acceptance. Working with locally available materials will also lower the costs, decrease the likelihood of supply disruptions, and may provide an opportunity for local economic growth.

One example of the development and application of appropriate technology is the "pot-in-pot" technology developed by Bah Abba from Nigeria [30]. Northern Nigeria is an impoverished region where people in rural communities eke out a living from subsistence farming. With no electricity, and therefore no refrigeration, perishable foods spoil within days. Such spoilage causes disease and loss of income for needy farmers who are forced to sell their produce daily. Nigerian teacher Mohammed Bah Abba motivated by his concern for the rural poor and by his interest in indigenous African technology sought practical, local solution to these problems. His extremely simple and inexpensive earthenware "Pot-in-Pot" cooling device is starting to revolutionize lives in this semi-desert area.

The innovative cooling system developed in 1995 consists of two earthenware pots of different diameters, one placed inside the other. The space between the two pots is filled with wet sand that is kept constantly moist, thereby keeping both pots damp. Fruit, vegetables and other items are put in the smaller inner pot, which is covered with a damp cloth and left in a very dry, ventilated place. The water contained in the sand between the two pots evaporates toward the outer surface of the larger pot where the drier outside air is circulating. The evaporation process causes a drop in temperature of several degrees, cooling the inner container. This design is now used to maintain the food supply, provide cool drinking water, and store vaccines and other medicines. As this example demonstrates, designs to meet the challenges of the developing world do not require the use of high-technology, energy and material intensive, expensive solutions. In fact, this example of appropriate technology demonstrates the effectiveness of simple designs that can be developed, manufactured, and sold locally using readily available materials for the benefit of the environment, economy, and society.

3. Conclusion

The nature of the design of products, processes and systems ranging from the smallest molecule to the largest metropolis will have major consequences for the future of humans and the biosphere. The Twelve Principles of Green Engineering is a tool for asking the right questions, considering the right factors, and building in the right parameters as design criteria. The examples presented in this chapter, while examples of notable and important innovations, represent merely a sampling of our present day accomplishments that need to be far exceeded by those designers that follow us in future generations. We should celebrate the fact that with increasing rapidity our creativity will be surpassed by even greater discoveries.

Without question, design for sustainability is one of today's most complex technical undertakings being engaged by engineers of all types. The Twelve Principles of Green Engineering will have unavoidable trade-offs in our designs that must be evaluated and will also exhibit beneficial synergies that must be understood and realized. All of these designs will be taking place in context of dynamic population, economies, scientific understanding, and changing ecosystems. As the old adage says, if you don't know where you are going, any road will take you there. The Twelve Principles provide a compass for ensuring that by taking certain engineering steps, while thinking systemically and with a consideration of life-cycle impacts, you will be moving toward a more sustainable outcome.

References

[1] P.T. Anastas, J.B. Zimmerman, Environ. Sci. Technol. 37 (2003) 94A–101A.
[2] Office of Management and Budget, Office of Information and Regulatory Affairs, 2003 Report to Congress on the Costs and Benefits of Federal Regulations Washington, DC, 2003.
[3] Report of The President's Commission on The Accident at Three Mile Island, October 1979.
[4] Report of the Presidential Commission on the Space Shuttle Challenger Accident, June 1986.
[5] Biotechnology Industry Organization, New Biotech Tools for a Cleaner Environment, June 2004.
[6] Cargill Dow, NatureWorks Technical Data Sheet, www.cargilldow.com.

[7] M. Henneberry, Paint and Coatings Industrial Magazine, 6 (2002).
[8] J.M. Symons, T.A. Bellar, J.K. Carsewell, J. DeMarco, K.L. Kropp, G.G. Robeck, B.L. Smith, A.A. Stevens, J. Am. Water Works Ass., 67 (1975) 634–647.
[9] J.J. Rook, Water Treatment Exam., 23 (2) (1974) 234–243.
[10] US Environmental Protection Agency, Integrated Risk Information System (IRIS). Office of Research and Development, 1997.
[11] E.D. Williams, R.U. Ayers, M. Heller, Environ. Sci. Technol., 36 (2002) 5504–5510.
[12] R.A. Sheldon, Chem. Ind., 1992, 903.
[13] M. Bullock, V. Dioumaev, Nature, 424 (2003) 530–532.
[14] H.C. Von Baeyer, *The Sciences*, 1 (2000) 12–15.
[15] http://www.physics.udel.edu/wwwusers/watson/scen103/intel-new.gif.
[16] N. Narayanan, P. Roychoudhury, A. Srivastava, Electron. J. Biotechnol., 7 (2) (2004) 167–179.
[17] J. Sidener, San Diego Tribune, May 17, 2004.
[18] L. Jacobson, Can Computers Fly on the Wings of a Chicken?, Special to the Washington Post, July 8, 2002, A07.
[19] Franklin Associates, Characterization of Municipal Solid Waste in the United States: 1998 Update, USEPA Office of Solid Waste report No. EPA530, July 1999.
[20] F. Ackerman, J. Soc. Philos. Technol., 2 (2) (1997) 1–7.
[21] www.natureworksllc.com.
[22] C. Green, AURI Agriculture Innovation News, 8, 4 (1999).
[23] Commission on Life Sciences, National Academies of Science, Biobased Industrial Products: Research and Commercialization Priorities, 2000.
[24] Water Science and Technology Board, National Research Council, Confronting the Nation's Water Problems: The Role of Research, 2004.
[25] http://water.usgs.gov/owq/map1.jpeg.
[26] K. Hooper, T. McDonald, Environ. Health Persp., 108 (5) (2000) 387–392.
[27] K. Matyjaszewski, Macromol.Symp. 152 (2000) 29–42.
[28] J. McAuley, Environmental Issues Impacting Future Growth and Recovery of Polypropylene in Automotive Design. In: Proceedings from Society of Plastics Engineers, Dearborn, MI, 1999, www.plasticsresource.com/recycling/ARC99/Mcauley.htm.
[29] I. Wernick, J. Ausubel, Industrial Ecology: Some Directions for Research prepared for the Office of Energy and Environmental Systems, Lawrence Livermore National Laboratory, ISBN 0-9646419-0-7 1997.
[30] N. Lubick, Desert Fridge, Scientific American, 1 (November) (2000).

Sustainability Science and Engineering: Defining principles
Martin A. Abraham (Editor)
DOI 10.1016/S1871-2711(05)01003-2

Chapter 3

Ethics of Green Engineering

P. Aarne Vesilind[a], Lauren Heine[b], Jamie R. Hendry[c],
Susan A. Hamill[d]

[a]*R. L. Rooke Professor of Engineering, Department of Civil and Environmental
Engineering, Bucknell University, Lewisburg, PA 17837, USA*
[b]*Director of Applied Science, GreenBlue Institute, 600 E. Water St., Charlottesville, VA
22902, USA*
[c]*Assistant Professor, Department of Management, Bucknell University, Lewisburg, PA
17835, USA*
[d]*Founder, Sustainable Labs, 5440 SW Buddington St., Portland, OR 97219, USA*

The green engineering revolution is today a visible component of the everyday operations of many manufacturing, service, and professional firms. There is, however, a great deal of confusion about what constitutes green engineering, and more importantly, there is widespread confusion as to when corporate and engineering green activities can be considered to be morally admirable. Doing the right thing for the wrong reason is a fortunate accident, but such actions have no moral worth. In this chapter, our objective is to clarify when, if ever, with regards to green engineering, an organization and its leaders behave morally for the *right* reason; i.e. they choose to do the right thing not because it is beneficial to the survival and profitability of the firm, but because they believe that doing the right thing will benefit the greater society and future generations.

We begin by describing the historical and political initiatives that have emerged to create the concept of green engineering and then use examples to show how most corporate green engineering programs are often simply self-serving actions that also happen to be beneficial in terms of global environmental pollution and resource use. Finally, we argue that true moral worth comes when a corporation decides to practice green engineering because doing so benefits others, and not because it is the profitable option.

1. Green engineering

The operational quest for sustainability is defined as green engineering, a term that recognizes that engineers are central to the practical application of the principles of sustainability to everyday life. The relationship between sustainable development, sustainability, and green engineering is shown below:

Sustainable   →   Green   →   Sustainability
Development   Engineering

Sustainable development is an ideal that can lead to sustainability, but this can only be done through green engineering.

Green engineering is still a nascent field and the underlying principles of green engineering are fluid. Most people are fairly confident that science and technology can support sustainability through the development of sustainable materials and processes, but we also recognize that technology is not a panacea for all global problems and that a higher level of technology will not guarantee sustainability. Technology can, however, provide innovative solutions when society challenges scientists and engineers to promote human and ecological well-being and to embrace systems concepts in their development activities. According to Grübler [1], "Through technology, humanity has for 300 years increasingly liberated itself from the environment. The job is not yet complete as billions of people continue to be excluded from the benefits of technology. The next immediate task is to assure their inclusion". Grübler argues that the task ahead is to progressively liberate the environment from adverse human interference.

> For this century-long journey, we will need more technology, not less. As the world's population grows to 10 billion or more, we recognize that nature cannot be shielded perfectly from human intervention any more than we can be shielded perfectly from nature. But technological change can relax our grip and lighten our tread on the natural world. Our choices are constrained by what already exists and the environmental legacy of the past. But over the long term the capacity for social choices to shape technology is endless.

Green engineering is emerging as an umbrella term that engages engineers from diverse disciplines to pave the way for sustainable co-existence between humans and the rest of the natural world. The goal therefore is sustainability, and green engineering is the way to achieve this goal.

2. Motivations for practicing green engineering

In order to understand the reasons why humans behave as they do, one must identify the driving forces that lead to particular activities. The concept of the driving force can also be used to explain engineering processes. For example, in gas transfer the driving force is the difference in concentrations of a particular gas on either side of an interface. We express the rate of this transfer mathematically as $dM/dt = k(\Delta C)$, where M is mass, t time, k a proportionality constant, and ΔC the difference in concentrations on either side of the interface. The rate at which the gas moves across the interface is thus directly proportional to the difference in concentrations. If ΔC approaches zero, the rate drops until no net transfer occurs. The driving force is therefore ΔC, the difference in concentrations.

Similarly, in engineering and business alike, driving forces spur the adoption of new technologies or practices. The objective here is to understand what these motivational forces are for adopting green engineering and business practices. We propose that the three diving forces supporting green engineering are legal considerations, financial considerations, and finally ethical considerations.

2.1. Legal considerations

At the simplest and most basic level, green engineering is practiced in order to comply with the law. For example, a supermarket recycles corrugated cardboard because it is the law — either a state law such as in North Carolina, or a local ordinance as in Pennsylvania. This behavior is, at best, "morality lite". Engineers and managers comply with the law because of the threat of punishment for non-compliance. The decision to comply with the law is thus largely an amoral decision. Complying with the law is not morally good nor morally bad, although not complying may be considered morally bad. So in this situation, managers and engineers choose to do "the right thing," not because it is the right thing to do — but simply because they feel it is their only choice.

We should point out that the vast majority of firms will comply with the law regardless of the financial consequences. They will not even bother to conduct a cost-benefit analysis because they assume that breaking the law is not worth the cost.

Occasionally, however, firms may prioritize financial concerns over legal concerns and the managers may determine that by adopting an illegal practice (or failing to adopt a practice codified in law) they can enhance profitability. In such cases they argue that either the chances of getting caught are low, or that the potential for profit is large enough to override the penalty if they do get caught.

For example, a concrete producer in Vermont typically overloads its trucks, exceeding the weigh limit set for the highways. Occasionally they get caught, but the fine is not enough to deter the practice. As one executive put it, "We simply build the cost of the fines into the cost of operations".

A more significant example is the violation of air emission requirements by power plants. In November 1999, the US Environmental Protection Agency sued seven electric utility companies — American Electric Power, Cinergy, FirstEnergy, Illinois Power, Southern Indiana Gas & Electric Company, Southern Company, Tampa Electric Company — for violating "the Clean Air Act by making major modifications to many of their coal burning plants without installing the equipment required to control smog, acid rain and soot" [2]. On August 7, 2003, "Judge Edmund Sargus of the US District Court for the Southern District of Ohio found that Ohio Edison, an affiliate of FirstEnergy Corp., violated the Clean Air Act's NSR provisions by undertaking 11 construction projects at one of its coal-fired plants from 1984 to 1998 without obtaining necessary air pollution permits and installing modern pollution controls on the facility" [3]. Given the number of violations, it seems obvious that the companies had calculated that breaking the law and possibly getting caught was the least cost solution and thus behaving illegally was "the right answer".

In some cases, private firms can take advantage of loopholes in tax law that inadvertently allows companies to pretend to be environmentally green while in reality doing nothing but gouging the taxpayer. An example of this is the great synfuel scam [4]. In the 1970s, the US Congress decided to promote the use of cleaner fuels in order to take advantage of both the huge coal reserves in the United States and the environmental benefits derived from burning a clean gaseous fossil fuel made from coal. Producing such synfuel from coal had already been successfully implemented in Canada and the United States government wanted to encourage our power companies to get into the synfuel business. In order to promote this industry Congress wrote in huge tax credits for companies that would produce synfuel and defined a synfuel as chemically altered coal, anticipating that the conversion would be into a combustible gas that could be used much as natural gas is used today.

Unfortunately, the synfuel industry in the United States did not develop as expected because cheaper natural gas supplies became available. The synfuel tax credit idea remained dormant until the 1990s when a number of corporations (including some giants like the Marriott hotel chain) found the tax break and went into the synfuel business. Since the only requirement was to change the chemical nature of the fuel, it became evident that even spraying the coal with diesel oil or pine tar would alter the fuel chemistry and that this fuel would then be legally classified as a synfuel. The product of these synfuel plants was still coal, and more expensive coal than raw coal at that, but the tax credits were enormous. Some companies formed specifically to take advantage of the tax break, often with environmentally attractive names like Earthco, made huge profits by selling their tax credits to other corporations that needed them. The synfuels industry presently is receiving a gift from the US taxpayer of over $1 billion annually, while doing nothing illegal, but also while doing nothing to benefit the environment.

2.2. Financial considerations

Decisions about the adoption of green practices are also driven by financial concerns. This level of involvement with "greening" is at the level promoted by the economist Milton Friedman, who stated famously, "The one and only social responsibility of business [is] to use its resources and engage in activities designed to increase its profits so long as it … engages in open and free competition, without deception or fraud" [5]. In line with this stance, the firm calculates the financial costs and benefits of adopting a particular practice and makes its decision based on whether the benefits outweigh the costs or *vice versa*.

Many companies seek out green engineering opportunities solely on the basis of their providing a means of lowering expenses, thereby increasing profitability. Here are some examples [6]:

- In one of its facilities at Deepwater, New Jersey, Dupont uses phosgene, an extremely hazardous gas. The gas was shipped to the plant in tankers. In an effort to reduce the chance of accidents, DuPont redesigned the plant to produce phosgene on site and to use almost all of it in the manufacturing process, avoiding costs associated with hazardous gas transport and disposal.
- Polaroid did a study of all of the materials it used in manufacturing and grouped them into five categories based on risk and toxicity. Managers were encouraged to alter product lines in order to reduce the amount of material in the most toxic groups. In the first 5 years, the program resulted in a reduction of 37% of the most toxic chemicals, and saved over $19 million in money not spent on waste disposal.
- Dow Chemical challenged its subsidiaries in Louisiana to reduce energy use, and sought ideas on how this should be done. Following up on the best ideas, Dow invested $1.7 million, and received a 173% return on its investment.

Other firms may believe that adopting a particular green engineering technology will provide them with public relations opportunities: green engineering is a useful tool for enhancing the company's reputation and community standing. If the result is likely to be an increase in sales, and if sales are projected to rise *more* than expenses so that profits rise, the firm is likely to adopt such a technology. The same is true if the public relations opportunities can be exploited to provide the firm with expense reductions, such as decreased enforcement penalties or tax liabilities. Similarly, green technologies that not only yield increased sales but at the same time decrease expenses are the perfect recipes for the adoption of green practices by a company whose primary driving forces are financial concerns. For instance:

- Dupont's well-publicized decision to discontinue its $750 million a year business producing chlorofluorocarbons (CFCs) was a public relations bonanza. Not only did DuPont make it politically possible for the United States to become a signatory to the Montreal Protocol on ozone depletion, but it already had alternative refrigerants in the production stage and were able to smoothly transit to these. In 1990, the US Environmental Protection Agency gave DuPont the Stratospheric Protection Award in recognition of their decision to get out of CFC manufacturing [7]. The fact that the decision also proved to be highly profitable for DuPont apparently did not matter to the judges.
- The seven electric companies sued by the EPA in November 1999 for Clean Air Act violations (mentioned above) heavily publicized their efforts to reduce greenhouse gas emissions. For example, American Electric Power (AEP) issued news releases on May 8, June 11, November 21, 2002 regarding emissions reduction efforts at various plants. Not coincidentally, the US government, which had sued during the Clinton administration but which was now operating under the Bush administration, was in the process of revising the portions of the Clean Air Act which the company had violated. Presumably, regulators were favorably impressed with the company's hard work; in August 2003, the EPA announced that it was dropping the suit and revamping that portion of the Act.

These examples clearly demonstrate bottom-line thinking: cases in which managers were simply trying to practice "good business," seeking ways to increase the difference between revenues and expenses so that profits would rise. As far as we can tell, these decisions were not influenced by the desire to "do the right thing" for the environment. Here we again have examples of amoral decision making. Businesses are organized around the idea that they will either make money or cease to survive. In the "financial concerns" illustrations, green practices were adopted as a means of making more money.

On occasion, though, managers are *forced* into considering the adoption of greener practices by the threat that not doing so will cause expenses to rise and/ or revenues to fall. For example, in October 1998, ELF (Earth Liberation Front) targeted Vail Ski Resort, burning a $12 million expansion project to the ground [8]. In the wake of this damage, the National Ski Areas Association (NSAA) began developing its Environmental Charter in 1999 with "input from stakeholders, including ... environmental groups" [9] and officially adopted the charter in June 2000 [10]. In accordance with the charter, NSAA has produced its Sustainable Slopes Annual Report each year since 2001 [11]. Apparently, the driving force behind the decision to adopt the Environmental Charter was largely a response to financial concerns rather than by the desire to treat the environment responsibly — it was an amoral decision. That is, NSAA was spurred to create the Environmental Charter by concerns about member companies' bottom lines: Further "ecoterrorist" activity could occur, thereby causing expenses to rise; and the ELF action may have sufficiently highlighted the environmental consequences of resort development to the point that environmentally minded skiers might pause before deciding to patronize resorts where development was occurring, thereby causing revenues to fall.

Similarly, for firms trying to do business in Europe, adopting ISO 14000 is close to a required management practice. The ISO network has penetrated so deeply into business practices that firms are nearly locked out if they do not gain ISO 14000 certification.

There is ample evidence that one of the reasons businesses participate in the quest for sustainability is because it is good for business. The leaders of eight leading firms that adopted an environmentally proactive stance on sustainability were asked in one study to justify the firms' adoption of such a strategy [12]. All companies reported that they were motivated first by regulations such as the control of air emissions, pretreatment of wastewater, and the disposal of hazardous materials. One engineer in the study admitted that: "The [waste disposal] requirements became so onerous that many firms recognized that benefits of altering their production processes to generate less waste".

The second motivator identified in this study was competitive advantage. Lawrence and Morell quote one director of a microprocessor company, who noted that "by reducing pollution, we can cut costs and improve our operating efficiencies". The company recognized the advantage of cutting costs by reducing its hazardous waste stream [12].

Another study, conducted by PriceWaterhouseCoopers, confirmed these findings [13]. When companies were asked to self-report on their stance on sustainable principles, the top two reasons for adopting sustainable development were found to be

1. Enhanced reputation (90%)
2. Competitive advantage (cost savings) (75%)

It is not clear if the respondents were given the option of responding that they practiced sustainable operations because this was mandated by law. If it had, there is no doubt that all companies would have publicly stated that they are, indeed, law abiding.

So it seems likely that the two primary driving forces behind the adoption of green business and engineering practices are (1) legal concerns and (2) financial concerns. Can we argue that such behavior is morally admirable simply on the basis that the outcomes (e.g., cleaner air and water) are morally preferable? We say no. In accord with Sethi [14], we argue instead that actions undertaken in response to legal and financial concerns are actually *obligatory*, in that society essentially demands that businesses make their decisions within legal and financial constraints. For an action to be morally admirable, however, the motivating force driving has to be far different in character.

2.3. Ethical considerations

The first indication that some engineers and business leaders are making decisions where the driving force may not be due to legal or financial concerns comes from several cases in American business. Although most business or engineering decisions are made on the basis of legal or financial concerns, some companies believe that behaving more environmentally responsibly is simply the right thing to do. They believe that saving resources, and perhaps even the planet, for the generations that will follow is an important part of their job. When making decisions, they are guided by the "triple bottom line." Their goal is to balance the financial, social, and environmental impacts of each decision.

A prime example of this sort of thinking is the case of Interface Carpet Company [15]. Founded in 1973, its founder and CEO until 2001 was Ray Anderson, now Chairman of the Board. By the mid-1990s, Interface had grown to nearly $1.3 billion in sales, employed some 6600 people, manufactured on four continents, and sold its products in over 100 countries worldwide. In 1994, several members of Interface's research group asked Anderson to give a kick-off speech for a task force meeting on sustainability: they wanted him to provide Interface's environmental vision. Despite his reluctance to do so — Anderson had no "environmental vision" for the company except to comply with the law — he agreed. Fortuitously, as Anderson struggled to determine what to say, someone sent him a copy of Paul Hawken's *The Ecology of Commerce* [16]; Anderson read it, and it completely changed not only his view of the natural environment, and his vision for Interface Carpet Company, but also his entire conception of business. In the coming years, he held meetings with employees throughout the Interface organization explaining to them his desire to see the company spearhead a sustainability revolution. No longer would they be content to keep pollutant emissions at or below regulatory levels: instead, they were going to strive to be a company that created zero waste and did not emit any pollutants *at all*. The company began to employ The Natural Step [17] and notions of Natural Capitalism [18] as part of its efforts to become truly sustainable. The program continues today, and although the company has saved many millions of dollars as a result of adopting green engineering technologies and practices, the reason for adopting these principles was not to earn more money but rather to do the right thing.

The example of Interface as a company with environmental vision is especially significant because they are not a consumer products company, but rather a materials producer. They do not sell their carpets to the individual consumer in carpet stores but rather produce commercial floor coverings. Thus their attitude toward sustainability would not be beneficial on the consumer level where individual purchasers might make decisions based on perceived "greenness" of the company. In fact, it is not easy for a carpet producer to choose green constituent materials in manufacture because carpets are generally made using many hazardous chemicals, including benzene, xylene, chlorine, and formaldehyde. Interface has been working intently on the problem since 1994, putting millions of dollars into research. Its website is filled with information about "flows of matter to and from the ecosphere".

Yet another example is that of Herman Miller, an office furniture manufacturing company located in western Michigan. Its pledge in 1993 to stop sending any materials to landfills by 2003 has resulted in the company's adoption of numerous progressive, but sometimes expensive, practices. For example, the company ceased taking scrap fabric to the landfill and began shredding it and trucking it to a firm in North Carolina that processes it into automobile insulation. This environmentally friendly process costs Herman Miller $50,000 each year, but the company leaders agree that a decision that is right for the environment is the right decision. Similarly, the company's new waste-to-energy plant has increased costs, but again company leaders feel it is worth the cost, as employees and managers are proud of the company's leadership in preserving the natural environment in their state [19].

Our point here is this: The decisions made by the leadership of Interface Carpet Company and Herman Miller were not morally admirable simply because they enabled these companies to reduce toxic emissions (among many other positive outcomes for the environment); they were morally admirable because the *driving force* behind those decisions was the desire to stop harming the Earth, to protect it so that future generations would be able to enjoy it as much as, or even more than, we do today. Conversely, in the cases of Dupont, Polaroid, and Dow Chemical cited earlier, the *driving force* behind their decisions to adopt green technologies was a desire to save the company money; the benefits to the Earth were simply a fortunate byproduct of those decisions.

In order to understand some of these deeper motivational factors for adopting green engineering practices one of the authors (Susan Hamill) asked engineers and business leaders for their *real* reasons for participating in sustainable operations. When pressed for their personal drivers for adopting green engineering in their work the respondents often said things like this:

- My son challenged me to leave the earth a better and more livable place for my grandchildren.
- I hope to do my part to ensure that the plane remains a nice place for my children and grandchildren to explore.
- Waste is ugly and engineers in general can't stand things that are not designed well. I also now have a child and I want her to have a future.

- I'd rather be useful than famous.
- Some of my personal drivers come from a deep respect and love of the natural world. I think most human beings enjoy the fresh air and water that are found in the remote places in nature. Unfortunately, this freshness is limited. I would like to be in a world where I don't have to worry about breathing clean air or drinking uncontaminated water.
- I think we should at a minimum be aware of what we are doing (make it a conscious decision rather than side effect of our thoughtless consumer life styles), and in the ideal world we would be doing everything we could to stop the current destruction and restore at least some of what we have already damaged.
- I am interested in the environment in general. I am also a student of energy efficiency, but mostly I have a love for the outdoors. Sustainable design has a certain beauty in economy. But what is more intriguing as a designer is the focus is less narrow than traditional engineering. There is an appreciation for context and the way in which processes are integrated. It approaches artistic beauty, particularly when all the competing design goals are met. It is a real high.
- I have been interested in the environment most of my life. I suppose I express that through my involvement with the electric car project. It's fun and lets me be a "gear head". It allows my passion to intersect with my need to "do no harm".

Based on these interviews, we have come to the conclusion that capitalist motives are inadequate to describe the driving forces that form the attitudes that an increasing number of engineers and industrial leaders have toward sustainability. Apparently some believe that green engineering is simply the right thing to do and that they owe something to future generations and thus want to conduct their business in a way that they use the least resources and cause the least damage to the environment. They believe that saving resources, and perhaps even the planet, for the generations that will follow is an important part of their job. When *ethical considerations* are the motivational driving forces for corporate decisions to adopt green engineering, these decisions have moral value.

3. Green engineering: a normative view

We identify three primary driving forces behind corporate decisions to adopt green engineering and business practices: legal, financial, and ethical considerations. Most firms do not even consider disobeying the law: legal concerns are their top priority. Financial concerns are nearly as high on the priority scale: managers consider it their duty to shareholders or owners to assure that the company makes an adequate profit, so they base decisions about green practices on a cost–benefit analysis of the likely consequences. These firms are not concerned with "doing the right thing" except inasmuch as "the right thing" means obeying the law and making money. In other words, these firms may decide to adopt green engineering and business practices strictly on the basis of legal and financial factors, without being significantly influenced by the desire to protect the natural environment. Only when the driving forces involve the desire to do good for all people do such decisions become moral in character.

This observation suggests that it might be possible to develop a normative model of green engineering. Such a normative view would ask the question: What *ought* to be the driving forces for adopting green engineering practices? Our proposed normative model is rooted in the work of developmental constructivist thinkers such as Piaget, Kohlberg, Rest, and others (e.g., Kohlberg [20], Piaget [21], Rest [22,23]) who noted that moral action is a complex process entailing four components: moral awareness (or sensitivity), moral judgment, moral motivation, and moral character. The actor must first be aware that the situation is moral in nature: that is, at the least, that the actions considered would have consequences for others. Second, the actor must have the ability to judge which of the potential actions would yield the best outcome, giving consideration to those likely to be affected. Third, the actor must be motivated to prioritize moral values above other sorts of values, such as wealth or power. And fourth, the actor must have the strength of character to follow through on decision to act morally.

Piaget, Kohlberg, and others (e.g., Duska [24]) have noted that the two most important factors in determining a person's likelihood of behaving morally — that is, of being morally aware, making moral judgments, prioritizing moral values, and following through on moral decisions — are age and education. These seem to be particularly critical regarding moral judgment: a person's ability to make moral judgments tends to grow with maturity as they pursue further education, generally reaching its final and highest stage of development in early adulthood. This theory of moral development is illustrated in Fig. 1.

Kohlberg noted that in the two earliest stages of moral development, which he combined under the heading "pre-conventional level," a person is primarily motivated by the desire to seek pleasure and avoid pain. The "conventional level" consists of stages three and four: in stage three, the consequences that actions have for peers and their feelings about these actions; in stage four, considering how the wider community will view the actions and be affected by them. Few people reach the "post-conventional" stage, wherein they have an even broader perspective: their moral decision making is guided by universal moral principles [25, 26] — that is, by principles which reasonable people would agree should bind the actions of all people who find themselves in similar situations.

We propose that the normative model of green engineering can be developed along the same lines. The moral need to consider the impact one's actions will

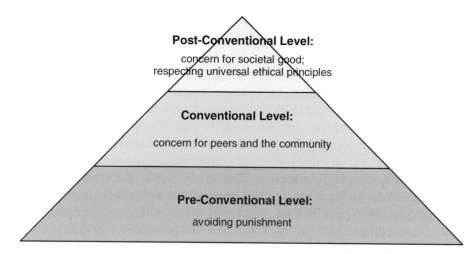

Fig. 1. Kohlberg's moral development model.

have on others forms the basis for the normative model we are proposing. Pursuing an activity with the goal of obeying the law has as its driving force the avoidance of punishment, and pursuing an activity with the goal of improving profitability is a goal clearly in line with stockholders' desires; presumably customers', suppliers', and employees' desires must also be met at some level. And finally, pursuing an activity with the goal of "doing the right thing," behaving in a way that is morally right and just, can be the highest level of green engineering behavior. This normative model of green engineering can be illustrated as Fig. 2.

There is a striking similarity between Kohlberg's model of moral development and our model of moral green engineering. Avoiding punishment in the moral development model is similar to a corporation staying out of trouble by obeying the law. The pre-conventional level and our legal concern level have similar driving forces.

At the second level in the moral development model is a concern with peers and community, while in our model the corporation undertakes green business practices in order to make more money for the stockholders and to provide a service or product for their customers that will in turn make the corporation more profitable. At this level, as in the previous one, self-centeredness and personal well-being govern decisions.

Finally at the highest level of moral development a concern with universal moral principles begins to govern actions, while for the corporate model, fundamental moral principles having to do with environmental issues control corporate decisions. In both of these cases the driving force or motivation is trying to do the right thing on a moral (not legal or financial) basis.

We suggest that moral green engineering occurs only when engineers and managers base their decisions about the adoption of green business and

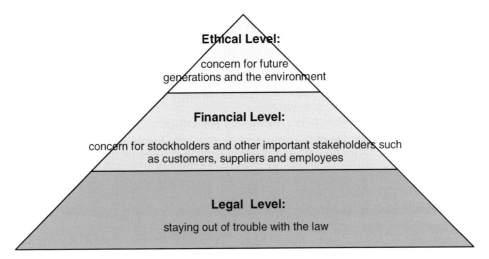

Fig. 2. The green engineering moral development model.

engineering principles on ethical considerations. That is, they recognize the broad impacts their decisions may have, and they act in such a way that their actions will be in the best interest of not only themselves, their companies, and their companies' direct stakeholders, but also the broader society and even future generations.

4. Conclusions

Green engineering will eventually lead the world to sustainability, but green engineering today occurs most often when doing the right thing also results in adherence to laws and regulations and in achieving greater profitability for the organization. This kind of green engineering, although often beneficial to society, is not morally admirable.

The true heroes of green engineering are those corporate leaders who believe deeply in the principles of green engineering and who try to work within these principles while still helping their corporations to be profitable. They enjoy working to promote sustainability, and do so not for show or profit, but because it gives them pleasure to do the right thing.

References

[1] A. Grübler, Technology and Global Change, Cambridge University Press, Cambridge, England, 1998.
[2] C. Lazaroff, US Government Sues Power Plants to Clear Dirty Air, Environment News Service. http://ens.lycos.com/ens/nov99/1999L-11-03-06.html.

[3] D. Fowler, Bush Administration, Environmentalists Battle Over 'New Source Review' Air Rules, Group Against Smog and Pollution Hotline, Fall 2003. Retrieved January 3, 2004 from http://www.gasp-pgh.org/hotline/fall03_4.html.

[4] D.L. Barlett, J.B. Steele, The Great Energy Scam, TIME 162 (October 2003) 60–70.

[5] M. Friedman, Capitalism and Freedom, University of Chicago Press, Chicago, 1962.

[6] K. Gibney, Sustainable development: A new way of doing business, Prism. American Society of Engineering Education, January 2003.

[7] S.B. Billatos, N.A. Basaly, Green Technology and Design for the Environment, Taylor & Francis, Washington, DC, 1997.

[8] J. Faust, Earth Liberation Who? Retrieved January 3, 2004 from the ABCNEWS.com website: http://more.abcnews.go.com/sections/us/DailyNews/elf981022.html.

[9] National Ski Areas Association (NSAA), Environmental Charter, http://www.nsaa.org/nsaa2002/_environmental_charter.asp.

[10] J. Jesitus, Charter promotes environmental responsibility in ski areas, Hotel Motel Manage. 215 (19) (2000) 64.

[11] National Ski Areas Association (NSAA), Sustainable Slopes Annual Report, http://www.nsaa.org/nsaa2002/_environmental_charter.asp?mode = s.

[12] A.T. Lawrence, D. Morell, Leading-Edge Environmental Management: Motivation, Opportunity, Resources and Processes, in: J.S.E. Post, D. Collins, M. Starik (Eds), Research in Corporate Social Performance and Policy, Supplement 1, JAI Press, Greenwich, CT, 1995.

[13] PriceWAterhouseCooper, Sustainability Survey Report, New York, August, 2002.

[14] S.P. Sethi, Dimensions of corporate social performance: an analytical framework, Calif. Manage. Rev. 17 (1975) 58–64.

[15] R.C. Anderson, Mid-Course Correction — Toward a Sustainable Enterprise: The Interface Model, The Peregrinzilla Press, Atlanta, 1998.

[16] P. Hawken, P. The Ecology of Commerce: A Declaration of Sustainability, Harper Business, New York, 1994.

[17] K.-H. Robért, Educating a nation: The natural step, In Context, no. 28, Spring 1991.

[18] A.B. Lovins, L.H. Lovins, P. Hawken, A road map for natural capitalism, Harvard Bus. Rev. 7 (1999) 145–158.

[19] L.T. Hosmer, Herman Miller and the Protection of the Environment, The Ethics of Management., McGraw-Hill Irwin, Boston, 2003.

[20] L. Kohlberg, The Philosophy of Moral Development, Vol. 1, Harper & Row, San Francisco, 1981.

[21] J. Piaget, The Moral Judgment of the Child, The Free Press, New York, 1965.

[22] J.R. Rest, Moral Development: Advances in Research and Theory, Praeger, New York, 1986.

[23] J.D. Rest, D. Narvaez, M.J. Bebeau, S.J. Thoma, Postconventional Moral Thinking: A Neo-Kohlbergian Approach, Lawrence Erlbaum Associates, Mahwah, NJ, 1999.

[24] R. Duska, M. Whelan, Moral Development: A Guide to Piaget and Kohlberg, Paulist Press, New York, 1975.

[25] J.A. Rawls, A Theory of Justice, Harvard University Press, Cambridge, MA, 1971.

[26] I. Kant, Foundations of the Metaphysics of Morals, translated by Lewis W. Beck, Bobbs-Merrill, Indianapolis, 1959.

Sustainability Science and Engineering: Defining principles
Martin A. Abraham (Editor)
DOI 10.1016/S1871-2711(05)01004-4

Chapter 4

Green Engineering Education

Robert P. Hesketh[a], Michael H. Gregg[b], C. Stewart Slater[a]

[a]Chemical Engineering, Rowan University, 201 Mullica Hill Rd., Glassboro,
NJ 08028, USA
[b]Engineering Education, Virginia Tech, Blacksburg, VA 24061, USA

1. Introduction

Green engineering was originally defined by the EPA as the design, commercialization, and use of processes and products that are feasible and economical while minimizing: generation of pollution at the source and risk to human health and the environment [1]. In a recent conference, this definition of green engineering was more broadly defined as transforming *existing engineering disciplines and practices to those that lead to sustainability. Green Engineering incorporates development and implementation of products, processes, and systems that meet technical and cost objectives while protecting human health and welfare and elevating the protection of the biosphere as a criterion in engineering solutions* [2]. As a result of this conference, nine green engineering principles were developed which engineers should follow to fully implement green engineering solutions:

1. Engineer processes and products holistically, use systems analysis, and integrate environmental impact assessment tools.
2. Conserve and improve natural ecosystems while protecting human health and well-being.
3. Use life cycle thinking in all engineering activities.
4. Ensure that all materials and energy inputs and outputs are as inherently safe and benign as possible.

5. Minimize depletion of natural resources.
6. Strive to prevent waste.
7. Develop and apply engineering solutions, while being cognizant of local geography, aspirations, and cultures.
8. Create engineering solutions beyond current or dominant technologies; improve, innovate and invent (technologies) to achieve sustainability.
9. Actively engage communities and stakeholders in development of engineering solutions.

As given in the first chapter of this book, the conference participants felt strongly that there is "a duty to inform society of the practice of green engineering". Green engineering, as stated in the preamble given above, comprises a set of tools for use by green engineers to achieve sustainability. In this regard, educational developments in both green engineering and sustainable development will be presented in this chapter.

The natural method for engineering professors to inform society is to first teach their students this subject in the classroom. If engineering students are taught how to incorporate green engineering into their work, then this will help its implementation by industry and lead us to a sustainable future.

The discipline embraces the concept that decisions to protect human health and the environment can have the greatest impact and cost effectiveness when applied early to the design and development phase of a process or product. In green engineering, risk assessment is applied to the design, retrofit, and optimization of feedstocks, waste streams, and unit operations in processes and products. The concept of risk-assessment takes into consideration the extent of harm a chemical and its use can pose to the environment. While traditional pollution prevention techniques focused on simply reducing as much waste as possible by treating all wastes as equal, risk-assessment methods used in pollution prevention can help quantify the degree of environmental impact for individual chemicals. With this approach to pollution prevention, engineers can design intelligently by focusing on the most beneficial way to minimize risk.

By applying risk-assessment concepts to processes and products, the engineer can:

• Quantify the environmental impacts of specific chemical on people and ecosystems
• Prioritize chemicals that need to be minimized or eliminated
• Optimize design to avoid or reduce environmental impacts
• Design greener products and processes.

This chapter highlights techniques to include green engineering in the chemical engineering curriculum. This may be through stand-alone courses, concepts in core courses such as thermodynamics or engineering economics, design projects, and as part of the assessment requirements for ABET Criteria 2000.

Green engineering can be alternatively defined as environmentally conscious attitudes, values, and principles, combined with science, technology, and engineering practice, all directed toward improving local and global environmental quality. Green engineering encompasses all of the engineering disciplines, and is consistent and compatible with sound engineering design principles. Green engineering focuses on the design of materials, processes, systems, and devices with the objective of minimizing overall environmental impact (including energy utilization and waste production) throughout the entire life cycle of a product or process, from initial extraction of raw materials used in manufacture to ultimate disposal of materials that cannot be reused or recycled at the end of the useful life of a product.

2. Strategy and motivation for teaching green engineering

The need to introduce green engineering concepts to undergraduate students is recognized to be increasingly important [3]. This need is driven in part in the USA by its Engineering Accreditation Commission Accreditation Board for Engineering and Technology (ABET) criteria 2000. In the general engineering criteria it states that engineers must have an ability to design a system, component, or process to meet desired needs within realistic constraints such as economic, environmental, social, political, ethical, health and safety, manufacturability, and sustainability. Discipline-specific criteria, such as in chemical engineering, further specify that engineers must have "ethics, safety and the environment" included in the curriculum. On a global perspective, several national professional engineering accreditation bodies from New Zealand and Australia require those engineering programs to produce graduates that have an understanding of sustainable technology and development [4–6]. Other programs, such as South Africa [7] and Ireland [8] have similar wording to that in the USA accreditation requirements. In the United Kingdom, chartered and incorporated engineers must "undertake engineering activities in a way that contributes to sustainable development" [9]. On the basis of this international consensus of accreditation bodies, the need to introduce green engineering and sustainability concepts in engineering education is a required outcome.

A possibly more significant driver for teaching green engineering to students is that their future employers in the chemical industry have moved toward a sustainable future. Major chemical companies such as BP [10,11], BASF [12], Dow [13,16], DSM [14], DuPont [15,16], GSK [17], Merck [18], and Rohm and Haas [19] have adopted a green engineering approach to achieve a sustainable future. Many of these companies are members of the Dow Jones Sustainability

Index (DJSI) World & DJSI STOXX which totals 300 companies from 22 countries and was launched in 1999 [20]. For example, the leaders in chemicals, energy and food & beverages, industrial goods & services, non-cyclical goods & services sectors in 2004 were DuPont, BP, Unilever, 3 M, and Procter & Gamble, respectively [21].

In addition, professional organizations have advocated issues in sustainable development such as AIChE's CWRT center [22], the American Chemical Society [23], the Chemical Industry Council Responsible Care program [24], and the American Society for Engineering Education [25]. Three of these societies, (ASCE, AIChE, ASEE), have formed "The Engineers Forum on Sustainability" [26]. One of the major tenets of this forum is to identify and distribute "information on engineering education programs that incorporate sustainability". The American Society of Civil Engineers (ASCE) stipulates in their code of ethics that, "engineers shall hold paramount the safety, health and welfare of the public and shall strive to comply with the principles of sustainable development in the performance of their professional duties" [27]. As in the consensus in accrediting organizations mentioned above, professional societies globally are promoting work in green engineering- and sustainable development-related areas.

The historic and probably most common method to introduce aspects of green engineering has been through a senior/graduate level elective course on environmental engineering, with emphasis on *end of the process* treatment. In the 1990s, courses were developed that focus on methods to minimize or prevent waste streams from existing chemical plants. These trends mirror those in industry, in which initial efforts were applied to waste treatment, whereas current efforts are aimed at reducing the total volume of effluent treated as well as the nature of the chemicals treated. Currently, courses in green engineering, environmentally conscious chemical process engineering, and engineering for sustainable development have replaced many of the environmental and pollution prevention courses. Most of these courses are in the senior year and are optional engineering courses. These stand-alone courses are excellent in providing detailed coverage of the subject and are needed in the engineering curriculum. Unfortunately, students may get the impression that green engineering is either optional or something taken up at the end of the design process, since this course is usually both optional and at the end of their undergraduate education. Additionally, students get the impression that only one professor, the one currently teaching the course, knows about this subject matter which reduces the importance of this subject in the students mind.

To emphasize that engineers should be using green engineering and sustainability throughout the design process, perhaps a better method is to introduce these concepts throughout the curriculum. Through this integrated approach, students see green engineering throughout their 4 years of engineering, thereby showing the high importance of this subject and reinforces the need to employ this subject in industry. A drawback with this method is that faculty may agree to add this material, but its coverage becomes diluted and sporadic throughout the curriculum. Both these methods of education need to be encouraged and further developed.

Engineering programs need to evaluate how they are training engineers in concepts of green engineering and may use either of the approaches or both the approaches. Two major issues should be considered in using either of these methods. First is the need for professors to recognize that green engineering should be taught to all engineering undergraduates. Second is that professors need to be shown that many of the subjects that they are currently teaching are part of green engineering. They need to see the link between what they currently teach and what they could be teaching within the context of green engineering.

In spite of the compelling motivation to incorporate green engineering materials at all levels of the chemical engineering curriculum, lack of familiarity with the green engineering subject matter, a stronger interest on the part of the instructor in his or her specialized research area, or simple inertia and lack of time can prevent faculty from addressing green engineering issues in core engineering courses. For example, a faculty whose expertise is in catalysis is likely to teach an undergraduate reaction engineering course. He or she may not have a clear understanding of how reaction engineering can be applied to green engineering and will be unlikely to add these concepts to their course. What is required by the engineering faculty are modules in green engineering which specifically relate to the subject matter that they are teaching.

This chapter contains examples of how green engineering is being incorporated in engineering education and provides references for detailed examples of this material.

3. Overview of green engineering and related courses in engineering curricula

Most engineering departments throughout the United States list at least one course in environmental training. Initially this course was on air and water pollution control and in many cases has expanded to a survey course on environmental engineering. In the 1990s, pollution prevention courses were added to engineering programs. An example of courses in pollution prevention can be found at the old National Pollution Prevention Center (NPPC) website [28]. This website is still on-line, but new material is being added through the Center for Sustainable Systems [29]. The new website has a list of 6 MS PowerPoint modules used in teaching green engineering and sustainability. This listing gives syllabi for courses added between 1988 and 1995; nearly all of the courses listed were designed as electives for graduate students or upper division undergraduates.

In a recent survey on pollution prevention [30], chemical engineering departments were asked for information on how they taught pollution prevention within their curriculum. The responses can be loosely classified into three categories:

- Programs in which pollution prevention is taught as a separate elective class (30%).
- Programs that offer a course in air pollution or waste treatment and include pollution prevention as a component within these elective courses (40%).
- Programs that do not provide any specialized training in pollution prevention but may include some material within the regular course sequence, usually, the senior design course (30%).

In nearly all cases, the courses are targeted at upper division undergraduate or graduate students and are elective courses. Only a small number of departments require all seniors to take a course in pollution prevention. Although the number of survey responses represents a minority of chemical engineering departments, these results would appear to be consistent with anecdotal information that many chemical engineering programs are now looking into ways in which *pollution* prevention can be incorporated into the graduate and undergraduate curriculum.

TWM Research located in Upper Montclair, New Jersey was commissioned to conduct a survey to determine what textbooks universities and professors were using to teach green engineering or pollution prevention courses in the 2002 academic year. Table 1 shows the results of the number of universities and departments teaching green engineering-related courses. In this survey we found that 37 universities had courses in either pollution prevention or green engineering and in many cases this course was offered to a number of departments in the engineering college. As shown in Table 2 these courses were mainly taught in a chemical engineering department with several courses taught by civil and/or environmental, mechanical, and environmental engineering science departments. This is to be expected, since the concept of pollution prevention more readily employed in chemical engineering than in other engineering disciplines.

Table 1
Universities teaching green engineering or pollution prevention courses

Universities offering green engineering or pollution prevention courses	37
Departments offering green engineering or pollution prevention courses	42
Total number of green engineering or pollution prevention courses offered	84

Table 2
Departments teaching green engineering or pollution prevention courses

Department classification	Number of courses offered in 2002/03 academic year
Chemical Engineering	25
Civil and/or Environmental Engineering	10
Mechanical Engineering	4
Environmental and/or Engineering Science	2
Engineering Technology	1

As in the previous survey, only a small number of chemical engineering departments are teaching a green engineering-related course; 25 chemical engineering departments out of 157, listed in the AIChE directory, have either a green engineering course or a pollution prevention course taught in 2002. This amounts to only 16% of all US chemical engineering departments. This shows that further efforts are needed to place green engineering into chemical engineering curriculums.

There are several recent articles on green engineering-related courses given in the senior and graduate years. A general description of green engineering education for chemical engineering is given in a paper by Shonnard et al. [31]. Notre Dame and West Virginia University have taught a chemical engineering course titled, "Environmentally Conscious Chemical Process Engineering," which was developed as a result of the consortium with the chemical engineering departments at the University of Notre Dame, West Virginia University and the University of Nevada at Reno [32]. This overall program includes the development of three new courses: (1) Environmentally Conscious Chemical Process Design, (2) Ecology and the Environment, and (3) Environmental Flows [33]. In addition, they are incorporating research results into instructional modules that are integrated throughout the chemical engineering curriculum, with a special emphasis on the design sequence. Previously, Abraham described possibly the only pollution prevention course that is required for all chemical engineering seniors at the University of Toledo [34]. Grant et al. [35] describes a senior/graduate elective taught at North Carolina State University that focuses on environmental management, while Simpson and Budd [36] describe a similar course developed at Washington State University.

Elective courses in life cycle analysis are being taught at several universities. In 1992, Chalmers University of Technology developed the first LCA course in Sweden. This course started as a graduate only course, but since 1996, the LCA course is given as an engineering elective course for all engineering students at Chalmers [37]. In addition to the LCA course, they also give "the 2 hour

lecture" orientation about LCA in the compulsory environmental course for
1st/2nd year students at Chalmers. Michael Gregg has also developed a course
for engineers at Virginia Tech [38] as has Jack Jeswiet at Queen's University in
Canada [39].

New courses in sustainable development are also being developed. Georgia
Tech has established an Institute for Sustainable Technology and Development
and through funding from General Electric has developed a series of courses in
this area [40]. These courses were developed by Jorge Vanegas; Introduction to
Sustainable Development; Case Studies in Sustainable Development, Designing
Sustainable Engineering Systems, and the Sustainable Problem Solving Laboratory. Numerous programs are also being developed worldwide in which sustainable development is integrated within engineering curriculums and will be
discussed below.

These courses are designed to provide, a selected set of students that are
interested in the environment, an excellent set of tools to tackle problems in
pollution prevention. When green engineering or pollution prevention is taught
as an elective course, the majority of students will graduate without the knowledge of how their engineering practice will affect society, ecosystems, and the
environment. In addition, by placing this subject at the end of their university
preparation, this tends to leave an impression with students that environmental
concepts are only considered after the engineering work is completed. Since one
of the precepts of green engineering is that it should be conducted at all levels of
engineering practice and design, we believe that it should be taught at all levels.
Instead of having only an optional course in environmental or green engineering, we believe that it is more appropriate to integrate green engineering concepts in a range of courses within an engineering *discipline*.

4. Green engineering-related texts and references

In the survey on green engineering-related courses there were a number of
textbooks used. Table 3 gives a listing of these textbooks.

This first textbook in Green Engineering is by Allen and Shonnard, which
originated as a result of a program developed by the US Environmental Protection Agency. In 1998, the US EPA initiated a program in green engineering
to develop a text book on green engineering; disseminate these materials and
assist university faculty in using these materials through national and regional
workshops coordinated with the American Society for Engineering Education
(ASEE), Chemical Engineering Division. The textbook titled, "Green Engineering: Environmentally Conscious Design of Chemical Processes" [47] by
Allen and Shonnard is designed for either a senior or graduate chemical engineering course and has a series of accompanying modules that can be

Table 3
Textbooks used for green engineering or pollution prevention courses

Textbooks
Allen and Shonnard: *Green Engineering: Environmentally Conscious Design of Chemical Processes* [47]
Bishop: *Pollution Prevention: Fund & Practice* [54]
Pierce, Weiner, and Vesilind: *Environmental Pollution & Control* [41]
Ganesan: Pollution Prevention: *Waste Management Approaches* [42]
Higgins: *Pollution Prevention Handbook* [43]
El-Halwagi: *Pollution Prevention through Process Integration* [55]
Graedel: *Streamlined Life-Cycle Assessment* [44]
Curran: *Environmental Life-Cycle Assessment* [45]
Miller: *Living in the Environment, Environmental Science, and Sustaining the Earth* [46]

employed throughout the curriculum. This book is divided into three major sections: (1) Chemical Engineer's Guide to Environmental Issues and Regulations (2) Environmental Risk Reduction for Chemical Processes, and (3) Moving Beyond the Plant Boundary. The first section provides an overview of major environmental issues, and an introduction to environmental legislation, risk management, and risk assessment. The second section contains tools for assessing the environmental profile of chemical processes and the design tools that can be used to improve environmental performance. These tools include release estimation approaches and pollution prevention strategies, total cost accounting, and green process design. This group of chapters begins at the molecular level, examines unit operations, and then proceeds to an analysis of process flow sheets. The final section contains the tools for improving product stewardship and improving the level of integration between chemical processes and other material processing operations. This textbook includes software tools that have been developed for green engineering with a brief summary given in Table 1. Additional information on this text can be found on the green engineering EPA website [48].

On the basis of the chemical engineering pollution prevention survey [30], the most popular textbook used in an advanced elective course on pollution prevention is the text by Allen and Rosselot [49], which is divided into three sections that describe macro, meso, and micro-scale pollution prevention. The recent text by Mulholland and Dyer [50] provides a practical guide for practicing pollution prevention in the chemical process industries. Freeman [51] has produced a comprehensive handbook referenced by many pollution-prevention educators. Other general texts include those by Rossiter [52], Theodore [53], and Bishop [54]. For those courses with an emphasis on mass integration, the text by El-Halwagi [55] is available. For case studies and pollution prevention problems, one can consider the compilation of problems by Allen [56]. Other resource texts on pollution prevention can be found on National Pollution Prevention Center for Higher Education website [28].

Texts in sustainable development are very useful in teaching green engineering. A book by Azapagic, Perdan, and Clift gives 11 case studies on sustainable development for engineers [57]. Other new texts that are useful in teaching sustainable development concepts to engineers include [58] background on life cycle assessment [59–61].

In the general area of design for the environment there are a number of texts that have been developed. In the field of electrical circuit manufacture a book by Lee Goldberg titled Green Electronics is used at the University of Texas-El Paso [62].

5. Green engineering throughout the curriculum

Within the engineering curriculum it is natural to weave the tools of green engineering throughout the curriculum. A number of examples of this method can be found at universities, such as Virginia Tech [63], University of Texas-El Paso [64,65], Carnegie Mellon [66], Berkeley, Georgia Tech, University of Windsor [38], Rowan University [67], and the University of Surrey. In this section, examples of programs that have incorporated green engineering into the curriculum will be given. In the following section, specific examples of green engineering in engineering and related courses are presented.

The College of Engineering at Virginia Polytechnic and State University has developed a green engineering concentration within their B.S. program. This concentration requires two green engineering core courses (Introduction to Green Engineering, Environmental Life Cycle Analysis), six credit hours of interdisciplinary electives and six credit hours of disciplinary engineering electives (which are required to contain at least 25% green engineering content). This program started with a steering committee that identified 70 courses throughout Virginia Tech's curriculum that had an identifiable environmental/green engineering/sustainable component; 56 of those courses were in engineering. As a result all of Virginia Tech's engineering departments offer courses devoted to or containing significant green engineering content. In chemical engineering at Virginia Tech, these courses are the core chemical engineering courses of Mass & Energy Balances, Separation Processes, Chemical Reactors, Chemical Engineering Laboratory, and Process and Plant Design. The Green Engineering Program also offers a one-credit hour not-for-graduation seminar series entitled *Green Engineering Lecture Series*. Lecturers for this series are engineers invited from industry, government, and academia to discuss their engineering experiences and the environmental ramifications of those activities. This incorporation on a college basis is an excellent example of integrating green engineering concepts throughout the curriculum.

Carnegie Mellon's Green Design initiative has produced a number of green engineering modules to give a basic introduction to environmental issues [68]. These modules can be found on their website [66]. A brief synopsis of each module is given in Table 4.

At the University of Texas — El Paso, Charles Turner is leading the development of an integrated program on sustainable development. In this program 6 engineering departments have contributed a total of over 30 courses with a content on sustainable development [69].

At the University of Surrey, Adisa Azapagić is introducing concepts on sustainability starting in the first year and continuing into the final year of chemical engineering [70,71]. At Surrey they believe that sustainability should not be an 'add-on' to the curriculum, instead it should be integrated systematically throughout the curriculum. In the first year all engineering students have a module on sustainable development. In this module an introduction to sustainable development is given followed by concrete examples of sustainable development in each of the engineering disciplines. These examples are reinforced with guest lectures from the industry. Tutorials are given throughout with discussions on sustainable industrial systems and an examination of products such as mobile phones and computers are assigned and a case study on wastewater systems is presented [57,72]. Throughout this module life cycle thinking is introduced to the first year engineering students. The chemical engineers have an additional course in the first year that expands on this multidisciplinary course. In this additional module, chemical engineering students are given an assignment to examine the sustainable activities of a chemical company of their choosing. In the second year at the University of Surrey, the chemical engineering students learn European and British legislation related to sustainability and examine the voluntary activities that companies are pursing toward achieving sustainability. The final part of this second year course is to employ life cycle assessment software to examine simple problems such as packaging of products. In the third year of coursework, students apply the IChemE sustainability metrics [73] to their design projects and in the fourth year participate in a multidisciplinary design project that incorporates aspects of sustainable development.

In the area of graduate education, more work is done on the masters level and the University of Surrey offers a PhD in Sustainable Development [74]. Virginia Tech is currently expanding its Green-Engineering Program to include graduate level courses.

At the University of Oxford, a similar program has been developed for chemical engineering [75]. The first year starts with a case study on sustainable energy. This is followed by a 1-week module on energy and the environment, which includes topics such as photovoltaics, fuel cells, climate change, incineration, and hydrogen economy. In the second year, all students follow a series

Table 4
Green design modules for the curriculum — Carnegie Mellon

Module/paper	Description	Use
Introduction to green design	ASEE paper presents general principles of environmentally conscious product/process/service design	Introduction
Radioactive waste management	This module is a historical perspective of the issues associated with radioactive waste management, the extent of the problem, and a discussion of how the problems have developed over time. Questions included for class discussion	Undergraduate humanities and social sciences course entitled science, technology, and policy
Disposition of personal computers	Complete case study with 4 student questions Teacher's Guide provided	Corporations, undergraduate history and policy courses, upper undergraduate/graduate-level environmental management
Reverse engineering for green design	PowerPoint presentation giving examples of reverse engineering a one-time use camera and a coffeemaker. Application of life-cycle analysis on these products is used to examine possible product improvements. The teacher's notes contain the MS PowerPoint notes which give more detail on the photographs of the 2 products	Capstone design course in mechanical engineering
Rechargeable battery management and recycling (PDF format)	NiCd batteries. Contains 8 questions which include a new LCA problem for students	Graduate level environmental management class
Life cycle assessment: asphalt vs. concrete	This case study describes life cycle assessment and then discusses the EIOLCA approach in detail. EIOLCA software (available free on the web) is required to assess the environmental implications of two alternative pavement construction materials	Upper undergraduate and graduate level course in environmental management
Full cost accounting	A course module on incorporating environmental and social costs into traditional business accounting systems. Contains example class projects and assignments	MBA courses, graduate and upper level undergraduate environmental engineering course

Table 4 (*continued*)

Module/paper	Description	Use
Semiconductor FCA case study	Full cost accounting case study of a semiconductor fabrication facility	Same as above
Case study — landfill power generation	Examines methane releases from landfills, which account for 4% of greenhouse gas emissions. Total cost accounting questions at end	Same as above
Case study — motor vehicle emissions testing	Total accounting case study to examine methods to reduce automotive emissions in the state of Pennsylvania	Same as above

of lectures exploring the concepts of sustainable development, with two case studies. In the third year, design projects are based on applying sustainable engineering to projects such as the design of a hydrogen plant for fuel cells, a refugee camp in an African country, the design of an offshore wind-farm, an autonomous roadtrain car system, and a solar-powered aircraft. In the fourth year, the students can take optional lectures in life cycle assessment and fuel cells.

There are several undergraduate education initiatives in engineering at the University of Cambridge [76]. These initiatives are detailed on the Department of Engineer's Centre for Sustainable Development webpage. Four modules are given to engineering students. The first module consists of eight lectures to first year students with the objective to introduce the economic, industrial, and social contexts faced by engineers. In the third year of study, students take 16 lectures on wastewater and solid-waste pollution. In the fourth year, students have the option to work on research projects and also to enroll in a sustainable development course for engineers.

At Rose-Hulman Institute of Technology, the civil engineering department has integrated sustainability through their curriculum [77]. Sustainability topics were traditionally covered by a single faculty member in the environmental and water resource engineering courses. This led to the impression that this topic was not important in other areas of civil engineering and was heavily reliant on one faculty member. At Rose-Hulman they are starting to integrate the following sustainability topics:

- Basic knowledge — the fundamental concepts of sustainable development and their implications on the engineering practice
- LEED — Green Building Rating System
- Geotechnical consideration of using recycled materials in earthwork
- Recycled materials in transportation (asphalt, etc.)

- Recycled materials in construction/deconstruction
- Natural resource management (currently focusing on water resources).

An example of an integrated green engineering program in aeronautical engineering, is given by Thomas Gally at Embry-Riddle Aeronautical University [78]. In this paper, he gives examples of how a curriculum could be infused with green engineering concepts. He discusses the various aspects of green aircraft design including energy conservation, engine emissions, life cycle analysis of materials, and noise pollution. Finally, he proposes a life cycle analysis of some very interesting philosophies of operation and marketing related to airlines. For example, Boeing makes a smaller, more energy efficient aircraft that will directly fly to a city, while an Airbus makes a large aircraft that would only fly to hub cities. Which of these two methods would reduce overall fuel usage?

An example of integrating sustainable development throughout a university can be found at Brabant University, Netherlands [79]. This initiative labeled, the CIRRUS project, started in 1999 within the Technology and Natural Sciences faculty of the Brabant University of Professional Education. The model that was developed requires that:

1. Each course, project, and other activities in the curriculum present the issues relevant for sustainability such as materials use, energy, design approaches, economics, business operation methods, etc.
2. An introductory course presents the overall concept of sustainability needs and creates the general framework for issues and details treated elsewhere.
3. Attitude, lateral thinking, interdisciplinary ability aimed at sustainability will have a large emphasis throughout all activities in the study and increasingly so toward the end.

The objective of this integration is to prepare graduates that are experts in their chosen field, but have the knowledge, competences and attitude needed for sustainable development practices. To achieve this implementation, 10 faculty members underwent extensive training in sustainable development and these faculty members trained the 250 remaining faculty members. This integration is currently ongoing.

Rowan University's chemical engineering program has integrated green engineering starting in the freshman year and continuing through the senior year. We believe that green engineering concepts can be readily coupled with what is currently being taught in a chemical engineering curriculum. At Rowan University our freshman year starts with multidisciplinary projects and case studies utilizing aspects of green engineering. In the sophomore year, the chemical engineering students are required to purchase the Allen and Shonnard text [47] and use this in conjunction with all of their chemical engineering courses. In our

Table 5
Integration of green engineering in the chemical engineering curriculum

Chemical engineering course	Green engineering topic
Freshman Engineering Clinic	Green engineering project drip coffee maker Introduction to environmental regulations Introduction to life cycle assessment
Sophomore Engineering Clinic	Life cycle assessment of a product Environmental regulations
Material and Energy Balances	Emissions terminology/calculations "Green" material and energy balances
Mass Transfer/Equilibrium Stage Separations	Mass separating agent Risk assessment
Material Science	Estimation of properties, EPA PMN case studies: polymers or electronic materials
Heat Transfer	Introduction to heat integration
Chemical Thermodynamics	Estimation of chemical properties
Separation Processes	Pollution prevention strategies Novel "Green" separation process integration
Chemical Reaction Engineering	Pollution prevention strategies Green chemistry
Process/Plant Design	Heat integration and mass integration Flowsheet analysis Life cycle assessment
Process Dynamics and Control	Pollution prevention modeling and control
Unit Operations Laboratory	Green engineering experiments
Design for Pollution Prevention	Heat and mass integration Process analysis
Senior Engineering Clinic/Senior Project	Real industrial projects in green engineering

multidisciplinary sophomore engineering clinic all engineering students participate in projects related to green engineering [80]. Table 5 gives a summary of green engineering activities that can be incorporated into a chemical engineering curriculum.

5.1. Examples of integration of green engineering in core courses

This section of the chapter will give examples of the integration of green engineering materials in existing engineering courses. Many of these materials are available online (www.rowan.edu/greenengineering) as part of the ASEE/EPA initiative underway at Rowan University [81]. The goal of the initiative at Rowan University is to provide materials that can be readily adapted to currently existing courses [67]. The materials that are being provided consist of instructor guides to assist in mapping green engineering topics into various chemical engineering courses and provide homework problems, in-class examples, and case studies for faculty to use. Green engineering modules prepared for core courses are available online [82]. This is a password-protected site in which students do not have access to the homework problem solutions.

5.1.1. Freshman engineering — projects and life cycle

As a result of the increasing importance of environmental issues, most universities have instituted environmental courses that can be taken by all university students to fulfill their humanities requirements. These courses typically have titles such as Man and the Environment or Environmental Ethics and have a goal of making students more aware of their actions in a global environment. In advising first-year engineers, these types of courses would be very useful to introduce green engineering concepts. In advising freshman engineers, these types of courses would be very useful to introduce green engineering concepts.

Many engineering colleges have now instituted a freshman-engineering course. These courses provide excellent opportunities to introduce freshman to the basic concepts of green engineering. Instead of employing a lecture style format, freshman could be introduced to green engineering through case studies and hands-on projects. For example, at Rowan University, students in our Freshman Engineering Clinic investigate commercial household products through reverse engineering. The students are very familiar with products such as coffee machines, computers, hair dryers, and common household toys, because they have been exposed to these items since birth. Hesketh et al. [83] have students who dissect coffee machines to find out how they work. They discover a large number of individual components and are asked to conduct a life cycle assessment of these materials. A similar module is given for both coffee machines and one-time-use cameras [66]. Other freshman engineering programs, such as the one at New Jersey Institute of Technology [84], use a case study approach in which students have to site and design a manufacturing facility that either uses or generates hazardous materials. In this example, students are asked to consider pollution prevention strategies in their process plant design.

In a recent paper, Kampe and Knott [85], give the ideas of how they introduce green engineering concepts to over 1200 first-year engineering students at Virginia Tech. In this course, they do not introduce green engineering as a separate subject. Instead, they expand traditional problems to include environmental issues. An example given in this paper involves analyzing life cycles of various orange juices (fresh or concentrated). A recent paper by Wiedenhoeft [86] shows

Table 6
Sustainable/green engineering projects for freshman at University of Texas—El Paso

Department	Sustainable/Green engineering problem
Civil Engineering	Life cycle of a disposable camera
Electrical Engineering	Role of remote sensing and signal processing in addressing the current, global concern about the impact of human activities on the Earth's resources and natural processes and how the technology aids in promoting sustainable development/engineering

how to introduce basic concepts of pollution prevention to freshman students. Another example in the material balance course by Rochefort [87] introduces pollution prevention using a material balance module developed by the Multimedia Engineering Laboratory at the University of Michigan [88]. The extent of green engineering concepts covered in the aforementioned courses depends on the student audience.

The freshman year is also a good place to introduce the concept of a life cycle. Some good examples that can be used at this level are materials found from various websites [89–91] and books [59,92]. At the University of Texas-El Paso each department has contributed a project on sustainable development (Table 6).

At Brabant University of Professional Education, in the Netherlands, students are given several cases on simple issues that are familiar to students [93]. The projects that have been run include shopping bags from paper or polyethylene; sustainability at home; and cleaner fuels for transport. The shopping bag case involves an LCA approach. Sustainability at home requires students to evaluate the consumption of energy and materials and household emissions and wastes that are generated. Students also must consider issues of the available environmental space and the ecological footprint of these activities. Here aspects as 'rights' and ethics, equity, economic development, and technical challenges appear. The cleaner fuels for transport case is based on a systems approach in which students consider traffic issues, current, and new fuel systems (e.g. H_2), and are asked if a cleaner fuel will resolve issues in sustainable development. At the end of these modules they are asked to view these problems from the perspectives of engineers, policymakers and business people.

5.1.2. Freshman – introduction to design
At Rose-Hulman Institute of Technology, Introduction to Design is a required freshman course featuring teams of three or four students working on real civil engineering design projects for real clients [77]. The topics in civil engineering that are covered are communication of engineering design, being a professional,

time management, ethics, sustainability, and engineering business management. In the sustainability module, students are asked to reflect on their role as engineers in sustainable development and the sustainability of their project design. Each year, the class makes a trip to a major city (either St. Louis, Chicago, or Cincinnati) to visit engineering projects and meet with engineers. They include a sustainable engineering project as one of the visits during this yearly trip.

5.1.3. Sophomore engineering

Further integration of green engineering can be applied across the disciplines at the sophomore year. At Rowan University, our Sophomore Engineering Clinic is a multidisciplinary sophomore design course with a major focus on communication skills. A recent project was analyzing the solid waste stream of the university. Chemical and civil engineering students worked in teams to assess and improve solid waste management strategies for the campus. The teams researched recycling technologies, performed a life cycle analysis for each major component of the waste stream, investigated fluctuations in the markets for various recyclable materials, and weighed economic, environmental, and social factors that impact recycling programs. One of the driving philosophies behind the project was to increase awareness of the lifetime and fate of common products and introduce the concept of product stewardship to engineers. This goal was easily accomplished within the context of a traditional chemical engineering course by applying life cycle analysis to any disposable item (e.g. food packaging, paper, beverage containers). This type of problem assignment could also be placed in a mass and energy balance course.

At Brabant University multidisciplinary projects in sustainability have been conducted with teams of engineers [94]. Examples of these projects include assessing the need for information on sustainable energy options in households and in small and medium enterprises, improving energy efficiency of a new home for the elderly, and a re-use plan for Xerox copiers.

By the sophomore year, a more detailed life cycle analysis can be undertaken. During this year, students will be learning about mass and energy balances so this is a good point to look at a detailed life cycle.

5.1.4. Material and energy balances

The introductory material and energy balance course is a logical place to put basic terminology and concepts of green engineering. A module for a material and energy balance course has been prepared [95] and is currently available [82]. The curriculum module developed [96] has 25 problems (with solutions) that can be used by an instructor for in-class examples, cooperative learning, home work problems, etc. Two to four problems have been developed for each main topic in material and energy balances and the majority of these problems have multiple parts. Most require a quantitative solution, while others combine both a chemical-principle calculation with a subjective or qualitative inquiry. The problems take a topic from a particular subtopic/topic (section/chapter) and then find a green engineering analog. Some cover specific terminology, principles, or calculations covered in both texts, such as in the calculation of vapor pressures of volatile organic compounds (VOCs) while others introduce concepts only covered in a green engineering text.

Unit conversions typically used in green engineering process calculations, methods of representing pollutant/emission concentrations, and various defining equations used in green engineering can easily be included. Overall "closing the balance" of a chemical manufacturing process, balances on recovery and reuse operations in green engineered processes, green chemistry in stoichiometry, combustion processes, and environmental impact fit nicely into the core fundamentals of material balances. Single and multiphase systems should include calculation of pollutant volatility using vapor pressure and condensation calculations (gas–liquid equilibrium) for vapor recovery processes. Representation of various forms of energy in a green engineering process can also be incorporated into the course such as recovery of energy in a process, energy use in green chemistry reactions, and energy of combustion processes.

The concept of occupational exposure is introduced by having students perform a unit conversion with a dermal exposure equation. In a similar way, workplace exposure limits are introduced in the context of calculating concentration using mole and mass fractions. This helps optimize time usage and course flow, since as prior papers on various subjects have pointed out "to put in X, you need to take out Y". By taking basic material and energy concepts and designing a problem to introduce a green engineering concept, a unique integration of concepts occurs.

Some problems have additional questions that require students to investigate the literature, go to a website, or perform a more qualitative analysis of the problem. For example, in the dermal exposure problem, the student must go to an EPA or related website to determine threshold limiting values and permissible exposure limits for other chemicals. The level of green engineering material is quite elementary since the objective is to give students some familiarity with concepts that would form the basis for more substantial green engineering problems in subsequent courses such as transport, thermodynamics, reactor design, separations, plant design, etc.

Table 7 presents an overall conceptual view of green engineering topics mapped to those in a material and energy balance course. The mapping is done in a very generic way so that an instructor can see the general outline of the topics taught in a material and energy balance course and some of the general areas of green engineering concepts. Not all of the concepts covered in a material and energy balance course have a green engineering analog and *vice versa*. That is why the EPA-supported Green Engineering Project has multiple modules

Table 7

Conceptual mapping of green engineering topics in a material and energy balance course

Green Engineering Topic	Material and Energy Balance Topic (follows Elementary Principles of Chemical Processes, Felder & Rousseau, 3rd)
How green engineering is utilized by chemical engineers in the profession	Chapter 1: What Some Chemical Engineers Do for a Living
Unit conversions typically used in green engineering process calculations	Chapter 2: Introduction to Engineering Calculations
Various defining equations used in green engineering	
Typical method of representing concentrations of pollutants in a process (%, fractions, ppm, etc.)	Chapter 3: Process and Process Variables
Overall "closing the balance" of a chemical manufacturing process	Chapter 4: Fundamentals of Material Balances
Balances on recycle operations in green-engineered processes	
Green chemistry in stoichiometry	
Combustion processes and environmental impact	
Use of various equations of state in green engineering design calculations for gas systems	Chapter 5: Single Phase Systems
Pollutant concentrations in gaseous form	
Representation and calculation of pollutant volatility using vapor pressure	Chapter 6: Multiphase Systems
Condensation calculations (gas–liquid equilibrium) for vapor recovery systems	
Liquid–liquid extraction balances for pollutant recovery systems	
Representation of various forms of energy in a green engineering process	Chapter 7: Energy and Energy Balances
Recovery of energy in a process — energy integration	Chapter 8: Balances on Nonreactive Processes
Use of heat capacity and phase change calculations	
Mixing and solutions issues in green engineering	
Energy use in green chemistry reactions, combustion processes	Chapter 9: Balances on Reactive Processes
Overall integration of mass and energy balances in green engineering on a overall plant design basis	
Use of various simulation tools and specifically designed software for green engineering design	Chapter 10: Computer-Aided Calculations
Representation of mass and energy flows for transient processes with green engineering significance	Chapter 11: Balances on Transient Processes
Industrial cases studies of green engineered manufacturing processes	Chapters 12–14: Case Studies

developed for other courses in the chemical engineering curriculum. The material in this module was developed to be used at the first semester sophomore level and therefore integrates green engineering concepts in a way that a student starting a chemical engineering program can readily understand.

5.1.5. Engineering thermodynamics and heat transfer

Mark Schumack at the University of Detroit Mercy used the unifying theme of energy conservation as a student learning aid in his engineering thermodynamics and heat transfer courses [97]. One of his objectives was to demonstrate the relevance of heat transfer and thermodynamics to important societal issues. The energy conservation theme was introduced using three methods: weekly homework assignments, a design project in heat transfer, and class discussions. Homework assignments from the texts were supplemented with definitions of terms, essays, and problems requiring engineering calculation. Of the 6–8 homework problems assigned per week, 1 or 2 were replaced with energy conservation assignments. Four of the special assignments were essays on energy conservation is an ethic, practical energy conservation suggestions, energy conservation and the poor, and the hydrogen economy. Homework problems included retrieving and analyzing energy utilization and reserves, quantification of the R-value, air- or argon-filled windows, and the validation of advertising claims of a reflective room light using compact fluorescent technology. The design project was on ground-coupled heat pumps for weather conditions in the state of Michigan.

5.1.6. Materials courses

In a paper by Kampe [98], a description is given of the incorporation of green engineering in a required Materials Science and Engineering course, which is also available as a technical engineering elective at Virginia Tech. The selection of a material-of-construction for engineering components or systems has environmental implications that should be considered in design. In some cases, the design objective and environmental stewardship are directly related and mutually compatible. This occurs in using a design objective that minimizes costs that include the waste generated or energy consumed. In many other cases, the primary design objective may not seem to have a relationship to environmental issues such as using an objective that seeks to maximize the structural efficiency (e.g. strength, size, lifetime) of some component or system. In reality, both situations have direct environmental relevance since their objectives may further rely on the manufacture of selected materials and their associated availability in a form suitable for the given application.

Green engineering represents a design philosophy or approach where the implications of a particular design or material selection are considered on a total lifetime basis. That is, while a certain material may offer advantages in terms of prior practice or in-service performance, it may additionally require substantial industrial and/or societal investment in terms of production, disposal, and public health. Thus, it is appropriate and ethical for engineers to consider such issues at the design stage of product development, since they are optimally positioned to make decisions in which environmentally responsible options can be considered and potentially implemented.

In one particular example, the green engineering approach is illustrated in which the environmental load associated with the selection of a specific material can be routinely assessed as part of the overall decision-making process used in engineering design. This presents the selection of a replacement material for asbestos insulation in habitable buildings. This case study seeks to resolve the public health issue of asbestos replacement that minimizes in-service heat losses, and considers the energy expenditures associated with the availability of certain candidate replacement materials.

Another example of integration in a materials course is given by Stephen Stafford at the University of Texas El Paso [64]. In his sophomore level Introduction to Materials Science and Engineering Course, required by all engineering disciplines, he has a Sustainable Engineering Emphasis using life cycle analysis. He highly recommends the use of the text by Pat Mangonon entitled *The Principles of Materials Selection for Engineering Design* [99], which deviates from typical materials texts by sections relating to Life of Component Factors, Cost and Availability, Codes and Statutory requirements, Risk Issues, and Performance and Efficiency. Materials selection based on life cycle assessment is also covered with case study studies in the areas of beverage container recycling and automotive materials design criteria for safety, crash worthiness, and recycle potential. Most of these topics can be easily incorporated into the course's topical outline. At least 2 weeks of the semester will be dedicated to a case study format and the presentation of open-ended engineering problem analysis.

5.1.7. Mechanics of materials

Again at Rose-Hulman Institute of Technology [77], this class has traditionally included behavior of elasto-plastic materials, composite beams and columns, and similar structural components, but focused on the properties of manufactured plastics, timber, and metals. Sustainability is now being introduced through several class assignments considering the material behavior of recycled materials, such as plastic residential decking. These assignments merely replace others dealing with materials that can be modeled as elasto-plastic, so there is no loss of content with the revision. Students will consider how the engineering behavior of these materials will affect design of a simple engineered system, and will be asked to reflect on the significance of sustainability to their work as designers.

5.1.8. Separation processes

Separation topics covered can be applied in a green engineering way in an overall role in pollution prevention such as in the reduction of byproducts, waste minimization, emissions reduction, etc. The choice of the proper mass-separating agent from a green engineering standpoint for the particular industrial separation is a key criteria to be presented. Ultimately, a separation course should present sound rationale for the "green" integration of separation technologies in a reuse/recovery mode where valuable material(s) may be recovered and reused in the overall process. These approaches should be applied in the discussion of design and application of the various separation methods to the system being purified, fractionated or concentrated. Separation processes courses also need to encompass a broad range of both traditional and novel unit operations such that a student can see the pros and cons in their application from a green engineering standpoint.

At Rowan University, a two-semester sequence of separations courses are taught with the first being equilibrium staged (distillation, extraction, absorption, etc.) and the second rate controlled (reverse osmosis, ultra/microfiltration, adsorption, crystallization, etc.). When each of these processes is discussed, a particular problem or case study can be employed showing an application for material recovery/reuse or related pollution prevention. For example, reverse osmosis applications in pollution prevention, reuse/recovery, and mass integration in a variety of manufacturing processes can be described. Reverse osmosis use in electroplating industry to recover and reuse purified water and recover and reuse concentrated plating metals is an excellent example from an environmental and economic standpoint since both separated streams can be reused. For a more advanced topic, students can investigate the integration of a novel technology, membrane pervaporation, with a traditional separation, distillation, in azeotropic separation. A good design case here is replacing the entrainer benzene in ethanol–water separation with the pervaporation technique, since the potential release of benzene in the environment is removed. In the above cases, students can perform calculations to quantify the environmental improvements.

A series of problems were developed to accompany a chemical engineering Separation Processes course that intends to integrate concepts of Green Engineering. The materials were developed around the "Green Engineering" text and can be used with any of the texts covering separation processes or mass transfer operations [82]. Some schools teach these topics in a mass transfer course covering some of the traditional separation processes of distillation, absorption/stripping. At many schools these equilibrium-staged processes are frequently grouped together in a course with suitable name. The problems can also be used to supplement elective or required courses in advanced separations or rate-controlled separations, such as the membrane-based operation problems developed for this module.

The recommended approach is for problems to be used for in-class examples, cooperative learning exercises, homework or exam problems. The problems are grouped according to the separation process. One to two problems have been developed for each separation process and the majority have multiple parts. Most require a quantitative solution, while others combine both a process theory, application, or design calculation with a subjective or qualitative inquiry. The problems take a topic from a particular separation process and then find a green engineering application or green engineering analog. To view the full problem set with solutions, the reader is referred to http://www.rowan.edu/ greenengineering

In some problems, additional questions are asked that require the student to investigate the literature, go to a website, or perform a more qualitative analysis of the problem. For example, the student may need to go to an EPA website to determine potential hazards related to the release of the chemical being separated or find more environmentally benign solvents to use in the operation.

It is envisioned that the instructor covering a particular separation process topic, such as liquid–liquid extraction column calculations, would be presenting some in-class examples or cooperative learning exercise. Instead of a generic type calculation that has no green engineering significance, the instructor could use the problem developed in this module. Not only would the student get the active experience of solving a concentration type problem, but a new concept from the Green Engineering text can be introduced. In this case, the student would learn about the concept of how a separation process is used in the reuse and recovery of materials (process mass integration) in a chemical manufacturing plant. The objective of most problem is to introduce this concept to the student taking information from the Green Engineering text so that additional lecture activity on the green engineering topic is not needed, although the instructor may wish to review the Green Engineering text section beforehand or assign reading of sections to students to accompany the problem.

If the problems are assigned as homework exercises, then the instructor can provide additional green engineering commentary in reviewing the solutions with the students. Alternately, the instructor may require the student to do additional reading from the green engineering text to accompany the homework exercise. It is encouraged that the instructor actively engage students in the learning process for both separation process concepts as well as the green engineering materials.

5.1.9. Electronics I and electronic devices

At the University of Texas El Paso, Electrical Engineering professor Benjamin C. Flores [65] has students in the junior level Electronic Devices and Electronics

I courses integrate aspects of sustainability. In Electronic Devices students are asked to design diodes or transistors to meet specified constraints. Students are required to describe how to fabricate the devices they design including the use and disposal of the wide variety of toxic chemicals employed. Additionally, they learn about the impact on the environment of the disposal or recycling of the chemicals in their processes.

In Electronics I, students examine the chemicals used and produced in the manufacture of electronic circuits on a printed board. They learn to understand the impact of the products they design on the environment. The students are required to read the chapter on green printed wiring board manufacture in the text by Lee Goldberg [62]. A laboratory activity includes the students implementing an electronic circuit on a copper printed board. They measure the volume of PCB etchant solution (ferric chloride $FeCl_3$) and potable water used to produce the board. Each student prepares a paper discussing precautions taken in using the PCB etchant solution, the effect of discarding $FeCl_3$ into the environment.

5.1.10. Chemical reaction engineering

The synthesis of a process design represents a hierarchical decision process, in which the choice of a particular component impacts all other process decisions. The central feature of most chemical processes is the conversion of raw materials into useful products. As a result, the reactor design is one of the central tasks in the synthesis of a chemical process. The selection of design characteristics, i.e. reactor type, conversion, temperature, use of solvent, etc., dictate many of the remaining process considerations associated with separations and recycle, heat exchange, and use of utilities. Thus, it is appropriate to consider the environmental impacts of a reactor design problem in the context of green engineering [100].

Numerous traditional topics of reaction engineering can be applied to green engineering. For example, in a parallel reaction scheme wherein one reaction leads to the desired product, the reaction temperature, the concentration of the reactant, or the reactor type can often be used to control the selectivity. Similarly, the incorporation of a heterogeneous catalyst can accelerate the rate of reaction or affect the reaction selectivity. An additional element of pollution prevention in reaction engineering is the development of new reactor separator configurations. We have developed a student project using a membrane reactor for the production of ethylene from ethane resulting in lower energy consumption requirements as well as higher conversion [101].

A final area in which pollution prevention can be emphasized in the reactor design class is the area of green chemistry. Here, one investigates whether a new reaction route can be identified that minimizes the possibility of worker or surrounding environmental exposure? Alternatively, the question could be asked, "can one of the products be used as a raw material for another feed stream?" A mapping of green engineering topics (from Allen and Shonnard Green Engineering text) into a Reaction Engineering course using the Fogler text [102] is shown in Table 8. It can serve as a guide to incorporate green engineering into other chemical engineering courses.

5.1.11. Design for pollution prevention or design for the environment

At Rowan University, a senior elective/graduate course in design for pollution prevention is offered every fall semester. In this class, students are exposed to advanced engineering design computing tools. The course is intended to provide the students with an understanding of current technology in the design field specifically molded for energy conservation, waste minimization, and pollution prevention at the source by process modification and pollutant interception. The students are first introduced to environmental regulatory law and the relation between the industrial activity and the environment. The rest of the semester is then devoted to develop the necessary skills to design and retrofit processes so the environmental impact is minimized. To accomplish this, students learn basic optimization theory from unconstrained optimization to linear and non-linear programming modeling. The course is computer-intensive as students are required to pose and solve optimization problems using commonly available software such as Excel (Mircosoft Corp, Redmond, WA) and specialized commercial packages such as GAMS (General Algebraic Modeling System, GAMS Development Corp, Washington, DC). The rationale behind the choice of these programs lies in the fact that practicing engineers are more likely to find Excel in the workplace than GAMS while graduate students and researcher may benefit from a more comprehensive optimization tool such as GAMS that can be applied to many fields. The course then covers topics in heat integration, heat exchanger network design, heat integration in distillation columns, and finally mass exchangers network systems.

5.1.12. Senior design courses

Chemical engineering process design. At the core of green engineering is environmentally conscious design of engineered products and processes. In chemical engineering, this is addressed throughout the curriculum, but culminates in a capstone design course in the senior year. The design of chemical processes proceeds through a series of steps, beginning with the specification of the input–output structure of the process and concluding with a fully specified process flowsheet. Traditionally, environmental performance has only been evaluated at the final design stages, when the process is fully specified. In green engineering, a three tier process can be employed at a variety of stages in the design process, allowing the process engineer more flexibility in choosing design options that improve environmental performance.

Table 8
Mapping of green engineering concepts to a reaction engineering course

Topic	Reaction engineering text source [102] (Fogler 1999)	Green engineering text source
In a reactor design project, two pathways are examined to produce cumene using Tier 1 Environmental Performance Tools	Chapter 1: Mole Balances	8.2 Tier 1 Performance Tools (pp. 200–215) 9.3 Pollution Prevention for Chemical Reactors 9.3.1 Material Use and Selection for Reactors
	Chapter 2: Conversion and Sizing Chapter 3: Rate Laws and Stoichiometry	
Membrane Reactors	Chapter 4: Isothermal Reactor Design	9.5 Pollution Prevention Applications for Separative Reactors (pp. 283–286)
Method of Half-Lives — the use of the half-lives method — example is the attack by OH radicals on hydrocarbons.	Chapter 5: Analysis of Rate Data	5.3 Estimating Environmental Persistence — 5.3.1 Estimating Atmospheric Lifetimes
Maximizing desired products and minimizing undesired products by: choice of reactor types, reaction temperature and pressure	Chapter 6: Multiple Reactions	9.3.2 Reaction Type and Reactor Choice 9.3.3 Reactor Operation (pp. 261–274)
Psuedo-steady-state approximation for the attack by OH radicals on hydrocarbons.	Chapter 7: Nonelementary Reaction Kinetics	5.3 Estimating Environmental Persistence — 5.3.1 Estimating Atmospheric Lifetimes
Heat Integration — minimizing energy inputs (only an introductory comment for reactor design)	Chapter 8: Energy Balances	10.1 & 10.2 Process Energy Integration
	Chapter 9: Unsteady-State Nonisothermal Reactors	
Green Chemistry: Alternative Pathways	Chapter 10: Catalysis	7.2.3 Synthesis Pathways
	Chapter 11: External Mass Transfer Chapter 12: Internal Diffusion Chapter 13: Residence Times Distributions Chapter 14: Models for Nonideal Reactors	

A hierarchy of tools has been presented in the text by Allen and Shonnard [47] for evaluating environmental performance throughout the design process and is summarized in Table 9. At the early stage, the designer is concerned with the environmental implications of major input and output materials, with issues of product use and recycle, and of waste treatment alternatives. As the design becomes more defined and detailed, issues of pollution prevention at the process level become important (in-process recovery and recycle, heat integration, and mass integration). In the final design stage, the engineered process is evaluated and perhaps optimized using not only economic objectives but also a comprehensive set of environmental objectives.

The "tier 1" environmental assessment is applied at the earliest stage of design when a large number of design alternatives, for example, raw materials and reaction pathways, is present and where only the most basic input and output information is available for raw materials, products, and by-products. This assessment is based on toxicological properties of each raw material, product, and by-product. The goal is to screen out chemicals and pathways having very high impact potentials. Following this, a preliminary process flowsheet, with reactors, separation units, storage vessels and stream flow rates, would be evaluated using a "tier 2" environmental assessment. Direct emissions from

Table 9
Green engineering hierarchical design summary [31]

Green engineering evaluation	Design stage	Tools
Tier 1	Pathway Development and Screening	Screening of alternatives using: Economic criteria Environmental criteria Toxicity criteria
Tier 2	Preliminary process flowsheet — specific process units and streams are defined	Emissions estimation Selection of appropriate: Mass separating agents Process units Processing conditions
Tier 3	Detailed design	Process integration methods (energy/materials) Multimedia environmental fate modeling Relative risk assessment using metrics

major process units and fugitive releases from numerous minor sources are estimated using emission factors. Targeted pollutants can then be identified from the process emissions (Toxic Release Inventory chemicals, greenhouse gases, ozone depleting substances, criteria pollutants, etc.) and listed per mass of product. In addition, energy consumption, total mass of materials, and water usage is also compiled. Material and energy flow profiles of chemical processes are sometimes available for use in these assessments.

Later, at the point of detailed process design, normally only 2 or 3 alternative flowsheets remain to be evaluated. A "tier 3" environmental assessment would add multimedia environmental fate and transport calculations to the emissions from the "tier 2" assessment. In addition, a number of impact indicators would be used to characterize the various environmental effects of each emitted chemical.

Construction technology. In a green design course at Indiana University – Purdue University Indianapolis (IUPUI) a module on the design of roofs is presented [103]. Green design and construction refers to architectural design and construction practices that take into consideration a number of issues related to the environment, including, but not limited to, energy savings in heating and cooling, environmental-friendly construction materials, wastewater, and placement on site. Despite the fact that only 3% of new buildings in the USA have some environmental-friendly features, there is an ever-increasing interest in green design and construction. This manifests itself in paying special attention to energy conservation both in the production of materials used in buildings as well as energy conservation during functioning of the building. The guiding principle in such undertakings has been consideration and development of green solutions/alternatives for diverse building components/parts that: have fewer

Table 10
Design for the environment course topics

I. Introduction to environmental issues, regulations and risk assessment (defining the problem, an overview of environmental issues, waste generation and management, exposures and risk assessment, environmental legislation, the roles and responsibilities of chemical engineers)

II. Assessing and improving the environmental performance of chemical processes (assessing environmental fate and impacts of chemicals, green chemistry, assessing the environmental performance of flowsheets, pollution prevention for unit operations, flowsheet analysis for pollution prevention (HAZOP and Pinch technology), total cost accounting

III. Life-Cycle Assessment and Industrial Ecology

IV. Case studies

---

adverse ecological consequences, enhance the aesthetic appearance, are cost competitive, and are economically feasible when compared with traditional methods.

Civil engineering design and synthesis. In the required senior level – capstone design course in civil engineering at Rose-Holman has, aspects of sustainability have been incorporated [77]. In this course, students work in teams on real projects submitted by corporate and government sponsors. Presently, alternatives considered for design often include sustainable technologies, such as wetlands, alternative energy sources, or recycling of available materials, and the students are aware that the alternatives are environment-friendly. However, inclusion and consideration of sustainability is still project-dependent, being driven by project needs or the client interests, rather than as a fundamental consideration. What is needed in this course is a requirement that all projects are evaluated using sustainability metrics.

5.1.13. Engineering clinic/industrial/research projects for juniors and seniors
In the last 2 years of the Rowan Engineering curriculum, students take a project-based course for all four semesters called Junior and Senior Engineering Clinic. In this clinic, student teams work on multidisciplinary research and design projects supported by industry, state or federal agencies [101,104]. Many of these projects investigate the use of new and innovative technologies to replace traditional unit operations [105]. All of these projects start with an assessment of the current process and predict the impact of the new technology on the economics of the process and examine reductions in generation of pollutants at the source and assess reductions of risk to human health and the environment. In many of the industrially funded projects there is a large reduction in risk and pollutant generation. Many of these projects have continued in the summer by involving students through a Research Experience for Undergraduates Program in Pollution Prevention funded by the National Science Foundation.

6. Elective/required courses in green engineering

The traditional method to add a new subject in engineering is to introduce the topic as an engineering elective course. As described previously this has been done at several universities. The advantage of this method is that students see a concentrated coverage of the topic for one semester. The disadvantage is that only a select number of students take these courses.

One very innovative approach is to have a required course for all engineers. This could be done in the first year as discussed above at the University of Surrey [70,71] or in the second year. At the University of British Colombia, Dr.

Crofton teaches a required course for civil engineers titled, "Introduction to Sustainable Development," which is delivered using an interactive web-based format. This course insures that all civil engineers begin to think of sustainable development early in their educational development.

Another approach to teaching this topic as a stand-alone course is through independent study. At the University of Minnesota — Duluth, Rich Davies teaches green engineering as an independent study course [106] and many of his students enroll for this course during their industrial co-op. The advantage of this situation is that the green engineering project for this course is linked to an industrial problem that they are currently working on. This is an innovative method to combine green engineering with industrial problems.

This section presents a short synopsis of a number of approaches to teaching green engineering as a stand-alone course. These three courses should give the reader an appreciation for the variety of topics and teaching methods employed at several universities. The first two examples use a method of delivery as an upper level engineering elective in design. The second method is to combine safety and green engineering into an upper level required course.

6.1. University of Texas – Austin

Professor David Allen [107] teaches several courses at the University of Texas at Austin in the green engineering area. In chemical engineering a course titled, "Design for Environment" and in the general education area for all students is "Engineering Sustainable Technologies [108]". The engineering course is a 3 hour upper level course that meets twice per week for 1 semester and is based on the Allen and Shonnard text [47]. Additional reference texts include Allen and Rosselot [49] and El-Halwagi [55]. The general outline of the course is shown in Table 10. The goal of this course is to present engineering tools that are used to improve the environmental performance of chemical products and processes. This includes assessment tools, such as life cycle assessment and risk assessment, as well as design tools such as heat and mass exchange network synthesis. On completion of this course students are expected to have:

1. An understanding of the important environmental issues, and how chemical processes and products impact the environment
2. A conceptual understanding of risk assessment
3. A familiarity with major pieces of environmental legislation
4. An appreciation of the ethical duties and responsibilities of engineers in environmental problem solving;
5. Ability to assess the environmental fate of chemicals
6. An ability to estimate exposures to chemicals

7. Ability to identify and evaluate the environmental impacts of materials and reaction pathways
8. Ability to assess the environmental performance of process flow sheets
9. Ability to improve the environmental performance of unit operations
10. Ability to improve the environmental performance of process flowsheets
11. Ability to evaluate potential economic benefits of superior environmental performance
12. Quantitative and qualitative understanding of product life cycles
13. Quantitative and qualitative understanding of the role of chemical processes in industrial material and energy flows.

6.2. Notre dame

At Notre Dame, Brennecke and Stadtherr [32] believe that a concentrated course will uniquely equip students with the active knowledge and the ability to implement pollution prevention that totally eliminate pollutants at the source. Their one-semester, senior level course has the following objectives: (1) to educate students on the real costs of operating processes that release pollutants to the environment (including costs of related to regulations, community, and potential future liabilities); (2) to provide students with an awareness of strategies for minimizing or reducing the environmental impact of a given chemical process; and (3) to give students the opportunity to work on the design of processes using new technologies that totally eliminate pollutants at the source.

The course includes four major components. First, an introduction to pollution prevention is given which includes management and maintenance procedures, as well as simple process modifications that can prevent pollution, especially through the reduction of fugitive emissions. In addition, this portion of the course includes an introduction to life cycle analysis and industrial ecology. Second, pertinent environmental regulations that impact the design and operating costs of chemical processes are presented. Third, an overview of new technology and current research efforts to develop alternative technologies that minimize waste or eliminate pollutants through solvent substitution, the use of different raw materials and intermediates, and the development of more selective catalysts and reactor systems is given. The fourth and key part of the course involves the development and comparison of chemical process designs that juxtapose conventional chemical processes with new, environmentally benign technologies. Some examples of these processes are given in the paper and include the decaffeination of coffee, extraction of soybean oil, and the production of *p*-nitroaniline.

6.3. Michigan technological university

Michigan Tech has a unique approach to teaching green engineering by combining it with a required safety course. Green engineering is taught within a combined senior-level semester-long course titled "CM4310 Chemical Process Safety/Environment". In this manner about twothirds of the lectures are on safety and the remaining third are on green engineering. In the green engineering portion of the course the following subjects are covered:

1. Introduction to environmental issues, Chapter 1
2. Major federal environmental regulations, Chapter 3
3. Environmental properties of chemicals, Chapter 5
4. Environmental releases, Chapter 8
5. Unit operations and pollution prevention, Chapter 9
6. Detailed flowsheet impact assessment, Chapter 11

The references to the chapters are based on the Allen and Shonnard text [47]. Goals of the course are to:

- teach fundamental concepts of safety and environmental issues related to chemical processing,
- present methods and software tools for assessing safety and environmental performance of process designs,
- provide methods to improve safety and environmental performance.

Ten weeks in the course are for process safety and five for green engineering. Course format for the green engineering portion uses lectures, weekly homework assignments, a writing assignment on an environmental issue, and a graduate student-run workshop to demonstrate the "tier 3" impact assessment software tool, EFRAT. During the workshop, using a simple process flowsheet as a case study, students link a process simulation output file with the environmental impact assessment software, estimate emissions from process units, and generate output tables of environmental impact indices for process units and chemicals. Grading for this course is based on end-of-chapter homework assignments, the writing assignment, and an examination.

This course has the advantage that all chemical engineering students that graduate from Michigan Tech will have an understanding of green engineering. The disadvantage of the course is that the students cover only a small portion of the material in a short period of time.

7. General education courses

Many universities are beginning to recognize that developing technological literacy should be a part of the general education of all university students. One possible approach to developing technological literacy would be to teach students the general concepts of mass and energy balances, applying these concepts in the analysis of coupled engineered and natural systems at local, regional, and global scales. One such course is offered at the University of Texas by the Chemical Engineering Department. The course begins with a brief description of biogeochemical flows (grand cycles) of the six elements (carbon, hydrogen, oxygen, nitrogen, phosphorus, and sulfur) that are the major constituents of living tissues, which accounts for 95% of the biosphere. Understanding these "grand cycles," which describe how the Earth's systems process materials, is critical to developing an understanding of global environmental changes. The course then focuses on engineered systems, noting that many of the grand cycles are now significantly affected by the flows generated by human activities. So, the grand cycles of material flows must include a description of materials and energy flows in both natural and engineered environments.

The general education course titled, "Engineering Sustainable Technologies" is open to non-engineers and meets the science and math requirements for general education. The focus of this course is on flows of materials and energy in engineered environments at local, regional, and global scales, and the interaction of those anthropogenic flows with natural cycles of materials and energy. The course begins with a brief description of biogeochemical flows (grand cycles) and then examines anthropogenic material flows at the national level, in industrial sectors, and for consumer products. The overall outline of the course is given below:

    I. Introduction to environmental issues, regulations and risk assessment
   II. Environmental partitioning
  III. Introduction to biogeochemical flows
  IV. Introduction to engineered material and energy flows
   V. Tools for characterizing material and energy flows — Life cycle assessment, Material and Waste flow analyses
  VI. Evaluation of life cycles
 VII. Detailed case studies.

The Textbook used by E.S. Rubin is titled, "Introduction to Engineering and the Environment".

This leads to analyses of anthropogenic material flows at the national level, in industrial sectors, and for consumer products. Students in the course gain a quantitative appreciation of the interactions between engineered systems and the natural environment, they are introduced to problem-solving techniques and they become familiar with some of the most widely applied engineering principles — mass and energy balances.

Other general education courses in aspects of sustainable development are, listed by Fiona Croft [109]. She gives the following examples:

Examples include: *Psychology and Sustainable Development* (Psychology, Hofstra University); *Strategies for Sustainable Development* (courses by this title are offered in the business schools at both McGill and University of Michigan); *Environmental Justice* (Sociology, Brown University); *Western Environmental Policy* (History, California Institute of Technology); and *Economy, Environment, and Community* (Economics, Tufts University); *Biotechnology, Nature and Society* (Biology, Tufts University); *Hanford Social and Environmental History* (Sociology, Washington State University); *Strategies for Environmental Management* (this course at the University of Michigan's Business School addresses sustainable technology development); and *Environmental Ethics* (alternative technology is among the issues considered in this Philosophy course at the University of Alberta).

At this time there are many more examples of courses in sustainability in general education courses at universities and it is suggested that these types of courses be examined and incorporated in engineering programs.

8. Green chemistry courses

Another area in which the basics of green engineering is being taught is through the science courses. There are a number of chemistry departments that have incorporated green chemistry throughout their programs[110]. The Green Chemistry Institute has sponsored a number of initiatives to develop green chemistry curriculum materials [111]. These can be obtained through the ACSs chemistry.org website under Green Chemistry Education. On this website are articles on how to introduce green chemistry into the curriculum, laboratory experiments and demonstrations, case studies based on the Presidential Green Chemistry Challenge Awards, as well as a video and DVD of presentations on green chemistry. One of these resources titled, "Going Green: Integrating Green Chemistry into the Curriculum," gives specific examples of this integration. Many of these examples focus on integrating green chemistry into core courses as well as having a standalone course in green chemistry. Organic chemistry is an excellent subject area to introduce green chemistry in both the lecture and lab sections. Several examples are given for new or modified experiments in this course. [112,113]. Incorporating green chemistry into core chemistry courses will positively influence student's opinions on chemistry as well as promote green engineering.

9. Conclusion

The engineer, as the designer of products and processes, also has a central role in designing chemical processes that have a minimal impact on the environment. We as educators can prepare our students to use the risk assessment tools of green engineering to design new processes and modify existing processes. As a result, green engineering could become a central component of the engineering curriculum. This chapter has reviewed several of the methods currently being used to teach green engineering within engineering programs. Examples are given in which green engineering is integrated throughout a curriculum as well as examples of stand-alone courses in green engineering. Many examples of the integration of green engineering in specific courses are also given.

(Proceedings of the American Society for Engineering Education can be obtained at http://www.asee.org/about/events/conferences/search.cfm)

References

[1] U.S. Environmental Protection Agency, Green Engineering web site, http://www.epa.gov/opptintr/greenengineering/index.html, 2005.

[2] Ritter, S. K., "A Green Agenda for Engineering: New set of principles provides guidance to improve designs for sustainability needs," July 21, 2003, 81 (29) Chemical & Engineering News pp. 30-32.

[3] N. Bakshani, D.T. Allen, In the States: Pollution Prevention Education at Universities in the United States. Pollution Prevention Review 3 (1) (1992) 97–105.

[4] Institution of Professional Engineers of New Zealand, Requirements for Initial Academic Education for Professional Engineers, http://www.ipenz.org.nz/ipenz/forms/pdfs/Initial_Academic_Education.pdf, December 2003.

[5] The Institution of Engineers, Australia, Manual for the Accreditation of Professional Engineering Programs, http://www.ieaust.org.au/membership/res/downloads/AccredManual.pdf 7 October 1999.

[6] Canadian Council of Professional Engineers, 2003 Accreditation Criteria and Procedures, http://www.ccpe.ca/e/files/report_ceab.pdf.

[7] Engineering Council of South Africa, Standards and Procedures System — Standards for Accredited University Engineering Bachelors Degrees, Document: PE-61 Revision — 1 Date: 20 July 2000, http://www.ee.wits.ac.za/~ecsa/pe/pe-61.pdf.

[8] The Institution of Engineers of Ireland, Accreditation Criteria for Engineering Education Programmes, http://www.iei.ie/uploads/common/files/IEI%20AccredCriteria.pdf, November 2003.

[9] Engineering Council UK, UK Standard for Professional Engineering Competence, Chartered Engineer and Incorporated Engineer Standard, http://www.engc.org.uk/documents/Eng_IEng_Standard.pdf, 2005.

[10] B.A. Kuryk, VP, Global Issues Management & Product Stewardship, BP, in the AIChE Critical Issues Series, Proceedings of the Global Climatic Change Topical Conference of the AIChE 2002 Spring Meeting, New Orleans, LA, 10–14 March 2002.

[11] M. T. Foley, Director External Affairs and Business Development, BP Solar in the AIChE Critical Issues Series, Proceedings of the Global Climatic Change Topical Conference of the AIChE 2002 Spring Meeting, New Orleans, LA, 10–14 March 2002.

[12] Press Release, Local and Foreign Chemical Companies put Sustainable Development into Action, 30 March 2004, http://www.basf.com.sg/apac/ViewPress.asp?articleid = 507& page = 1.

[13] Dow Public Report Update 2000–1999 Results: Sustainable Development, http://www.dow.com/about/2000pr/pg2.htm2005.

[14] DSM Sustainability, http://www.dsm.com/en_US/html/sustainability/sustainability_home.htm,7 2005.

[15] DuPont Sustainable Growth, http://www2.dupont.com/Our_Company/en_US/glance/sus_growth/sus_growth.html, 2005.

[16] Chemical Companies Embrace Environmental Stewardship, Millenium Special Report Chem. Eng. News, 6 December, 77 (49) CENEAR 77 (49) (1999). 55–63 ISSN 0009-2347 http://pubs.acs.org/hotartcl/cenear/991206/7749sustain.html.

[17] Sustainable Prices for Medicines and Chemical Processing at GlaxoSmithKline (GSK) http://www.gsk.com/financial/reports/halfyear2001/review/downloads/gsk_hy_review_2001.pdf and annual reports http://www.gsk.com/financial/reports/ar/pdf_excel/report/report.pdf.

[18] Merck, http://www.merck.com/overview/speech/051401.htm.

[19] R. Gupta, Rohm and Haas 2000 EHS and Sustainability Annual Report, http://www.rohmhaas.com/EHS/index.html2005.

[20] Press Release, Zurich, 4 September 2003, Results of the DJSI Review 2003, http://www.sustainability-indexes.com/djsi_pdf/news/PressReleases/DJSI_PR_030904_Review.pdf.

[21] Dow Jones Sustainability Indexes (DJSI) — Annual Review 2003, Zurich, Switzerland, 4 September 2003, http://www.sustainability-indexes.com/djsi_pdf/publications/Presentations/DJSI_PRT_030904_Review.pdf.

[22] Center for Waste Reduction Technologies — http://www.aiche.org/cwrt/.

[23] www.acs.org.

[24] American Chemistry Council Responsible Care http://www.americanchemistry.com/.

[25] ASEE Board of Directors, ASEE Statement on Sustainable Development Education, http://www.asee.org/about/Sustainable_Development.cfm, 2005.

[26] The Engineers Forum on Sustainability, ASCE, AIChE, and ASEE, http://www.asee.org/resources/organizations/aboutefs.cfm, 2005.

[27] American Society of Civil Engineers ASCE Code of Ethics. Retrieved January 10, 2003 from http://www.asce.org/inside/codeofethics.cfm.

[28] National Pollution Prevention Center. University of Michigan, http://www.umich.edu/~nppcpub/resources/.

[29] Center for Sustainable Systems, University of Michigan, http://css.snre.umich.edu/

[30] R.P. Hesketh, M.A. Abraham, "Pollution Prevention Education in Chemical Reaction Engineering", in: M.A. Abraham, R.P. Hesketh (Eds), Reaction Engineering in Pollution Prevention, Elsevier Science, New York, 2000.

[31] D.R. Shonnard, D.T. Allen, N. Nguyen, S.W. Austin, R. Hesketh, Green Engineering Education Through a US EPA/Academia, Environmental Science and Technology 37 (23) (2003) 5453–5462.

[32] J.F. Brennecke, M.A. Stadtherr, A course in environmentally conscious chemical process engineering, Computers and Chemical Engineering 26 (2002) 207–318.

[33] J.F. Brennecke, J.A. Shaeiwitz, M.A. Stadtherr, R. Turton, M.J. McCready, R.A. Schmitz, W.B. Whiting, "Minimizing Environmental Impact of Chemical Manufacturing Processes," Proceeding of the 1999 Annual Conference of the American Society for Engineering Education, Charlotte, NC (1999).

[34] M.A. Abraham, A Pollution Prevention Course that Helps Meet EC Objectives, Chem. Eng. Edu. 34 (3) (2000) 272.

[35] C.S. Grant, M.R. Overcash, S.P. Beaudoin, A Graduate Course on Pollution Prevention in Chemical Engineering, Chem. Eng. Edu. 30 (4) (1996) 246.

[36] J.D. Simpson, W.W. Budd, Toward a Preventive Environmental Education Curriculum: The Washington State University Experience, J. Environ. Edu. 27 (2) (1996) 18.

[37] M. Tillman, H. Baumann, After 10 years and 300 Students Our LCA Teaching Experience, Conference Proceedings of Engineering Education in Sustainable Development, Delft, the Netherlands, October 24–25, 2002.

[38] M.H. Gregg, Environmental Life Cycle Analysis for Engineers, Proceedings of the Conference of American Society of Engineering and Education, Session 3251, 2002.

[39] J. Jeswiet, An Approach to Education in Life cycle Engineering — 2002. CIRP International Manufacturing Education Conference, Eschede, 2002, pp. 87–95.

[40] General Electric Fund Project, http://www.sustainable.gatech.edu/curriculum/ge_report.php2005.

[41] J.J. Peirce, R.F. Weiner, P.A. Vesilind, "Environmental Pollution and Control," Butterworth-Heinemann; 4th edition, October 1997, ISBN 0750698993.

[42] R.R. Dupont, L. Theodore, K. Ganesan, Pollution Prevention: The Waste Management Approach to the 21st Century, Lewis Publishers, Inc., December 1999 ISBN: 1566704952.

[43] T.E. Higgins, Pollution Prevention Handbook, CRC Press, March 1995 ISBN: 1566701457.

[44] T.E. Graedel, Streamlined Life-Cycle Assessment, Pearson Education POD; June 10, 1998.

[45] M.A. Curran, Environmental Life-Cycle Assessment, McGraw-Hill Professional Publishing, July 1998.

[46] G.T. Miller, Living in the Environment : Principles, Connections, and Solutions, 13 edition, Brooks/Cole, January 2 2003.

[47] D.T. Allen, D.R. Shonnard, Green Engineering: Environmentally Conscious Design of Chemical Processes, Prentice-Hall, Englewood Cliffs, NJ, 2001.

[48] U.S. Environmental Protection Agency, Green Engineering website - http://www.epa.gov/opptintr/greenengineering/index.html2005.

[49] D.T. Allen, K.S. Rosselot, Pollution Prevention for Chemical Processes. Wiley,, New York, 1997.

[50] K.L. Mulholland, J.A. Dyer, Pollution Prevention: Methodologies, Technologies, and Practices, AIChE Press, New York, 1998.

[51] H.M. Freeman, Industrial Pollution Prevention Handbook, McGraw-Hill, New York, NY, 1995.

[52] A.P. Rossiter, Waste Minimization through Process Design, McGraw-Hill, New York, 1995.

[53] L. Theodore, Y.C. McGuinn, Pollution Prevention, Van Nostrand Reinhold, New York, 1992.

[54] P.L. Bishop, Pollution Prevention: Fundamentals and Practice, McGraw-Hill, New York, 2000.

[55] M.M. El-Halwagi, Pollution Prevention through Process Integration, Academic Press, New York, 1997.

[56] D.T. Allen, N. Bakshani, K.S. Rosselot, Pollution Prevention: Homework and Design Problems for Engineering Curricula. American Institute of Chemical Engineers, American Institute for Pollution Prevention, and Center for Waste Reduction Technologies, New York, 1992.

[57] A. Azapagic, S. Perdan, R. Clift, Sustainable Development in Practice: Case Studies for Engineers and Scientists, Wiley, New York, 2004 ISBN: 0470856092 or 0470856084.

[58] C. Soares, Environmental Technology and Economics: Sustainable Development in Industry, Elsevier Science & Technology Books, 2000 ISBN: 0750670800.

[59] T.E. Graedel, Streamlined Life-Cycle Assessment, Prentice-Hall, Englewood Chiffs, NJ, 1998 ISBN: 0136074251.

[60] G. Sonnemann, Integrated Life-Cycle and Risk Assessment for Industrial Processes, Lewis Publishers, 2003 ISBN: 1566706440.

[61] B.S. Blanchard, W.J. Fabrycky, Systems Engineering and Analysis, Pearson Education, 1998 ISBN: 0131350471.

[62] L. Goldberg, Green Electronics/ Green Bottom Line, Newnes, 1999 ISBN: 0750699930.

[63] K. Gibney, Combining Environmental Caretaking with Sound Economics, Sustainable Development is a New Way of doing Business, Prism, 1999.

[64] S.D. Turner, W.-W. Li, A. Martinez, Developing Sustainable Engineering across a College of Engineering. Session 1402, Proceedings of the 2001 Annual Conference of the American Society for Engineering Education, (2001).

[65] Sustainable Engineering Initiative, The University of Texas at El Paso http://www.utep.edu/green/objectives.htm2005.

[66] Green Design Institute, Carnegie Mellon University, http://www.ce.cmu.edu/GreenDesign/ 2005.

[67] R.P. Hesketh, C.S. Slater, M.J. Savelski, K. Hollar, S. Farrell, A Program to Help in Designing Courses to Integrate Green Engineering Subjects, Int. J. Eng. Edu. 20 (1) (2004) 113–128.

[68] C. Hendrickson, N. Conway-Schempf, H. Scott Matthews, F.C. McMichael, Green Design Educational Modules and Case Studies, Session 2451, Proceedings of the ASEE Conference, St. Louis, 2000.

[69] C.D. Turner, W.-W. Li, A. Martinez, Developing Sustainable Engineering Concepts in a College of Engineering, Proceedings of the 2001 Virginia Tech College of Engineering Green Engineering Conference.

[70] S. Perdan, A. Azapagić, R. Clift, Teaching Sustainable Development to Engineering Students, Int. J. Sust. High. Edu. 1 (3) (2000) 267–279.

[71] Structure of the Chemical Engineering Programmes at the University of Surrey — Flowchart: http://www.surrey.ac.uk/eng/ug/cpe/images/flowchart-1.jpg.

[72] S. Perdan, A. Azapagić, Sustainable Engineering Design: An Interactive Multimedia Case Study, Int. J. Sust. High. Edu. 4 (1) (2003) 33–43.

[73] Institution of Chemical Engineers (IChemE) Sustainable Development Progress Metrics, http://www.icheme.org/sustainability/metrics.pdf, released 16th May 2002.

[74] The Centre for Environmental Strategy, University of Surrey http://www.surrey.ac.uk/eng/ pg/ces/phd.htm2005.

[75] Sustainable Development Activities at University of Oxford, Chem. Eng. http:// www.eng.ox.ac.uk/chemeng/sustsum.html, 2005.

[76] The Department of Engineering Centre for Sustainable Development, Undergraduate Teaching, http://www-g.eng.cam.ac.uk/sustdev/teaching.html, 2005.

[77] M. Robinson, K. Sutterer, Integrating Sustainability into Civil Engineering Curricula, Session 2615, Proceedings of the 2003 Annual Conference of the American Society for Engineering Education, 2003.

[78] T. Gally, A Systemic Approach to Teaching Sustainable Development in an Aerospace Engineering Program, Session 2202, Proceedings of the 2002 Annual Conference of the American Society for Engineering Education, 2002.

[79] J. Venselaar, N. Roorda, T. Severijn, Integrating Sustainable Development in Engineering Education The Novel CIRRUS Approach, Proceedings of the Conference of Engineering Education in Sustainable Development, Delft, the Netherlands, October 24–25, 2002, http://www.projectcirrus.net/engels/cirrus_e.html.

[80] B. Sukumaran, J. Chen, Y. Mehta, D. Mirchandani, K. Hollar, A Sustained Effort for Educating Students about Sustainable Development, Session 1793, Proceedings of the 2004 Annual Conference of the American Society for Engineering Education, 2004.

[81] US Environmental Protection Agency Office of Pollution Prevention and Toxics and Office of Prevention, Pesticides, and Toxic Substances CX 83052501-0 titled Implementing Green Engineering in the Chemical Engineering Curriculum, October 7, 2002.

[82] Green Engineering Website for integrating green engineering throughout the curriculum. www.rowan.edu/greenengineering, 2005.

[83] R.P. Hesketh, K. Jahan, A.J. Marchese, C.S. Slater, J.L. Schmalzel, T.R. Chandrupatla, R.A. Dusseau, Multidisciplinary Experimental Experiences in the Freshman Engineering Clinic at Rowan University, Proceedings of the 1997 Annual Conference of the American Society for Engineering Education (1997).

[84] E. Golub, D. Hanesian, H. Hsieh, A.J. Perna, The Siting and Design of a Manufacturing Facility Using Hazardous Materials, Proceedings of the 1999 Annual Conference of the American Society for Engineering Education, Charlotte, NC, 1999.

[85] J.C.M. Kampe, T. W. Knott, Exposing First-Year Students to Green Engineering, Session 3251, Proceedings of the 2002 Annual Conference of the American Society for Engineering Education, 2002.

[86] R.V. Wiedenhoeft, Historic Background for Colorado School of Mine's Nature and Human Values Course, Proceedings of the 1999 Annual Conference of the American Society for Engineering Education, Charlotte, NC, 1999.

[87] W.E. Rochefort, A Traditional Material Balances Course Sprinkled with Non-Traditional Experiences, Proceeding of the 1999 Annual Conference of the American Society for Engineering Education, Charlotte, NC, 1999.

[88] S. Montgomery, http://www.engin.umich.edu/labs/mel/meBalances.html, 2005.

[89] D.T. Allen, An Industrial Ecology: Material Flows and Engineering Design, in: Dr. Adisa Azapagic, Dr. Slobodan Perdan, Prof. Roland Clift (Eds), Sustainable Development in Practice: Case Studies for Engineers and Scientists, Wiley, New York, 2003.

[90] D. Allen, University of Texas — Austin, Green Product, http://www.utexas.edu/research/ceer/che302/greenproduct/start_here.htm.

[91] EPA LCA Homepage, http://www.epa.gov/ORD/NRMRL/lcaccess/index.htm, 2005.

[92] M.A. Curran (Ed.), Environmental Life Cycle Assessment, McGraw-Hill, New York, 1996.

[93] J.J. Hageman, J.J. van der Boom, J. Venselaar, Integrating Sustainable Development in Engineering Education: The Case for Chemistry and Chemical Engineering, Conference Proceedings, Engineering Education in Sustainable Development, Delft, the Netherlands, October 24–25, 2002, http://www.projectcirrus.net/engels/cirrus_e.html.

[94] L. Dejong, L. van Beek, T. Severijn, J. Venselaar, Multidisciplinary Projects as Learning Tool for Sustainable Approaches: Experience and some Critical Assessment Conference Proceedings, Engineering Education in Sustainable Development, Delft, the Netherlands, October 24–25, 2002, http://www.projectcirrus.net/engels/cirrus_e.html.

[95] C.S. Slater, R.P. Hesketh, Incorporating Green Engineering into a Material and Energy Balance Course, Chem. Eng. Edu. 38 (1) (2004) 48–53.

[96] C.S. Slater, Green Engineering Project — Material and Energy Balance Course Module, June 2003, http://nebula.eng.rowan.edu:82/Home.asp.

[97] M. Schumack, Incorporation of an Energy Conservation Theme into Thermal Science Courses, Session 2533, Proceedings of the 2002 Annual Conference of the American Society for Engineering Education, 2002.

[98] S.L. Kampe, A Method to Incorporate Green Engineering in Materials Selection & Design, Session 1625, Proceedings of the 2002 Annual Conference of the American Society for Engineering Education, 2002.

[99] P. Mangonon, The Principles of Materials Selection for Engineering Design, 1st edition, Prentice-Hall, Englewood Cliffs, NJ, 1998 ISBN: 0132425955.

[100] R. Smith, E. Petela, Waste Minimization in the Process Industries: Part 2. Reactors, Chem. Eng. 509 (1991) 17.

[101] S. Farrell, C.S. Slater, R.P. Hesketh, M.J. Savelski, K.D. Dahm, Membrane Projects with an Industrial Focus in the Curriculum, Chem. Eng. Edu. 37 (2003) 68.

[102] H.S. Fogler, Elements of Chemical Reaction Engineering, 3rd ed, Prentice-Hall, Englewood Cliffs, NJ, 1999.

[103] E.M. Sener, P. Baty, Green Design and Construction: An Example: Commercial Green Roofs, Session 1621 Proceedings of the 2003 Annual Conference of the American Society for Engineering Education, (2003).

[104] J.A. Newell, S. Farrell, R.P. Hesketh, C.S. Slater, Introducing Emerging Technologies into the Curriculum through a Multidisciplinary Research Experience, Chem. Eng. Edu. 35 (2001) 296.

[105] S. Farrell, R.P. Hesketh, C.S. Slater, Exploring the Potential of Electrodialysis, Chem. Eng. Edu. 37 (2003) 52.

[106] ChE 3791 Independent Study — Green Engineering, University of Minnesota — Duluth, http://www.d.umn.edu/~rdavis/courses/che3791/Green/greeneng.htm, 2005.

[107] Chemical Engineering Department Course Offerings, David Allen, allen@che.utexas.edu, http://www.utexas.edu/research/ceer/che341/, http://www.utexas.edu/research/ceer/che302/greenproduct/start_here.htm, 2005.

[108] Course homepage for Engineering Sustainable Technologies, David Allen, University of Texas at Austin, http://www.utexas.edu/research/ceer/che302/index.html, 2005.

[109] F.S. Crofton, Educating for Sustainability: Opportunities in Undergraduate Engineering, J.Cleaner Prod. 8 (2000) 397–405.

[110] T.J. Collins, Introducing Green Chemistry in Teaching and Research, J. Chem. Edu. 72 (1995) 965–966.

[111] D.L. Hjeresen, D.L. Schutt, J.M. Boese, Green chemistry and education, J. Chem. Edu. 77 (12) (2000) 1543–1544 1547.

[112] E.S. Santos, I.C.G. Garcia, E.F.L. Gomez, Caring for the Environment while Teaching Organic Chemistry J. Chem. Edu. 81 (2) (2004) 232–238.

[113] S.M. Reed, J.E. Hutchison, Green Chemistry in the Organic Teaching Laboratory: An Environmentally Benign Synthesis of Adipic Acid, J. Chem. Edu. 77 (12) (2000) 1627–1629.

PART II:
DEVELOPING THE PRINCIPLES

Sustainability Science and Engineering: Defining principles
Martin A. Abraham (Editor)
© 2006 Published by Elsevier B.V.
DOI 10.1016/S1871-2711(05)01005-6

Chapter 5

Systems Thinking

Walter Olson

Department of Mechanical, Industrial and Manufacturing Engineering, The University of Toledo, Toledo, OH 43606, USA

Engineer processes and products holistically, use systems analysis, and integrate environmental impact assessment tools". From the Sandestin Declaration of Green Engineering Principles

Every object on the earth is both a system unto itself and also part of other systems. The smallest grain of sand on a remote beach is a system because the internal interactions of its molecules and atoms and its potential future to be airborne or moved by the waves of the seas. Would the grain of sand exist if it were not for an ocean that supports waves and violent storms, a landmass that supports the deposition of the sand and a material such as quartz rock that permitted the grain of sand to be?

Furthermore, the systems of the earth are highly nonlinear. Thus, seemingly unimportant events result in major actions of the system. Weather forecasters are intensely aware of these phenomena. Weather models that are representative of global weather systems result in different forecasts with only the smallest changes to the input variables. Because of the nonlinearities that exist (whether known or unknown), no part of the system can be removed without making the system less.

This fundamental recognition that the earth is a system drives the need for green engineers to use systems thinking in their pursuit of their discipline. A basic tenet of green engineering is that one should not harm others or the environment by an engineering solution to a problem. But, how does one know? There are many examples where well-meaning engineers attempted to find solutions only making problems worse.

Oxygenated gasoline reduced local air pollutants emissions. A simple and economical solution added methyl tert-butyl ether (MTBE) to the gasoline to solve the problem. MTBE would eventually enter and contaminate ground waters through leaks and spillages. Why? While engineers that treat airborne problems have a discipline that is separate from those treating water borne problems, the earth is a system that does not separate air events from those of water [1].

Engineers knew of petroleum reserves offshore of the North Slope of Alaska for many years. To solve the problem of recovery, a simple solution was to build causeways and gravel pads out from the coast into the brackish shore waters to support petroleum retrieval. However, after executing this solution, five species of fish were at risk: arctic cisco, least cisco, arctic char, arctic cod and fourhorn sculpin. Why? The engineers did not know of the systems of life dependent on the intercoastal flows of brackish water [2].

Bangladesh is a country with most of its population in areas lacking potable surface waters. Wells were drilled by the WHO in Bangladesh to provide reliable fresh water. Only later it was that there arsenic in the aquifer would lead to a national calamity. Why? The ground chemical system is not isolated from systems of water [3].

In each case, the solution users discovered hazards after design, engineering and execution were completed. In each case, the engineers at the time of design would support the claim that the design was the best. Why then did harm come?

We believe that the answer to the questions above is that engineers considered the solution of the problem in isolation from the existing systems in place. Engineers control and manipulate a subset of system interactions to the benefit of man. To address a problem, engineers are trained to focus and concentrate on the immediate observed phenomena. This has the positive effect of reducing the number of variables needed "to solve a problem". However, if the engineer performs a reduction in neglect of all other operating systems, the interactions of the solution with the other operating systems are also neglected. In many cases, the interactions are inconsequential and have only minor wider effects. However, as the examples demonstrate, significant problems can occur because interactions were not looked for, and therefore not known at the time of design.

Systems thinking broaden the engineer's ability to identify potential interactions and to evaluate their consequences. Systems thinking offer green engineers the ability to understand the structure that their product or process exists in and interacts with. This expands the appreciation of their work impacts and provides opportunities to make their work more efficient. In addition, systems thinking also give the engineer information about where they may have problems with their product or process.

In the following sections, the reader is introduced to several common definitions used in describing systems. This is followed by informal and formal methods of performing systems thinking and systems analysis.

1. Definition of a system

The science systems and their interactions rightfully belong to the field of "Systemology" [5]. However, most engineers would better recognize the field by the term "Systems Analysis".

A system is a collection of objects that receive inputs from the external environment and/or from other objects of the system, and transforms the inputs to outputs. The focal point of a system is usually an action verb that captures the essence or gist of a process.

1.1. Objects, inputs and outputs

At the most basic but also highest level of a system, there is but one object with inputs and outputs. This is represented graphically as shown in Fig. 1. The object placed in the block could represent a city, a manufacturing plant, a building, a test tube, a human being, a single cell, or almost anything. Similarly, the inputs could represent visitors to the city, steel coils to a manufacturing plant, heat to the building from the sun, air to the test tube, money to the human being, nutrients to the cell, or others things that enter an object of the system. The outputs might be carbon dioxide produced by the city, wastewater from the manufacturing plant, heat produced within the building to the external environment, color of the chemicals in the test tube, words spoken by the human, and motion of a cell or other things that leave the object. Usually, the inputs and outputs are placed on directed arcs or paths with the arrows showing the type of interaction.

While engineers are most interested in the inputs and outputs of an object, there may exist a class of entities that are required by the object but are not in themselves transformed. These take the function of enablers or catalysts and are often called mechanisms. These need to be identified and placed on the diagram as a special type of input. Depending on the graphical standard that is used for the project, mechanisms are indicted by the position of the arrow entering the bottom of the object block (IDEF0 standard [6]) or alternatively by color-coded or dashed arrows. (Note: IDEF is the name of a widely used systems analysis technique. IDEF0, IDEF1, IDEF1X, IDEF2, etc., refer to different phases and structures of the technique. IDEF0, the basic diagramming and information gathering system was originally developed in the 1970s by SofTech, Inc.).

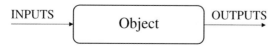

Fig. 1. Elemental system.

Additional special types of arrow that added to the block are constraints that limit the operation of the object in the block. Constraints again depend on the graphical standard used. Constraints are often identified by an arrow entering the top side of the object block (under the IDEF0 standard) or, alternatively, by color coding or dashing of the arrow. A system can be very large or very small. The size of a system depends on the person defining it. A system can be a set of real objects, a set of conceptual objects or a combination of both. Typically models are created that are conceptual systems that represent real systems. In some cases engineers construct conceptual models that interface to a real system but provide the engineer ways to modify the conceptual system and observe the results without affecting the real system. This can be used to great advantage to control a real system if the conceptual system is a reasonably close approximation of the real system. Interactions may also be abstract or concrete depending on the use of the analysis.

A system can be part of a larger system. A system can have smaller systems within itself. In such cases, the smaller systems are termed subsystems. It is correct to call the encompassing systems a super-system although this is rare. An example of a system with subsystems is Fig. 2 for a building climate control system. While the climate control decomposes into eight subsystems, each of the

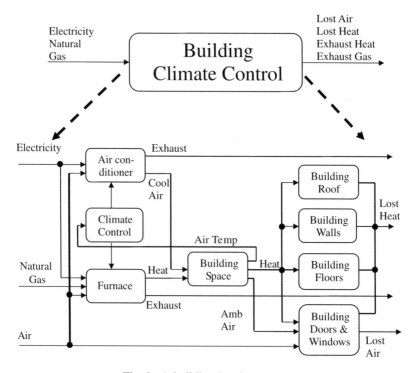

Fig. 2. A building heating system.

subsystems could further be decomposed. The building climate control system would be a subsystem of the overall building. A somewhat mathematical way of thinking of systems is that a system is a function that transforms inputs variables to output variables. Often times, the mathematical function is called a transfer function.

1.2. System types

Systems are termed closed if all of the inputs and outputs are contained within the system; that is to say, the system does not interact with its external environment in any way. Otherwise, a system is termed open. In reality, there is no such thing as a closed system. However, engineers often model their design problem as a closed system to avoid the complexities associated with environmental effects. It is not recommended that close systems be constructed, as it tends to limit one's views regarding the overall impacts.

Systems are termed static if they do not change with time. Otherwise, the system is dynamic. Somewhat archaic terms also used are scleric and non-scleric systems. Engineers are mainly concerned with dynamic systems. Of particular interest to the engineer are the transient or short-term performance of the systems and its long-term performance (so-called steady-state performance if that term is appropriate). Many dynamic systems exhibit a property of reaching a steady-state condition where the outputs may vary within well-defined bounds and in a well-defined manner. Such systems are often termed stable to differentiate the system from those that do not have bounded outputs. Some systems of interest show both properties of stability and instability depending on the inputs to the system. A common example is a microphone that has been set close to a speaker system. For low volume inputs uninfluenced by the speakers, the microphone and speakers perform well. However, if the input of speaker to the microphone becomes too large, the speakers begin to obnoxiously squeal demonstrating a lack of acceptable steady-state performance.

Systems may also be defined as feedback if the outputs of the system also become inputs to the same system. The speaker microphone combination above was a feedback system. Otherwise, the system is termed open loop. The build climate control system of Fig. 2 is open loop, where the microphone and speaker system as described above was a closed feedback system. Most real systems are feedback systems.

1.3. System architectures

Systems in analysis will have a characteristic architecture or structure. Common architectures include hierarchal, star, ring and mesh forms named after the appearance of their graphical representations. Certain hybrid systems will have a combination of the basic types. It is also possible for allotropes to exist where a system appears as one type for a given purpose and analysis, but also appears as another type for a different purpose and analysis. Each type of system has advantages and disadvantages in communications and control.

The shape of a hierarchal system is that of a tree with a major single stem, branches and at the most extreme extremities, leaves. In a hierarchal architecture, there are defined levels with defined range of interactions. The objects are organized into a tree graphical structure with levels of higher responsibility near to the main stem and lesser activities removed to the leaves. A typical system encountered in green engineering that is largely hierarchal is the production of chemicals from biomass as shown in Fig. 3.

Natural hierarchal structures occur in genetics where there is a parent and child relationship. In green engineering, this structure is often extended to feedstocks where one "stem" feedstock results in a number of products often at tertiary levels removed from the stem. Hierarchal structures are favored where there are command and control functions with a division of labor. As a result, most corporate structures are hierarchal. This permit each level given certain responsibilities and to execute a certain level of work. Communication is

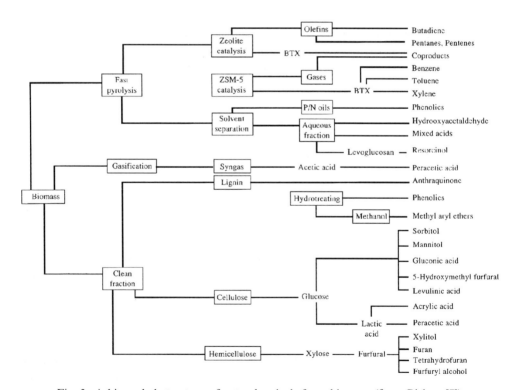

Fig. 3. A hierarchal structure of petrochemicals from biomass (from Bishop [7]).

favored from the top or stem to the leaves. It is less efficient in the reverse flow from leaves to stem. In true hierarchal structures, communication can only flow along the branch lines with no lateral communication between different branches. This can lead to serious communication problems in a complex, well-defined structure and reduce response time where problems exist. To overcome these limitations, many corporations have instituted lateral lines of communication to form a limited mesh under the formal name of "matrix management".

A star architecture has a center with paths out to each of the surrounding entities. A classical example of a star structure is borne in the cliché "*All roads lead to Rome*". Distribution centers are built on a star architecture. Star structures favor a complex center with many services upon which the outer elements can draw on when needed. This structure is highly efficient when the outer elements do not need full utilization of a resource. For example, one could model a gas station based on a star structure. The limitations of star structures are the resources constraints. Where the demand for a common resource exceeds the ability of the center to provide them, the star structure fails. As with the hierarchal structure, it favors flow of communication outwards but hinders communication in the reverse flow.

The ring structure is really an evolution of a linear structure where in the pure form, each node communicates at most with two others. When the linear form is closed and each node communicates with exactly two others, the structure is a ring. Rings/linear structures are extremely effective in representing systems where each object is accessed sequentially. As a result, most process flow sheets fall in this category. In green engineering, engineers will frequently encounter ring/linear structures in planning of projects. Ring/linear structures are ineffective in communication since a number of nodes must be traversed to relay a message. However, providing critical internal rings subordinated to the main structure can overcome this problem. The Plan, Do, Check, Act (PDCA) cycle shown in Fig. 4 is a good example of a directional ring structure.

The mesh structure represents a system where every node is, in its pure form, connected to every other node. In most meshes, there may be missing connection lines. This structure favors communications in any/every direction. However, it is extremely difficult to understand, control and to allocate resources in a mesh structure. A map of the United States is a mesh structure if one were to consider the cities at nodes and the roads and highways paths. It is largely because of mesh structures that engineers practicing green engineering must use "systems thinking". Green engineers will encounter a mesh structure when conducting life cycle assessment and analysis. An example of a mesh, discussed later, is Fig. 11 of the processes leading to a bar of soap.

It is obvious that the green engineer will encounter systems that are not pure architectures described above. Most of the systems in existence today are hybrids that combine two or more the architectures above. When architectures are

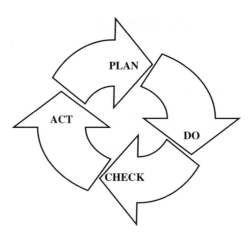

Fig. 4. Plan-do-check-act (PDCA).

willfully combined, the objective is usually to overcome a limitation that may exist in the pure system. Many of the systems in existence evolved without willful design but because of the efficiency of a particular architecture takes on its characteristics. Unless known forehand that a particular architecture exists, engineers are advised to consider a system to be a mesh until this can be disproved.

2. Plan, do, check and act

The fundamental method of systems thinking is one that should be familiar to all engineers: PDCA. This method often goes by the term "continuous improvement". Iterations of this simple process in systems analysis are the underlying basis of every other process or procedure used. Therefore, engineers must understand this process fully.

Essentially, planning is thinking ahead. Engineers must always plan what they are about to do. Failure to plan results in lost motion, following unproductive avenues and potentially has the impact of causing failure in what the engineer would do to accomplish. Planning consists of setting objectives or goals for the objective, determining the activities, setting objectives for those activities, determining the resources (time, labor, capital and tools) available and allocating those resources to the activities. Planning may also require information gathering, problem investigation and information reduction. At the beginning of engineering task, the freedom of an engineer are greatest; however, the information is the least. To determine appropriate actions, engineers need information. At some point however, information gathering must cease and the engineer will need to construct a sequence of actions that solve a problem. Plans need not be written and in most cases will not be. However, if the actions are complex, of high impact or of lengthy duration, the plan should be committed to a durable form.

During the "Do" phase of PDCA, the emphasis is activity execution. For the engineer, this means insuring that resources are available in timely manner, the sequence of activities leads to project completion and that the activities accomplish the assigned objectives. Actions may be sequential or concurrent. Where the action does not depend on resources or outcomes of a previous action, it is a candidate for a concurrent activity executed at the same time as other activities. However, it is most often that actions do depend on other activities for some part of the resources or on a particular outcomes. In this case, the activities must complete before continuing on to another activity. In some cases, only part of the previous activity need be performed before beginning the new activity in which case the activities may proceed concurrently but staggered in starting times. The engineer must always be aware during this phase of excessive resources consumption. If an activity is consuming excessive resources, the engineer must determine why and correct the situation either through better control of the activity or by finding additional resources. In some cases, the engineer will halt the activity because of discovering that the activity was poorly planned, based on false information or false assumptions, or will not meet the objectives intended. This usually results in an emergency replanning of the project, particularly if the activity were critical to project completion. In the opposite situation, an activity that finishes early and consumes too few resources is also problematic. Assuming that the activity was completed successfully, the remaining resources need to be redistributed and schedules for remaining activities need to be readjusted. Taking these potential situations together, the focus of the engineer is during execution is managing, monitoring quality of performance and immediate problem resolution.

The purpose of checking is to ensure that the project meets the defined objectives. Does the project do what it was intended to do? The engineer evaluates the project performance against the standards set during the planning phase. Checking is a determination of how well a plan and the execution met the overall goals and whether or not further action is warranted. Checking may require a formal evaluation where each of the project objectives has measurable performance objectives that are evaluated and documented. The engineer should look for unintended consequences. These may be favorable or unfavorable impacts created by the project. Favorable impacts may create opportunities for additional improvements. Unfavorable impacts may require corrective action. Checking provides a solid basis for establishing goal accomplishment as well as an identifying opportunities and situations for further work.

The "Act" phase is really a decision of whether or not the project is complete. During this phase, if the project is deemed successful during the "Check" phase, "Act" means to document the project and to publish the results so that it may be emulated elsewhere. Exceptional efforts lead to awards and credit during this phase. If the project was not successful, "Act" means to decide what further objectives are needed to perform the PDCA cycle yet again. An important activity at this point is to address project impacts. The focus of "Act" is to create goals and objectives for future improvement activity.

The next few sections of this chapter present both an informal and a formal analysis procedure. However, particularly at the introduction of a problem, it is often better to informally work with the objects of systems for clarifying one's understanding of a problem. Therefore, an informal method is described first before presenting the formal methods.

3. Informal systems thinking

The objective of informal systems thinking is to organize ones thoughts, to create an atmosphere of maximum learning about the problem situation and to create alternatives of action. Whether or not the informal method presented here leads to a formal systems analysis is up to the users, their requirements and their resources. Further discussion of informal systems analysis is in Checkland [4].

The starting point for any type of systems analysis is the problem situation. Initially, it is imperative that the person or persons tasked with studying and solving the problem put the problem into words and terms that they understand. An easy way to do this is graphically as a picture can often simplify and clarify complex terms. To illustrate, assume a manufacturing plant manager asks an engineer to solve the problem of excessive costs in metal cutting fluids. While the example is based on actual events, the information that could identify people or firms has intentionally been altered to protect their identities.

The first drawing made by the engineer is very simple (Fig. 5). The main point is to capture the problem. To emphasize informal, our drawing has cloud-like forms. These are not required. What is required is that the understanding of the problem by the engineer.

Next, if the plant manager made any other statements, these are captured. For example, the plant manager may have indicated a desire to start with system 5.

Fig. 5. Starting point of an informal systems analysis: the problem.

At this time, the engineer should jot down any knowledge that she/he may have about the problem (Fig. 6). This process will formulate questions that need answering. These are jotted down and used for initiating the investigation phase of the problem solving.

During the investigation phase, the engineer attempts to collect all relevant information regarding the problem. As the sketch evolves, the engineer may use these to help communicate and correct information by showing these to the people who have been consulted. The new data are added to the drawing in relationship to the other information. As the information develops, it is not unusual to need to redraw the sketch to provide room or to realign information. However, the old sketches should be saved unless it has been totally reproduced in a new sketch.

In the investigation of the example, the engineer discovered the major factors that must be added to the costs are for the cutting fluid handling system, its filters, makeup water and disposal costs. In addition, there were additional costs that could not be quantified for increased house keeping issues. While the literature on cutting fluids listed the major functions (lubrication, cooling of the work piece and tool and additional corrosion resistance,) it appeared that another plant (same firm in a different country) had switch to dry cutting. In addition, on another system, the engineer noted that a water valve was broken in the "ON" position for a period of over 2 months, which caused an extreme dilution of the cutting fluid on a similar line. No quality defects occurred on the product of that line. This data has been added to the sketch (Fig. 7).

Once the engineering is certain that all pertinent available information has been placed on the diagram in relationship to the other components of the sketch, the next step is reduction. In this effort, the engineer attempts to reduce the sketch to the elements critical to the problem. However, none of the sketches to date are discarded. They will be used later.

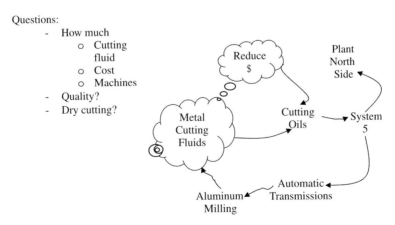

Fig. 6. Expanding on the problem based on a priori information.

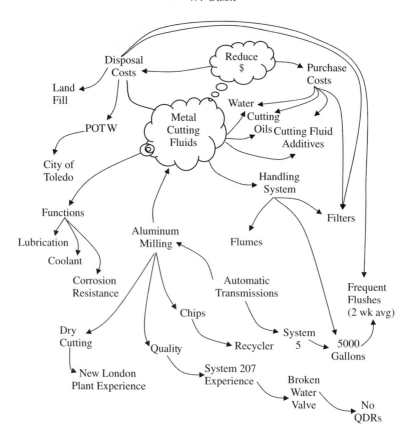

Fig. 7. Problem investigation.

During reduction, the engineer identifies the elements needed to solve the problem at hand, e.g. to reduce costs (Fig. 8). Once completing this task, several options will usually be apparent. The engineer selects the solution path (or possibly several best paths, if time and resources are available) to detail and complete with procedures for implementation, the resources needed and benefits resulting. In the example, System 5 has frequent flushes of 5000 gallons of water, cutting oils and additives that contribute to both purchase and disposal costs. To reduce these costs, the approaches are to reduce the frequency of the flushes, to reduce the size of the system, or to eliminate the system altogether. It appears that dry cutting may be the best alternative based on the New London Plant Experience.

There are a number of unknowns, which can be answered by a visit to the New London Plant. This resulted in Fig. 9, which now includes the information learned at the New London Plant. Dry cutting does result in quality problems; however, care needs to be taken to handle dust and chip collections. Where previously, chips were collected in the flumes of the cutting system, a separate system is needed. At New London, there was sufficient space in the plant to

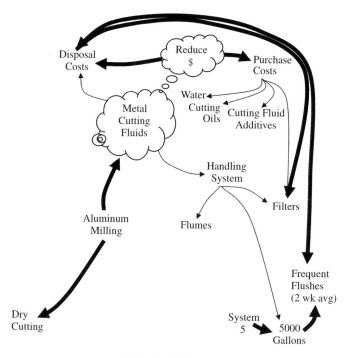

Fig. 8. Reduction.

install a chip-trolley collection system. These additions result in additional costs that essentially are a wash in cost between the wet and the dry cutting systems.

Since the dry cutting approach does not necessarily reduce costs the engineer rejects this approach. Returning to Fig. 8, the engineer focuses on the frequent flushes of the system. Further investigation (Fig. 10) reveals that the major cause for the frequent flushing of the system is foul odors caused by bacteria growth in the fluid system. A solution is to add sodium hydroxide as a biocide. This is rather inexpensive and lengthens the need to flush the system to approximately 60 days. This solution is adopted.

To complete the informal systems thinking process, the engineer must now perform one last critical step. The engineer returns to the previous drawings and notes how the solution that has been found affects the items on those sketches. Looking at sketch 3, when the system is flushed the liquids are disposed of in the public owned treatment works (POTW) of the city of Toledo. Therefore, the engineer must contact this facility to determine if the additional sodium hydroxide and the reduced frequency of flushing will have undue impacts. Fortunately, it did not. As a result the engineer declared task complete and went on to other tasks.

In recapitulation, the informal method is a set of sketches that track the progress of information and the relationships between the information as the project moves through a series of stages. There is a definition stage, an

W. Olson

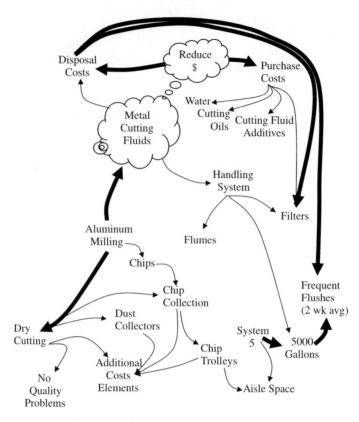

Fig. 9. Results of the new London plant visit.

information gathering stage, a reduction stage, a solution stage and a final evaluation stage. These stages may be repeated until a solution results that has acceptable overall impacts. Formal drawings are not required! The simple sketches serve as thought organization tools, communications vehicles and aids to memory. The process used is essentially the same as the formal system techniques below but eliminates the formalities and standardization needed when a large project team needs to be involved in the solution process. It is also not readily suitable to automation. In the following sections, the formal methods of system analysis are defined and described.

4. System analysis process

The following eleven steps outlines a rational systems analysis process:

- define the goal or objective,
- set system boundaries,

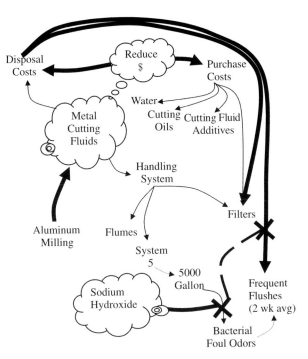

Fig. 10. Solution found!

- define system requirements,
- identify system objects, inputs and outputs,
- develop the system structure,
- collect data for the system parameters,
- build a mathematical model,
- test the model,
- analyze sensitivity,
- use system,
- evaluate the system.

These 11 steps incorporate the PDCA method both in the list entirety and in the individual performance of the steps.

4.1. Define the goals and objectives of green engineering systems

The first step in systems analysis is to define the goals and objectives. What is it that must be achieved? Reduction and elimination of pollution, improving the use of energy and materials throughout a product life cycle and prevention of hazards to humans and the environment are all possible starts to green engineering goals. However, these statements are usually too broad to be of much use except as overall green design philosophies. Goals should contain a quantifiable measure that assist in determining whether the goal has been met. Statements of goals might be:

- Use materials in an automobile design that are 95% recyclable.
- Eliminate the use of 1,2,3 tri-chloro-benzene in the manufacture of the vehicle windshield.
- Improve the vehicle highway gasoline consumption to 42 miles per gallon.

In defining goals of green engineered systems, it is important to understand that the primary functions of the product or process must be performed. Green engineering is rarely the sole goal of the system itself. For example, designing a green automobile requires focusing on the primary goal of providing mobility to a certain number of passengers under a set of a number of performance standards. Failure to meet the primary goal results in a failure to provide a green engineered product. Green engineering goal setting must be in synchronization with setting the product goals. However, once the green engineering goals have been established, they should not be considered subordinate to any other goals of the product.

4.2. Set system boundaries

Setting system boundaries is extremely important if the analysis will be timely, meaningful and useful. It is impossible to consider every possible interaction in a green engineered system. The success of green engineering is directly related to how large the system is considered to be. Setting a system boundary too large results in an analysis that cannot be completed. Setting a system boundary too small results in an analysis that has neglected the important interactions. As an initial start, a good rule of thumb is to consider system boundaries that contain the environment three levels removed from the desired system. Then this is modified as needed in order to meet the goal of the analysis. The system architecture will have a great influence on where the boundary is set.

A commonly used exercise to demonstrate complexity of products is to set the boundaries for the analysis of a bar of soap (Fig. 11). Soap, while a rather simple product, is still complex enough to make the boundary question difficult. There is no correct answer to where the boundary should be set; however, the boundary must be inclusive enough to include the objectives of the analysis. Boundary setting is a method of reducing the information to manageable dimensions for the project team.

4.3. Define requirements of the system

Definition of the requirements is the technical analysis of the goals and objectives. The goal is effective in stating what must be accomplished; it is not effective stating

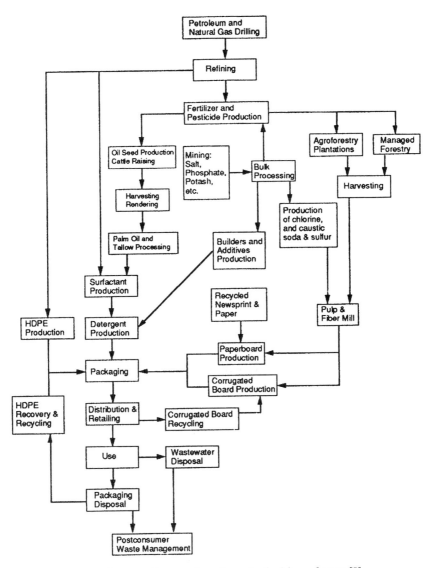

Fig. 11. Process diagram for a hypothetical bar of soap [8].

what activities need to take place to meet the goal. Defining the requirement expands the goal into subtasks that must be performed. The requirement definition should include actionable tasks and the metrics that they will be judged by.

For example, to reach the goal of 95% recyclability in an automobile, it is necessary to know the material in each component, its weight and its end of life properties. In addition one will need to know how to extract the material from the automobile. Where a material is not recyclable, it will have to be identified by functional performance and the potential alternative materials that it could be made of. Then a material selection process would be necessary. Alternatively, one could develop a method for recycling the existing material. This might involve itemizing the impediments and proposing possible solutions.

4.4. Identify system objects, inputs and outputs

The objective of this step is to identity all of the conversions that take place in the system under consideration. These are the objects of a system. At this time, it is also helpful to identify the things that the object transforms. This and the next step are closely related and in most analyses occur concurrently. Often, an analyst will identify an object and then ask the questions what comes before and what comes after to populate the system.

It is not uncommon at this stage to find that multiple systems are at work. For example in the improvement of recyclability of the automobile, one may find that there are information systems as well as material systems that apply to the problem at hand. Care needs to be taken to identify each system and treat its objects separately.

4.5. Develop the system structure

This step is really an organizational step. The task is to put the system in a logical form to aid in analysis. In some cases, this step alone will provide the information needed to meet the goal of the analysis. For clarity and ability to understand, this step is normally performed graphically. As a result, CAD tools that assist flowcharting may be used to benefit in this step.

One can start this step with any object; however, the work is often easier if one first considers the type of architecture the system is expected to expand into. For hierarchal and linear forms, it is usually best to start with the object that is at the highest level and expanding the chart toward the extremities. For star and mesh type structures, it is best to start with a centermost node and expand outwards. It is also desirable if a minimum number of paths cross each other to avoid confusion. A common recommendation is to condense the objects to between 5 and 9 to enhance clarity of understanding. Then each individual object can be expanded on a separate chart if necessary.

At the termination of this step, the analyst should review the goals and the boundaries that were previously set. The boundaries may have to be changed if it is obvious that the current structure will not meet the needs. It may also be possible to truncate or prune the structure if, based on the experience of the analyst, the additional nodes will not significantly change the results. The objective is to perform a valid analysis but reduce the effort in order to obtain it.

4.6. Collect the data for the system parameters

Environmental systems are data dependent for both analysis and prediction. The structure of the previous step guides data collection. For most green engineering systems, an analysis needs the flows of materials, amounts of materials, efficiencies, environmental impacts and information regarding the transformation process. In most cases, there will be multiple sources for the data. If the analysis is conducted in a corporate setting, data exists in the engineering divisions, plant operations and in purchasing. Process flow sheets, if they exist, are particularly useful. Often, minor quantities of secondary materials are not noted in flow sheets. Therefore, actual purchases will supplement the flow sheet data. Government databases, that can be used for properties include those of EPA and NIST. Mapping data may be obtained from USGS.

Data quality will vary depending on the source. The analyst should note where data quality is a problem. Databases will also be updated periodically to reflect corrections and new data. Therefore, the analysis should note the source and time of collection. Where possible, the data should be characterized by its certainty, usually with a measure of its spread and statistical distribution. Certain elements of data may be missing because either the data were never taken or there may have been an error in the data taking. In these cases, where it is necessary to proceed with the analysis without the data, an estimate will be used. In this case, the analysis must make clear what assumptions were made in arriving at the value. Later, the sensitivity analysis should test this estimate.

The analysis at this stage should consider the blocking of the data to provide separate data for both model building and model testing. This will be further discussed under testing of the model.

4.7. Build a mathematical model

Formal systems analysis is based on the concept that a model of an actual system can be built, which reflects the important properties of the systems. The model is used to both understand how the system functions and predict system response under a variety of inputs. A mathematical model is constructed to represent the system. There are a number of different mathematical models possible. The type of model used will depend on the goals of the analysis. The most rigorous models are a set of mathematical equations. However, for most systems, it may be very difficult or impossible to arrive at the equations necessary. A more common situation is to model the individual objects mathematically and then linked by their input and output paths.

In some cases, transfer functions can be written for the objects, which permit an analytical solution. However, in most green engineering cases, a Monte Carlo numerical simulation might be more appropriate. There are a number of computer tools available to assist in the analysis. Mathematica and Maple can be used to advantage for developing the mathematical relationships. MatLab and Simulink are powerful tools for performing continuous variable simulations. A simulation language, Modelica, has been developed to aid modeling. Modelica has been incorporated into Dymola, which eases the creation of models in Modelica. ARENA, Witness, ProModel and TaylorII are simulation tools that may be used for discrete event simulations.

4.8. Test the model

A model is a simplification of the real system that it represents. Therefore, the model must be tested to determine if the model results are comparable to the conditions in the real system. If the model does for the desired range of inputs specified by the problem requirements, the analyst moves on to the next step. However, if the model does not represent the real systems, the analyst must return to previous steps to understand why there are differences.

There are several different ways to test a model. Where the actual system is accessible to modification, the model and system conditions are set identically and run with the outputs of each compared. Usually this is performed only with small systems with very well-defined boundaries. Unfortunately, environmental systems rarely fit this category.

Another acceptable way of testing is to reserve data for testing purposes. To be effective the amount of data reserve should be as large as possible: it should never be less than 10%. In predictive models, the data block is often either the first data or last data in time sequence. Most models look to the future. Therefore, the model is use to predict the data of the latest test data block. Where back-casting is an option, one used the model to predict conditions earlier than the data used for model building.

4.9. Sensitivity analysis

Sensitivity analysis provides a measure of the risk in making decisions based on the model. All real systems have stochastic variances in the variables. Sensitivity analysis provides the analyst with information on stable the model is when subjected to the data extreme values. During data collection, data were evaluated according to its quality. Where data is suspect or estimated, the anticipated range of data should be tested in the model. If the model varies too greatly because of these tests, the analyst must either find better data or develop a less sensitive model.

Mathematically simple models lend themselves to performing sensitivity analysis by adding to each variable a symbolic delta, and determining how this variable is mathematically transformed. For models that are more complicated and particularly those involving algebraic manipulations of random variables, this technique cannot be used. In simple cases using random variables, convolutions of the distributions may be computed. However, most analysts resort to a technique known as Monte Carlo simulation. To perform a Monte Carlo simulation, the model is run a large number of times with the value of each variable under consideration sampled from a statistical distribution of suitable type using a random number generator. The number of runs needed will depend on the complexity of the model and the number of variables. Usually a very large number of runs, often in the tens of thousands are needed to validate a model. Then the variations of the runs are analyzed using standard statistical techniques.

Sensitivity analysis identifies those variables which have the greatest impact on the decision-making process. It is often possible after the sensitivity analysis to return to the model building stage and develop a simpler and more efficient model based only on the significant variables entering the decision.

4.10. Use the model

This is the payoff of systems analysis. The model is a window into the future on how a real system will perform given the input conditions. Thus, it is used by green engineers to arrive at an optimum set of starting conditions that have the least environmental impact. Furthermore, models provide engineers information useful in allocating resources. Using the model, the green engineer can build the actual system if it does not already exist.

4.11. Evaluate of the system

Consistent with continuous improvement, the green engineer needs to evaluate the system created. Does it perform to meet the system goals efficiently? Are there opportunities that have been revealed? Are there problems that must be resolved? Has the system become obsolete? Are there alternatives that promise better results? All of these questions should enter the engineer's mind at this stage of system analysis. This then leads to a new project or program and invokes the procedure above again.

5. Summary

Green engineering requires that engineers in finding solutions to problems consider the impacts of the solution on the external environment. The full range of impacts cannot be appreciated without considering the engineer task as part of an environmental system. Systems analysis provides the engineer with the ability to recognize the problem in its overall context. By examining the relationships of the project with the environment, the engineer can avoid making undue environmental impacts.

The philosophy of systems thinking is that it is a learning process. The engineer begins with what is already known, diagrams these for study, communication and verification of information, expands the information based on targeted objectives, solves the problem and then determines if the solution was adequate. It is an iterative process, which expands the engineer's vision and knowledge about project task. Plan, Do, Check and Act establishes a concise but thorough method for resolving problems and for continuous improvement. Systems thinking can be performed informally as well as formally. Systems analysis is the only tool known that insures that the engineer thinks about a problem holistically while solving a specific problem.

References

[1] P.R.D. Williams, MTBE in California Drinking Water: An Analysis of Patterns and Trends, Environ. Forensics 2 (2001) 75–85.
[2] P.C. Craig, Fish Use of Coastal Waters of the Alaskan Beaufort Sea: A Review, J. Am. Fish. Soc. 113 (3) (1985) 265–282.
[3] Md.M. Karim, Arsenic in Groundwater and Health Problems in Bangladesh, Water Res. 34 (1) (2000) 304–310.
[4] P. Checkland, Systems Thinking, Systems Practice, Wiley, New York, 1999.
[5] B.S. Blanchard, W.J. Fabrycky, Systems Engineering and Analysis, 3rd ed., Prentice Hall, Upper Saddle River, NJ, 1981, p. 9.
[6] Federal Information Processing Standards Publication 183: Integration Definition for Function Modeling (IDEF0), US National Institute of Standards and Technology. December 21, 1993. Available at http://www.idef.com/Downloads/pdf/idef0.pdf.
[7] P.L. Bishop, Pollution Prevention: Fundamentals and Practice, McGraw-Hill, New York, 2000, p. 366.
[8] Life Cycle Design Manual, Environmental Requirements and the Product System, EPA/600/R-92/226, United States Environmental Protection Agency, Washington, DC, 1993.

Sustainability Science and Engineering: Defining principles
Martin A. Abraham (Editor)
DOI 10.1016/S1871-2711(05)01006-8

Chapter 6

Systems and Ecosystems

J.A. Russell[a], W.H. Peters[a], N.N. Craig[a], B.C. Coull[b]

[a]*Department of Mechanical Engineering, University of South Carolina, 300 Main Street, Columbia, SC 29208, USA*
[b]*School of the Environment, University of South Carolina, 901 Sumter Street, Room 702G, Columbia, SC 29208, USA*

1. Introduction

In this chapter, we discuss ecosystems from two points of view: that of simple systems and that of complex systems. We offer a general overview of ecosystems and ecosystem structure from both view points; discuss the fundamental laws that govern ecosystems, the universal behaviors of ecosystems, and how ecosystem design can inform human system design. Lastly, we provide a preliminary table of cross boundary definitions in an effort to begin a dialogue that will allow engineers to compare and contrast ecosystems and human systems in terms of structure, components, and functions.

2. An overview of ecosystems as simple systems

We begin with an overview of ecosystems and ecosystem structure. So, what is an ecosystem? Allaby proposes the following definition [1]:

> An ecosystem is "a discrete unit that consists of living and non-living parts, interacting to form a stable system. Fundamental concepts include the flow of energy via food-chains and food-webs, and the cycling of nutrients biogeochemically. Ecosystem principles can be applied at all scales. Principles that apply to an ephemeral pond, for example, apply equally to a lake, an ocean, or the whole planet. In Soviet and central European literature 'biogeocoenosis' describes the same concept".

Biogeocoenosis can literally be translated as life and earth functioning together.

The non-living parts of natural systems consist of the atmosphere, lithosphere, and hydrosphere. The living components of ecosystems are organisms. All organisms can be divided into two groups, autotrophs and heterotrophs. This is a functional division based on the manner by which organisms obtain nutrition. The autotrophs or "self-feeders" provide the energy basis for all other life; they receive their energy flow from inorganic sources, while the heterotrophs or "other-feeders" receive their energy flow from the autotrophs [1].

Autotrophs, also known as producers, can be divided into two groups, the photosynthesizers and the chemosynthesizers. The photosynthesizers use electromagnetic energy in the form of sunlight. Chemosynthesizers use chemical energy that is usually in the form of reduced mineral compounds [2]. The autotrophs use the energy input to organize simple inorganic compounds such as carbon dioxide and water into more complicated biological compounds such as hydrocarbons, lipids, and proteins [3]. Although the chemosynthesizers are believed to be the first life on the planet [4], they are now far outnumbered by the photosynthesizers and contribute little to the overall energy flow of the biosphere [5].

The heterotrophs can also be divided into two main groups, the consumers and the decomposers. Consumers receive their energy flow by ingesting living or dead organic matter. Decomposers (fungi and most bacteria) receive their energy flow by absorption. Decomposers break down organic matter into inorganic matter, absorbing the nutrients they need while releasing the remaining material for re-use by the producers [6].

Figure 1 is a simplified representation of material and energy exchanges in an ecosystem highlighting the interactions between autotrophs and heterotrophs. The autotrophs capture energy (sunlight) and use it to drive a reaction between carbon dioxide (CO_2) and water (H_2O) that creates hydrocarbons ($C_6H_{12}O_6$) and diatomic oxygen (O_2). The energy rejected as low-quality heat by the autotrophs due to photosynthetic losses and other metabolic activity is represented as the respiration energy flow. The autotrophs carry out an energy conversion process where sunlight is converted into chemical energy and stored as biomass. This created biomass is referred to as net primary productivity (NPP) and can be defined as the energy fixed or assimilated by autotrophs minus autotrophic respiration [7]. The total energy fixed by autotrophs prior to their respiration losses is known as gross primary productivity (GPP).

The biomass created by the autotrophs is consumed either directly or indirectly by heterotrophs. The chemical energy stored in the biomass is released through oxidation and used to drive heterotrophic metabolism, which is the reverse of the biomass creation and is carried out through the combination of hydrocarbons and diatomic oxygen with the subsequent release of energy and carbon dioxide. These photosynthesis-oxidation reactions set up a balanced material exchange of carbon dioxide and diatomic oxygen between the autotrophs and heterotrophs. Heterotrophs also reject low-quality heat energy

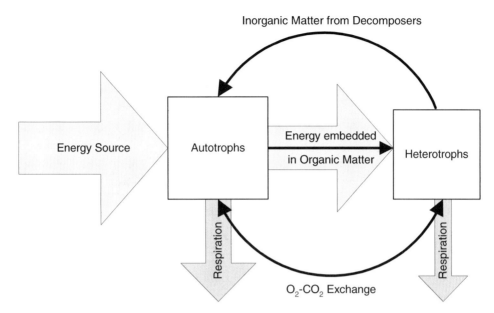

Fig. 1. Simplified energy flow and material cycling through an ecosystem.

through respiration. Heterotrophic created biomass is referred to as secondary productivity (SP) and can be defined as the energy assimilated by heterotrophs minus heterotrophic respiration [6].

All life is made up of the same basic elements: carbon, hydrogen, oxygen, nitrogen, phosphorous, and sulfur. The carbon, hydrogen, and oxygen are cycled between autotrophs and heterotrophs by the carbon dioxide and diatomic oxygen exchange. The nitrogen, phosphorous, and sulfur are cycled as well. The heterotrophs obtain nitrogen, phosphorous, and sulfur from the tissues of autotrophs. The autotrophs obtain these same elements through the work of a specialized group of heterotrophs, the decomposers. Decomposers take any dead tissue (autotrophic or heterotrophic) and break it down into simple inorganic compounds. The decomposers absorb the nutrients they need and the remaining matter (containing nitrogen, phosphorous, and sulfur) is available again for the autotrophs [6].

Figure 2 provides a graphic of these processes showing the decomposers separated from the other heterotrophs. For clarity, the heterotrophs that function by ingestion (herbivores, carnivores, omnivores, scavengers, and detritivores) are referred to as consumers and those that function through absorption (fungi and bacteria) are referred to as decomposers. The autotrophs that function through photosynthesis are labeled producers.

Thus, we have a simple systems view of ecosystems, which adequately describes the general structure and functions of ecosystems and some of the basic properties and behaviors of ecosystems. Erwin Schrödinger is one of many thinkers including Newton and Descartes who have attempted to gain a more detailed understanding

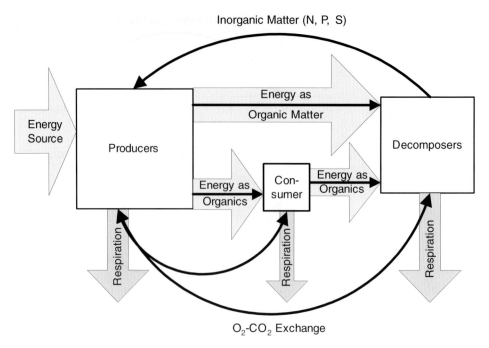

Fig. 2. Simplified energy flow and material cycling among producers, consumers, and decomposers.

of living systems through the fundamental physical laws of reductionist science [8]. In his book *What is Life?*, Schrödinger addresses these fundamental laws [9]:

"What then is that precious something contained in our food which keeps us from death? That is easily answered. Every process, event, happening — call it what you will; in a word, everything that is going on in Nature means an increase of the entropy of the part of the world where it is going on. Thus a living organism continually increases its entropy — or, as you may say, produces positive entropy — and thus tends to approach the dangerous state of maximum entropy, which is death. It can only keep aloof from it, i.e. alive, by continually drawing from its environment negative entropy — which is something very positive as we shall immediately see. What an organism feeds upon is negative entropy. Or, to put it less paradoxically, the essential thing in metabolism is that the organism succeeds in freeing itself from all the entropy it cannot help producing while alive".

Schrödinger tells us that living things must continually feed on negative entropy to avoid that state of maximum entropy, death. In our modern world of supermarkets and restaurants, we humans can easily forget that our survival depends on natural systems which are the ultimate source of negative entropy as well as breathable air, drinkable water, and other natural resources and services [6]. This is why natural systems or ecosystems must be taken into account in any discussion of sustainability. The quotation from Schrödinger is a restatement of the second law of thermodynamics and is one of the fundamental laws of ecosystems. The other fundamental laws include the conservation of mass and the conservation of energy. The earth is a thermodynamically closed system, therefore, according to the law of conservation of mass, the only material available to ecosystems and human systems is that which we find around us. Also, according to the law of conservation of energy, the sun is the only major source of negative entropy available on the earth. These are three of the fundamental laws governing ecosystems.

In summary, ecosystems can be seen as systems that create and maintain structures made up of cycling materials using energy obtained by the capture, conversion, and degradation of solar energy operating within the physical and thermodynamic constraints laid out by the laws of conservation of mass and energy and the second law of thermodynamics. However, the biosphere has not always been this cycling system of producers, consumers, and decomposers. As appealing as the concept of a stable cycling ecosystem is, it is an incomplete picture in both present and historical aspects.

3. An overview of ecosystems as complex systems

The basic introduction to ecosystems above has focused on energy flows and material cycles and the basic groups of organisms that make these cycles possible. This is a useful frame of reference from which to view ecosystems; in fact it is the basis for the field of Industrial Ecology. However, it is an incomplete description. Issues of scale, both temporal and spatial, cannot be addressed from the simple system frame of reference. Temporally, ecosystems display both cyclic (diurnal and annual) and evolutionary change. Spatially, the simple systems view ignores issues of scale, categorizing organisms as large as whales and as small as bacteria into neat groups for ease of discussion. Even in simple Cartesian coordinate terms there are huge variations in the biomes that make up the biosphere. In addition to well-defined biomes, biome interface areas such as estuaries display novel behavior that is not addressed when viewed as simple systems. These spatial and temporal variations and the complex dynamics that are associated with "real world" ecosystem behavior, particularly behavior associated with the dynamics of coupled human and natural systems, have thus far evaded scientific elucidation. The study of these complex ecosystem dynamics has recently become synonymous with the term biocomplexity.

Biocomplexity has been defined by Michener and colleagues [10] as:

> "properties emerging from the interplay of behavioral, biological, chemical, physical, and social interactions that affect, sustain, or are modified by living organisms, including humans".

Biocomplexity is an emerging field which recognizes the limitations of viewing ecosystems only from the traditional simple systems view and is attempting to study ecosystem from the complex systems frame of reference. Biocomplexity, rather than attempting to simplify systems so that they can be studied with traditional reductionist techniques, recognizes that the special evolutionary and emergent properties of living systems can only be understood by studying the systems in question as wholes including all of their complexity. However, as a new field it can as yet offer few laws or quantitative methodologies. The special properties of ecosystems recognized in biocomplexity studies include emergent and evolutionary behavior, unpredictable behavior brought about by non-linear dynamics, hierarchical system organization, and interactions spanning multiple system levels or spatiotemporal scales [10–13].

Thus, ecosystems are bounded as simple physical systems by traditional physical laws, but display seemingly inscrutable and enigmatic complex behaviors. Two ecosystem properties that are often noted are stability and resilience. Ecosystems are stable systems, e.g. they tend to return to a state of equilibrium after they are disturbed. Ecosystems are also considered to be resilient, e.g. they maintain their integrity when perturbated [14]. Though ecosystems do display stability and resilience within a certain range of perturbation or forcing, focusing on these properties alone is very misleading particularly when making decisions about what level of disturbance is allowable for ecosystems. This is due to the fact that all ecosystems change over time; a fact which makes it difficult to determine what the normal state of an ecosystem is [15]. Ecosystems as complex systems are comprised of many interconnected components [16]. Ecosystems display non-linear responses to linear forcings and also display discontinuities or bifurcation points [14].

Therefore, ecosystems display both stability and uncertainty. Ecosystems display the properties of stability and resilience, but also non-linear responses, discontinuities, and bifurcation points. What does this disparate picture of ecosystems mean to us as engineers who endeavor to design sustainable systems?

Whether one philosophically considers human systems to be a separate system or a subsystem of natural systems, we are in fact physically dependant on natural systems for our flow of negative entropy. Therefore, natural systems must act as a source for some human system needs. Because we share the same closed thermodynamic systems (the earth), natural systems must act as a sink for human system wastes. The properties of stability and resilience allow us to use natural systems as a limited source and sink. However, the non-linear and evolving properties of natural systems require that monitoring and caution be used to insure that human systems are not over sourcing or over sinking natural systems. Currently, humans appropriate as much as 40 percent of the NPP of the biosphere [7]. The NPP is the negative entropy that the producers supply that determines and maintains the structure of all ecosystems and keeps all of the higher trophic levels (including humans) alive. Forty percent is a massive amount for one species to consume, and whether or not it is a sustainable amount remains to be seen. Human systems, use of ecosystems as sinks is equally massive with current flows of materials and elements rivaling the size of the natural system's grand cycles [17].

If these negative trends are to be prevented from increasing, we must engineer human systems differently and perhaps natural systems can serve as a model for change. At the beginning of life, natural systems were not systems at all, but merely consumers feeding off of a primordial stockpile of nutrients. The crisis of a lack of food was solved by the development of the producers (photosynthesizing bacteria) that took advantage of the more permanent energy source from the sun. The next crisis was the oxidation of the planet caused by the abundance of producers. Oxygen was a toxin that threatened to extinguish life and inflame the planet. This crisis was solved by the development of respiration and thus a balanced system of producers and consumers was developed as summarized by Margulis and Sagan [18].

> "In far-from-equilibrium systems, waste products necessarily accumulate. But what may be garbage to one is dinner or building materials for another".
>
> "Earlier life forms tell us by their examples that long-term survival involves not so much halting pollution as transforming pollutants".

Therefore, if human systems are to grow or perhaps even stay the same size, we cannot depend on natural systems as exclusive sinks and sources. We must develop our own systems of producers, consumers, and decomposers. We must also determine what our current and future impacts on biological systems are and attempt to determine what level of impact is acceptable. Carrying out research in this manner requires cooperation among different branches in science and a requirement for researchers who are trained to cooperate and function in an interdisciplinary environment [11]. This interdisciplinary effort requires the development of a common language that will allow researchers from various disciplines to carry out meaningful dialogues [12].

The last section of this chapter attempts to use natural systems as a model to develop a language that will allow us to discuss human systems in terms of our current understanding of ecosystem structure and function.

4. Creating a common language

In order for a meaningful comparison of natural and human systems to be carried out, a common language needs to be developed [12,13,19]. The following is an attempted beginning of a common language or a language with shared meaning. Definitions were gathered from literature in biology and ecology. Where possible generalized definitions were crafted that accurately represent the concepts for both biotic and anthropic systems. When generalized definitions did not make sense, specific definitions were developed for anthropic systems and the term was changed to denote the incompatibility. Table 2 contains standard definitions of ecological terms in the left column. Table 1 is presented to define

Table 1
Glossary of acronyms

Acronyms	Definitions
ANRES	Anthropic non-renewable energy subsidy
AR	Anthropic respiration
ARES	Anthropic renewable energy subsidy
ATP	Adenosine triphosphate (metabolic energy carrier)
B	Biomass
GARES	Gross anthropic renewable energy subsidy
GPP	Gross primary productivity
HANPP	Human appropriation of net primary productivity
NPP	Net primary productivity
NTP	Net technological productivity
PNPP	Potential or pristine net primary productivity
R	Respiration
R/B	Respiration biomass
RNPP	Remaining net primary productivity
TNPP	Total net primary productivity
TPES	Total primary energy supply

Table 2
Ecological definitions and generalized equivalent definitions

Ecological/biotic definitions	Technological/anthropic or general definitions
Climax community: the final or stable community in a developmental series, in theory, the climax community is self-perpetuating because it is in equilibrium with itself and with the physical habitat. For climax communities GPP/R = 1. For autotrophic succession, GPP/R > 1, but moving toward unity (example disturbed land transitioning to forest). For heterotrophic succession, GPP/R < 1 but moving toward unity (e.g. bacteria in a sewage pond) [6].	*Anthropic climax community:* a sustainable community where in general, (HANPP + ARES)/(AR) = 1
Net primary productivity (NPP): the amount of energy left after subtracting the respiration of primary producers (mostly plants) from the total amount of energy (mostly solar) that is fixed biologically. NPP provides the basis for maintenance, growth, and reproduction of all heterotrophs (consumers and decomposers); it is the total food resource on the Earth [7].	*Anthropic renewable energy subsidy (ARES):* the renewable energy fixed or assimilated by human systems minus storage and conversion losses, (GARES- losses)

Table 2 (*continued*)

Ecological/biotic definitions	Technological/anthropic or general definitions
Gross primary productivity (GPP): the total rate of photosynthesis, including the organic matter used in respiration during the measurement period. This is also known as "total photosynthesis" or "total assimilation" [6].	*Gross anthropic renewable energy subsidy (GARES):* the total rate of renewable energy fixed or assimilated by human systems (e.g. solar or wind energy)
Net community productivity: the rate of storage of organic matter not used by heterotrophs (that is, net primary production minus heterotrophic consumption) during the period under consideration, usually the growing season or a year [6].	*Net community productivity:* the rate of storage of energy not used by heterotrophs (that is, net primary production minus heterotrophic consumption) during the period under consideration
Secondary productivities: the rates of energy storage at consumer levels. Since consumers use only food materials already produced, with appropriate respiratory losses, and convert to different tissues by one overall process, secondary productivity should not be divided into "gross" and "net" amounts. The total energy flow at heterotrophic levels, which is analogous to gross production of autotrophs, should be designated as "assimilation" and not "production" [6].	*Secondary productivity:* the rates of energy storage at consumer levels. The energy assimilated by heterotrophs minus heterotrophic respiration
Consumer: in the widest sense, a heterotrophic organism that feeds on living or dead organic material. Two main categories are recognized: (a) macroconsumers, mainly animals (herbivores, carnivores, and detritivores), which wholly or partly ingest other living organisms or organic particulate matter; and (b) microconsumers, mainly bacteria and fungi, which feed by breaking down complex organic compounds in dead protoplasm, absorbing some of the decomposition products, and at the same time releasing inorganic and relatively simple organic substances into the environment. Sometimes the term "consumer" is confined to macroconsumers, microconsumers being known as "decomposers". Consumers may then be termed "primary" (herbivores), "secondary" (herbivore-eating carnivores), and so on, according to their position in the food chain. Macroconsumers are also sometimes termed phagotrophs or biophages, while microconsumers correspondingly are termed saprotrophs or saprophages [1].	*Consumer:* see Heterotroph

Table 2 (*continued*)

Ecological/biotic definitions	Technological/anthropic or general definitions
Decomposer: a term that is generally synonymous with "microconsumer". In an ecosystem, decomposer organisms (mainly bacteria and fungi) enable nutrient recycling by breaking down the complex organic molecules of dead protoplasm and cell walls into simpler organic and (more importantly) inorganic molecules, which may be used again by primary producers. Recent work suggests that some macroconsumers may also play a role in decomposition (for example, detritivores, in breaking down litter, speed its bacterial breakdown). In this sense "decomposer" has a wider meaning than that traditionally implied [1].	*Example:* anaerobic digesters
Detritivore: a heterotrophic animal that feeds on dead material (detritus). The dead material is most typically of plant origin, but it may include the dead remains of small animals. Since this material may also be digested by decomposer organisms (fungi and bacteria) and forms the habitat for other organisms (e.g. nematode worms and small insects), these too will form part of the typical detritivore diet. Animals (i.e. the hyena) that feed mainly on the products (exuviae, e.g. dung), of larger animals, are termed scavengers [1].	*Example:* electrical utilities
Detritus: litter formed from fragments of dead material (e.g. leaf litter, dung, molted feathers, and corpses). In aquatic habitats, detritus provides habitats equivalent to those which occur in soil humus [1].	Same
Detrital pathway (detritus food chain): most simply, a food chain in which the living primary producers (green plants) are not consumed by grazing herbivores, but eventually form litter (detritus) on which decomposers (microorganisms) and detritivores feed, with subsequent energy transfer to various levels of carnivore (e.g. the pathway: leaf litter → earthworm → blackbird → sparrow hawk). Detritus from organisms at higher trophic levels than green plants may also form the basis for a detrital pathway, but the key distinction between this and a grazing pathway lies in the fate of the primary producers [1].	Same

Table 2 (*continued*)

Ecological/biotic definitions	Technological/anthropic or general definitions
Food chain: the transfer of energy from the primary producers (green plants) through a series of organisms that eat and are eaten, assuming that each organism feeds on only one other type of organism (e.g. earthworm → blackbird → sparrow hawk). At each stage much energy is lost as heat, a fact that usually limits the number of steps (trophic levels) in the chain to four or five. Two basic types of food chain are recognized: the grazing and detrital pathways. In practice, these interact to give a complex food web [1].	Same
Food web: a diagram that represents the feeding relationships of organisms within an ecosystem. It consists of a series of interconnecting food chains. Only some of the many possible relationships can be shown in such a diagram and it is usual to include only one or two carnivores at the highest levels [1].	*Food web:* a diagram that represents the energy exchanges of organisms within an ecosystem. It consists of a series of interconnecting food chains.
Herbivore: a heterotroph that obtains energy by feeding on primary producers, usually green plants [1].	Same
Autotroph: an organism capable of synthesizing its own food from inorganic substances, using light or chemical energy [20].	*Autotroph:* an organism capable of producing useful energy using inorganic energy sources such as light, chemical, wind, water, wave, geothermal, or nuclear energy (i.e. a producer)
Heterotroph: an organism that is unable to manufacture its own food from simple chemical compounds and therefore consumes other organisms, living or dead, as its main or sole source of carbon. Often, single-celled autotrophs (e.g. Euglena) become heterotrophic in the absence of light [1].	*Heterotroph:* an organism that receives its energy flow from another organism (i.e. a consumer)
Omnivore: a heterotroph that feeds on both plants and animals, and thus operates at a range of trophic levels [1].	*Omnivore:* a heterotroph that feeds on both autotrophs and heterotrophs, and thus operates at a range of trophic levels
Carnivore: any heterotrophic, flesh-eating animal [1].	*Carnivore:* a heterotroph that feeds only on other heterotrophs
Producer: in an ecosystem, an organism that is able to manufacture food from simple inorganic substances (i.e. an autotroph, most typically a green plant) [1].	*Producer:* see Autotroph

Table 2 *(continued)*

Ecological/biotic definitions	Technological/anthropic or general definitions
Assimilate: the portion of the food energy consumed by an organism that is metabolized by that organism. Some food, or in the case of a plant some light energy, may pass through the organism without being used [1].	*Assimilate:* the portion of input energy that an organism metabolizes
Ecosystem: a term first used by A. G. Tansley (in 1953) to describe a discrete unit that consists of living and non-living parts, interacting to form a stable system. Fundamental concepts include the flow of energy via food chains and food webs, and the cycling of nutrients biogeochemically. Ecosystem principles can be applied at all scales. Principles that apply to an ephemeral pond, for example, apply equally to a lake, an ocean, or the whole planet. In Soviet and central European literature "biogeocoenosis" describes the same concept [1].	*Anthropic system or technosystem:* a discrete human created system, consisting of producers and consumers. For example a country, state, or city
Grazing pathway (grazing food chain): a food chain in which the primary producers (green plants) are eaten by grazing herbivores, with subsequent energy transfer to various levels of carnivore (e.g. plant (blackberry, Rubus) → herbivore (bank vole, *Clethrinomys glareolus*) → carnivore (tawny owl, Strix aluco) [1].	Same
Metabolism: the total of all the chemical reactions that occur within a living organism (e.g. those involved in the digestion of food and the synthesis of compounds from metabolites so obtained) [1].	Same
Respiration: oxidative reactions in cellular metabolism that involve the sequential degradation of food substances and the generation of a high-energy compound, ATP (adenosine triphosphate) in aerobic respiration with the use of molecular oxygen as a final hydrogen acceptor; ATP, carbon dioxide, and water are the products thus formed [1].	*Respiration:* the expenditure of assimilated energy to do work
Respiration–biomass ratio (R/B ratio): the relationship between total community biomass (i.e. standing crop) and respiration. With larger biomass, respiration will increase but the increase will be less if the individual biomass units or organisms are large (reflecting the inverse relationship between size and metabolic rate). Natural communities tend toward larger organisms and complex structure, with low respiration rates per unit biomass [1].	*Anthropic R/B ratio:* Total energy expenditure divided by the biomass of the human population. *Note:* this definition is problematic due to the biomass energy disconnect in technosystems. Anthropic systems include an internal metabolism which sustains human life (e.g. the food we eat) and an external metabolism which could be called technical energy

Table 2 (*continued*)

Ecological/biotic definitions	Technological/anthropic or general definitions
Metastable system: a physical, chemical, or biological system that has temporarily stabilized at a higher than normal energy level, usually due to some outside inducement. A metastable system may be anything from an atom in which one or more electrons have skipped up to higher-level shells to an ecosystem such as a grassland in which the species composition has been changed by intense grazing pressure. Metastable systems are always in danger of collapse [4].	Same
Biome: a biological subdivision that reflects the ecological and physiognomic character of the vegetation. Biomes are the largest geographical biotic communities that are convenient to recognize. They broadly correspond with climatic regions, although other environmental controls are sometimes important. They are equivalent to the concept of major plant formations in plant ecology, but are defined in terms of all living organisms and of their interaction with the environment (and not only with the dominant vegetation type). Typically, distinctive biomes are recognized for all the major climatic regions of the world, emphasizing the adaptation of living organisms to their environment, e.g. tropical rain-forest biome, desert biome, tundra biome [1].	*Technome:* an area of the technosphere delineated by the type of technological organism that dominates *Candidates:* urban, suburban, commercial, industrial, agricultural, and recreational technomes
Biosphere: the part of the Earth's environment in which living organisms are found, and with which they interact to produce a steady-state system, effectively a whole-planet ecosystem. Sometimes it is termed "ecosphere" to emphasize the interconnection of the living and non-living components [1].	*Anthroposphere or technosphere:* a subset of the biosphere that includes humans and their creations
Energy subsidy: besides solar energy, an ecosystem may also be supported by external sources of energy that lessen its internal self-maintenance. This is called energy subsidy or auxiliary energy and is responsible for higher outputs (productivity) of an ecosystem [21].	*Technical energy subsidy:* any energy other than HANPP and NTP that maintains the technosystem at an elevated energy/metastable state
Human appropriation of net primary productivity (HANPP): can be defined as a geographic area's potential NPP which is "used directly, co-opted, or forgone because of human activities" [7].	Same

the acronyms used in Table 2. The right column contains a definition that has been generalized to allow the definition to be applied to industrial systems while still maintaining the original biological meaning. The following definitions are by no means rigorous or final, but are meant as a starting point to begin a dialogue between natural or biotic systems and human or anthropic systems.

References

[1] M. Allaby (Ed.), A Dictionary of Ecology, 2nd ed., Oxford University Press, Oxford, 1998.
[2] M.B. Rambler, L. Margulis, R. Fester (Eds), Global Ecology: Towards a Science of the Biosphere, Academic Press, Boston, 1989.
[3] R.E. Ricklefs, The Economy of Nature, 2nd ed., Chiron Press, New York, 1983.
[4] W. Ashworth, The Encyclopedia of Environmental Studies, Facts on File, New York, 1991.
[5] H. Lieth, R.H. Whittaker (Eds), Primary Productivity of the Biosphere, Springer-Verlag, New York, 1975.
[6] E.P. Odum, Basic Ecology, Saunders College Publishing, New York, 1983.
[7] P.M. Vitousek, P.R. Ehrlich, A.H. Ehrlich, P.A. Matson, Human Appropriation of the Products of Photosynthesis, BioScience 36 (6) (1986) 368–373.
[8] F. Capra, The Web of Life, Random House, New York, 1996.
[9] E. Schrödinger, What is Life?, Cambridge University Press, Cambridge, 1967.
[10] W.K. Michener, T.J. Baerwald, P. Firth, M.A. Palmer, J.L. Rosenberger, E.A. Sandlin, H. Zimmerman, Defining and Unraveling Biocomplexity, BioScience 51 (12) (2001) 1018–1023.
[11] A. Covich, Biocomplexity and the Future: The Need to Unite Disciplines, BioScience 50 (12) (2000) 1035.
[12] R. Colwell, Balancing the Biocomplexity of the Planet's Living Systems: A Twenty-First Century Task for Science, BioScience 48 (10) (1998) 786–787.
[13] K.L. Cottingham, Tackling Biocomplexity: The Role of People, Tools, and Scale, BioScience 52 (9) (2002) 793–799.
[14] D. Ludwig, B. Walker, C.S. Holling, Sustainablity, Stability, and Resilience, Conserv. Ecol. 1 (7) (1997).
[15] G.A. De Leo, S. Levin, The Multifaceted Aspects of Ecosystem Integrity, Conserv. Ecol. 1 (3) (1997).
[16] B.R. Allenby, W.E. Cooper, Understanding Industrial Ecology from a Biological Systems Perspective, Total Quality Environ. Manage. (Spring) (1994) 343–354.
[17] B.R. Allenby, Earth Systems Engineering, J. Ind. Ecol. 2 (3) (1999) 73–93.
[18] L. Margulis, D. Sagan, What is Life?, University of California Press, Los Angeles, 1995.
[19] J.A. Russell, Evaluating the Sustainability of an Ecomimetic Energy System: An Energy Flow Assessment of South Carolina, PhD. Dissertation, University of South Carolina, Columbia, 2004.
[20] The American Heritage College Dictionary, 3rd ed., Houghton Mifflin, Boston, 1993.
[21] W.A. Nierenberg, Encyclopedia of Environmental Biology, Vol. 1, Academic Press, New York, 1995.

Sustainability Science and Engineering: Defining principles
Martin A. Abraham (Editor)
© 2006 Published by Elsevier B.V.
DOI 10.1016/S1871-2711(05)01007-X

Chapter 7

Life Cycle Based Sustainability Metrics

Gregory A. Keoleian, David V. Spitzley

*Center for Sustainable Systems, School of Natural Resources and Environment,
University of Michigan, 440 Church St., Ann Arbor, MI 48109-1041, USA*

1. Introduction

Sustainability challenges confronting society in the 21st century include global climate change [1], declining fossil resources [2], persistent organic pollutants [3], freshwater scarcity [4], ecosystem degradation [5], biodiversity loss [6], over-population [7], and limited access to basic human necessities particularly in developing countries [8]. Ultimately, natural resource depletion and pollution are driven by material and energy flows associated with goods and services. The life cycle of a product system, which includes the material and energy flows across materials production, manufacturing, use and service, and end-of-life management stages, is a logical framework for understanding and improving the link between production and consumption activities and natural systems. It has become clear that significant changes in the production and consumption of goods and services are essential for maintaining the planet's life support system, which is increasingly threatened. Life cycle-based sustainability models and metrics play a key role in guiding the transformation of technology, consumption patterns, and corporate and governmental policies for achieving a more sustainable society.

Life cycle modeling represents a unique sustainability assessment framework for at least four reasons:

(1) The Life cycle of a product system encompasses all processes for addressing societal needs including materials production through end-of-life management.

(2) The Life cycle links production and consumption activities.
(3) The Life cycle boundary enables a comprehensive accounting of sustain-
    ability performance including environmental, social, and economic metrics.
(4) Metrics can be used by key stakeholders that manage and control the life
    cycle supply chains to guide their improvement.

A wide set of analytical methods and tools have been developed around a life
cycle system boundary. Table 1 provides a list of life cycle based techniques and
examples of metrics that have emerged over the last three decades.

These tools yield a wide array of metrics that can contribute to the understanding and assessment of environmental, social, and economic sustainability of
goods and services. Life cycle methods serve to help operationalize the broader
concepts of sustainable development as articulated in the Brundtland Commission definition: development which "...meets the needs of the present without
compromising the ability of future generations to meet their own needs" [9].

The objective of this chapter is to review the range of life cycle methods and
metrics for evaluating the sustainability of products and technology. This review
will highlight the relevant aspects of sustainability that each method addresses.
In addition to analyzing these tools, this chapter will demonstrate the application of life cycle models and metrics for diverse sectors including transportation, buildings, renewable energy, and consumer products. Strengths and
limitations of these methods and metrics for assessing sustainability will also be
discussed. The authors envision that life cycle metrics and indicators will continue to evolve in the decades ahead and in the process provide more explicit
meaning to the term sustainability.

2. Life cycle assessment

Life cycle assessment (LCA) is an analytical technique for assessing the potential environmental burdens and impacts associated with a product system from
the generation of the raw materials to the ultimate management of material

Table 1
Life cycle-based methods for sustainability metrics development

Method	Example metrics
Life cycle assessment	Ore consumption (kg), global warming potential (kg CO_2 equivalent)
Life cycle energy analysis	Energy consumption (MJ), net energy ratio
Life cycle cost analysis	Private costs ($), social costs ($), total costs ($)
Life cycle optimization	Optimal service life (years)
Life cycle sustainability matrix	Population obesity (%), rate of land conversion (ha/year)

remaining at the end-of-life [10]. LCA provides metrics that can be used to measure progress toward environmental sustainability. The stages in a product life cycle are shown in Fig. 1.

This method results in an environmental profile that measures environmental performance at each life cycle stage. LCA, which has undergone significant development over the last three decades [10–14], can be considered the most advanced method for assessing sustainability. The four components of LCA including goal and scope definition, inventory analysis, impact assessment, and interpretation, also serve as an important foundation of other life cycle-based methods as well as other sustainability assessment methods. These four steps and their relationships to each other are shown in Fig. 2.

Goal and scope definition establishes the objectives of the analysis, intended audience for study results, system boundaries, allocation rules, nature of the data to be collected, specific metrics to be evaluated, and peer review requirements. A critical activity within goal and scope definition is the determination and specification of a functional unit for the system under study. The functional unit describes the fundamental objective of the system and provides the basis for

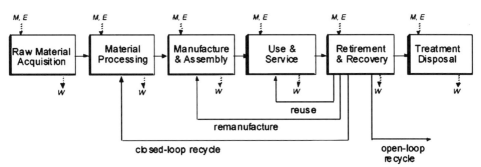

M, E   material and energy inputs for process and distribution
W   waste (gaseous, liquid, solid) output from product, process and distribution
⟶   material flow of product component

Fig. 1. Product life cycle stages.

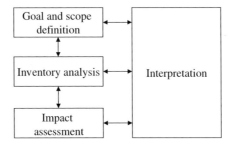

Fig. 2. Life cycle assessment activities [10].

the scope of the study. The functional unit should be measurable, meaningful to the intended audience, and relevant to data collection. Examples of functional units include 120,000 miles of operation for a five-passenger mid-sized automobile, 1 kWh of delivered electricity from a 1 GW baseload power plant, and 100,000 L of delivered carbonated beverages. LCA done for the purpose of evaluating alternative systems must compare systems on the basis of equivalent function as captured in the functional unit.

The life cycle inventory (LCI) analysis step is focused on the collection and analysis of data on the input and output flows associated with the system under study. In this step, system boundaries and allocation rules identified in goal and scope definition are applied. The inventory step and the specific challenges associated with data collection, allocation, and system boundary definition are discussed in detail in Section 2.1.

Life cycle impact assessment (LCIA) serves to translate the myriad input and output flow data compiled during inventory analysis into meaningful information regarding the environmental effects of the system. Impact assessment involves categorization of flows, characterization of impacts, and may also include normalization and weighting of results. Current practice in LCIA, standard methodologies, challenges, and limitations are discussed in detail in Section 2.2.

Interpretation considers the full study results in the context of the stated objectives, potential limitations, uncertainties in data, and the intended audience. In this step, opportunities for system improvement are discussed and results are placed in the appropriate context for the intended application. LCA results have been used to support policy deliberations [15], as an input to design improvement [16], and in support of product labeling [17].

2.1. Inventory analysis

LCI analysis is an accounting of the material and energy inputs and outputs between a system under study and the environment (termed elementary flows). Accounting in LCI considers flows across a series of individual life cycle stages. Activities conducted in LCI form the core of LCA. Typical metrics tracked in LCI include biotic and abiotic resource inputs, air pollutant emissions, water pollutant emissions, solid waste (hazardous and non-hazardous), recycled materials, products, and co-products. The challenges in conducting an LCI study relating to system boundaries and data collection are not unique to LCA. Successfully overcoming these challenges can provide the foundation for other sustainability assessment frameworks.

Several organizations have provided useful guidance for conducting a LCI analysis. The International Organization for Standardization (ISO) 14041 standard and the technical report ISO TR 14049 provide the internationally accepted code of practice for LCA [18,19]. In addition, the US Environmental Protection Agency (US EPA) [12], and the Society of Environmental Toxicology and Chemistry (SETAC) [14] provide additional clarification and guidance.

Although system boundaries and allocation procedures must be specified as part of the goal and scope definition, they directly impact the data collection procedure and are discussed here for clarity. In reality, the process of defining system boundaries and necessary allocations is an iterative process involving balancing study goals with data collection and analysis feasibility. System boundaries should be defined to combine a series of interrelated activities and operations into a comprehensive network supporting a specific function. Ideally, all system inputs and outputs should be captured as elementary flows to or from the environment. Elementary flows are described in a state of natural occurrence, i.e. no additional processing or transformation is performed outside of the system boundaries. In practice, system boundaries are defined using cut-off rule based criteria such as environmental relevance, mass contribution, and energy contribution.

Defining system boundaries can be especially problematic when some of the operations involved in the system produce multiple products [20]. In this situation, general guidance, including that codified in ISO 14041, on defining system boundaries recommends allocation according to one of four approaches — avoiding allocation, system expansion, causal allocation, or technical allocation.

Allocation of operational burdens should be avoided by attempting to subdivide the operation into smaller unit processes. When this is not possible, the system boundaries should be expanded to include additional system functions and related processes and/or co-products. In comparative studies, alternative systems should also be expanded to include these additional functions. For example, the waste management practices for some products may result in the generation of electricity in waste to energy plants.

Under a system expansion approach this electricity should be included in the system boundary and any alternative systems would need to include operations required for equivalent electricity production. When system expansion is not possible, operational burdens should be allocated between product outputs according to defined causal relationships between the products. If a painting operation produced multiple painted products, allocation could be based on the relative surface area of each product produced.

Allocation of environmental burdens between products can also be done on the basis of physical or technical information not directly attributed to operating burdens. The most common example of this is allocation based on product economic data.

As decisions regarding data allocation and system boundaries are reached, a record of these system definitions should be made. This record generally takes the form of a process flow diagram. LCI flow diagrams should provide details sufficient to effectively communicate unit processes studied and relationships in system modeling. These diagrams serve to guide the data collection process and verify data completeness.

One of the greatest challenges in conducting a LCI is the data collection process. Data limitations exist for several reasons including proprietary concerns, aggregation across more than one product system, and lack of a consistent tracking system. Analysts are more successful in gathering data when organizations holding the data are collaborating with the study. When this is the case, specific primary data should be collected following a well-documented procedure (see, for example LCI data collection forms provided in Annex A of ISO 14041) [18].

Frequently, LCI data collection will require information on operations and activities for which primary data are unavailable. In this case, analysts rely on existing LCI databases and other literature sources to compile data. Published databases designed for use in LCI are available for sale from several distributors. In addition, several organizations have conducted LCI studies which are publicly available at no cost. Both types of data sources contain information on common unit processes and systems associated with the production of frequently referenced materials. Publicly available LCI data sets are listed in Table 2.

Within these data sets as well as other published LCI studies, the accuracy of energy data tends to be greater than for air pollutant emissions and water pollutant emissions. Air pollutant emissions and water pollutant emissions can vary widely between databases due to differences in regulatory limits, technology, and measurement practices.

Case study: Life cycle inventory for a complex system — A typical North American Car. A majority of life cycle inventory studies in the 1970s through 1990s investigated relatively simple product systems, such as packaging, with a limited number of parts and materials involved. When the full system life cycle of materials extraction, processing, transport, forming, handling, use, and end-of-life is considered, even a relatively simple system becomes complex. Starting with a complicated system only serves to magnify data collection and modeling challenges.

Nevertheless, a team from the University of Michigan, in partnership with Ecobalance, Inc., and cooperating with Chrysler (now DaimlerChrysler), Ford, General Motors, the Aluminum Association, the American Iron and Steel Institute, and the American Plastics Council, conducted a LCI study on a complete North American automobile [21]. The material variety, product and process complexity, and scope of supply chain posed significant challenges for LCI. Automobiles typically contain over 20,000 individual parts. Collection of specific inventory data for each part was time and resource prohibitive.

In order to reduce the system complexity while maintaining an accurate LCI model of the system, the vehicle was subdivided into six systems, 19 subsystems, and 644 discrete parts and components composed of 73 materials. An example

Table 2
Selected examples of available life cycle inventory data sets

Source	URL	Data age	Geographic focus	Description
US LCI database project	http://www.nrel.gov/lci/	2004	United States	58 process modules describing the production and use of common materials and fuels
The eco-inventory of Packaging (BUWAL 250)	http://www.umwelt-schweiz.ch/buwal/	1998	Switzerland	Commodity packaging materials from production through conversion, distribution and disposal
Database for environmental assessment of materials	http://www.nims.go.jp/ecomaterial/ecosheet/ecosheet.htm	2000	Japan	Data on alloys, alloying elements, steelmaking processes, and social stocks
Association of Plastic Manufacturers in Europe	http://www.apme.org	1989–2004	Europe	Material production and forming data for commodity and engineering plastics
International Iron and Steel Institute	http://www.worldsteel.org/lci.php	1999–2000	Global	Material production and forming for 14 steel grades/types
LCI report for the North American aluminum industry	http://www.aluminum.org	1995	North America	Various processes within the aluminum product life cycle including information on primary and secondary aluminum
LC Acess	http://www.epa.gov/ORD/NRMRL/lcaccess/	n/a	Global	Searchable listing of available data sources

n/a = not applicable.

of this hierarchy is shown in Fig. 3. Small parts with similar materials and processing (e.g. screws, bolts, and other fasteners) were aggregated into single components in order to simplify the modeling. Data collection focused on 13 specific manufacturing facilities with representative processes, homogenous output, and strong relevance to the study.

The results of this modeling provide the LCI profile of a generic North American family sedan. The life cycle energy profile of this system is shown in Fig. 4, while LCI results for selected metrics are shown in Table 3. The results of this analysis reaffirmed the importance of the use phase as the major determinant of life cycle energy performance. The importance of the use phase is also

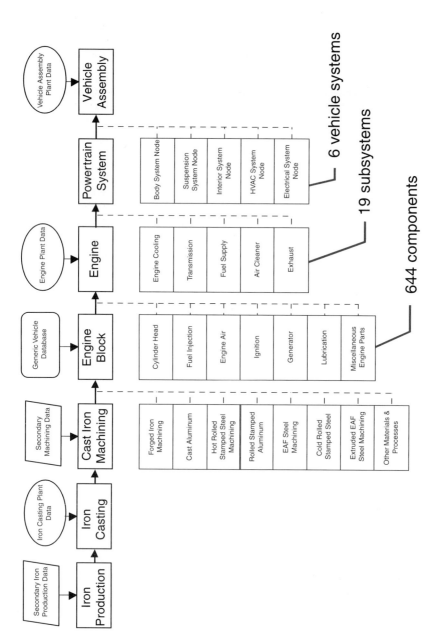

Fig. 3. Hierarchical structure example — manufacturing stage, engine block and related systems, and subsystems [21].

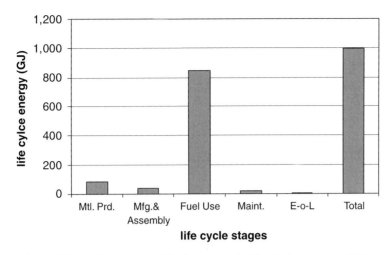

Fig. 4. Life cycle energy profile for a generic North American vehicle.

Table 3
LCI results for a generic North American vehicle

Inventory item	Life cycle performance	Use phase (%)
Energy (MJ)	995,000	87
Carbon dioxide (g)	61,300,000	88
Carbon monoxide (g)	1,940,000	97
Non-methane hydrocarbon (g)	259,000	92
Nitrogen oxides (g)	256,000	91
Solid waste (kg)	4380	25

Source: Sullivan et al., data corrected to final results.

apparent in airborne emission metrics. Other metrics, such as waterborne effluents (not shown in Table 3) and solid waste, exhibited much less dependence on the use phase.

2.2. Impact assessment

LCI analysis provides a useful framework for tracking and quantifying the material and energy inflows and outflows related to a product or process system. However, in order to characterize the environmental and societal effects of the system, LCIA is required. The purpose of LCIA is to assess a system's LCI analysis results to better understand their significance with respect to selected impact categories, such as resource depletion, human health, and ecological health. The procedures of LCIA are less standardized and more complex than those of LCI, and require more value judgments [23].

Several organizations have contributed to the standardization of LCIA. Noteworthy examples include the ISO 14042 standard and Technical Report 14047 [24,25], the SETAC Working Group on LCIA. More recent examples include the United Nations Environment Program (UNEP) — SETAC Life Cycle Initiative [26], and the US EPA through efforts to develop and disseminate the Tool for the Reduction and Assessment of Chemical and Other Environmental Impacts (TRACI) [27,28].

According to ISO 14042, LCIA consists of two required steps — classification and characterization — and three optional activities — normalization, grouping, and weighting. Classification consists of assigning LCI results (e.g. kg NO_x emissions) to specific impact categories of interest (e.g. photo-oxidant formation and acidification).

Common examples of LCIA categories include depletion of abiotic resources, depletion of biotic resources, impacts of land use, land competition, loss of biodiversity and life support function, greenhouse effect, stratospheric ozone depletion, human toxicity, eco-toxicity, smog formation, acidification, and eutrophication. Less common categories include odor, noise, radiation, waste heat, and casualties. The second required operation in LCIA, characterization, consists of calculating overall impact results by category. This operation typically requires the use of characterization factors that convert a specific inventory result to overall impact category totals (e.g. kg SO_2 equivalent acidification per kg NO_x emitted). Selected impact assessment categories and units of characterization are shown in Table 4.

Normalization is an optional element of LCIA in which category results are evaluated relative to a common standard in an attempt to enable comparison against a common baseline (e.g. impacts attributable to the system relative to total regional impacts).

Grouping, also an optional activity, involves sorting impact categories into groups sharing a common theme. A typical example of grouping involves ranking categories by priority — low, medium, or high.

The final optional activity in LCIA is weighting. Weighting consists of converting category results to a common scale using factors designed to reflect the relative importance of each category. Weighting factors are frequently used to calculate a single numerical score based on LCIA results to facilitate comparisons between systems.

Generally, LCIA is accomplished through the application of established characterization factors to LCI results. For example, automobiles have been shown to emit 850 and 164 kg of CO and NO_x, respectively, over a 10-year life cycle [35]. Characterization factors for impacts categories, including photochemical smog formation and acidification, relevant to North America are available in the TRACI software package from US EPA and are shown in Table 5 [28]. In this example, multiplying the reported emissions by the appropriate

Table 4
Selected impact categories and units of characterization

Impact category	Example characterization units	Sources of methodology discussion
Land use	ha-year	[29,30]
Loss of biodiversity	metric tons of net primary production/ ha, metric tons of gross primary production/ha	[31]
Global warming	kg CO_2 equivalent	[1,32]
Ozone depletion	kg CFC-11 equivalents	[27,32]
Human toxicity	kg benzene equivalent,[a] kg toluene equivalent,[b] DALY[c]	[27,33]
Eco-toxicity	Ecological toxicity potential relative to 2,4-D	[27,34]
Smog formation	kg O_3 equivalent, g-NO_x equivalents/ m	[28,32]
Acidification	kg SO_2 equivalents, mol H^+ equivalents	[28,32]
Eutrophication	kg NO_x equivalents, kg N equivalent	[28,32]

[a]For carcinogen impacts.
[b]For non-carcinogen impacts.
[c]For particulate (criteria) air pollutants, DALY = disability adjusted life years.

Table 5
Example LCIA calculations based on automobile emissions

Substance (i)	Inventory result (e_i) [35]	Photochemical smog characterization factor (cf_i) [36][a]	Acidification characterization factor (cf_i) [36][a]	Photochemical smog (I_i)	Acidification (I_i)
CO	850 kg	0.017 g NO_x eqv./ m/kg	n/a	11 g NO_x eqv./m	n/a
NO_x	260 kg	1.2 g NO_x eqv./m/ kg	40 H^+ mol eqv./kg	310 g NO_x eqv./m	10,000 H^+ mol eqv.
Category Total (I)	—	—	—	320 g NO_x eqv./m	10,000 H^+ mol eqv.

n/a = not applicable.
[a]US national characterization factors.

characterization factors and summing the results within each impact category accomplishes the fundamental components of LCIA. The relationship between impact category total (*I*), substance characterization factor (*cf$_i$*), and emissions of substance *i* (*e_i*) is as follows:

$$I = \sum cf_i \, e_i \qquad (1)$$

Table 6
Global warming potential characterization factors [1]

Substance	Global warming potential (kg CO_2 eqv./kg)
Carbon dioxide (CO_2)	1
Methane (CH_4)	23
Nitrous oxide (N_2O)	296
Hydrofluorocarbons (e.g. HFC 134a)	1300
Perfluorocarbons (e.g. CF_4)	5700
Sulfur hexafluoride (SF_6)	22,200

Table 5 presents the impact results for life cycle emissions of two air pollutants for a mid-sized automobile.

The use of characterization factors, such as those shown above, results in an assessment of impact indicators or "mid-points." The term mid-point indicates that these factors characterize impacts at an intermediate point between the source and a final observable effect. For example, greenhouse gases are emitted into the atmosphere and trap heat re-radiated from the Earth's surface leading to what is known as the greenhouse effect and ultimately, to global climate change. Observable effects include significant changes in temperature, precipitation, and sea level. The global warming potentials given in Table 6 are the characterization factors used for calculating the global warming impact. Global warming potentials are based on the radiative forcing (heat-absorbing ability) of each greenhouse gas as well as the decay rate of each gas relative to carbon dioxide over a 100-year time horizon. These factors do not provide any specific indication of the ultimate effects on sea levels or other end points as a result of emissions. Thus, the term mid-point is used to describe these impacts.

Some researchers have proposed alternative methods for impact assessment that begin with the end-points of interest and work backward in what is known as a top-down approach [37]. The top-down approach begins with the identification of end-points and the associated societal values, and then works toward emissions to derive characterization factors. The top-down approach is fundamentally consistent with ISO 14042 goals, but poses challenges for many impact categories due to the complexity of relationships and difficulty in forecasting end-point effects.

One example of the use of an end-point characterization factor is the evaluation of human health effects from criteria air pollutant emissions in TRACI [27]. Characterization factors for these effects were calculated in three stages. First, emissions for specific regions (US states) were modeled to determine expected changes in particulate matter concentrations resulting from each emission and the associated population exposures. The second stage translated concentration exposures into specific morbidity and mortality effects according to published concentration–response functions.

Finally, these morbidity and mortality effects are expressed in DALYs as a measure of expected combined years of life lost and years lived with disability. This procedure takes advantage of a series of well-documented relationships to relate the end-point of interest (DALYs) to emissions of a limited number of substances (NO_x, PM_{10}, $PM_{2.5}$, total suspended particulates, and SO_x). The top-down approach may ultimately result in more robust impact assessment, however, limitations in currently available data restrict the application of end-point factors. The bottom-up approach is the focus of this discussion as it is the more common approach, provides impact results sufficient for decision makers in most situations, and minimizes uncertainty relative to a top-down approach [38].

Uncertainties in impact assessment continue to pose a challenge to the effective use of LCIA in sustainability metrics and decision making. Sources of uncertainty in LCIA are inherent in the methodologies used to derive impact assessment characterization factors. Characterization factors provide a linear relationship between inventory results and quantified impacts. In reality, many impacts exhibit significant non-linearities with increasing environmental loadings. For example, soils may be buffered against acidification, economic factors will influence rates of consumption of abiotic resources and their reserve base, and plants and animals may exhibit the ability to absorb substances below a threshold with no observable effect.

Uncertainties in LCIA characterization factors and their application generally result from temporal and regional scaling. Uncertainty introduced through temporal scaling relates to the timing of the loading and the time horizon for impact evaluation. A common example of an environmental impact sensitive to temporal scaling is smog. Rush hour emissions are more likely to cause smog than emissions that occur overnight. However, existing characterization factors rarely distinguish between emissions at different points in time.

In addition to occurrences at various points in time, environmental impacts also occur at various spatial scales. Global impacts include climate change, ozone depletion, and resource depletion; regional or local impacts include acidification, photochemical smog, eco-toxicity, and human health. Regional differences have little or no effect on global characterization factors, but factors for regional impacts are heavily influenced by spatial differences. For example, unique geographic features in the Los Angeles basin led to specific transport phenomena that influence photochemical smog formation. Clearly, characterization based on phenomena observed in other regions would introduce uncertainties if applied to emissions in the Los Angeles basin.

Recent efforts, such as those by US EPA in the development of TRACI, have focused on the development of regionally appropriate characterization factors. Bare et al. reported that the use of regionally appropriate characterization factors can reduce uncertainty in impact assessment results for impacts such as acidification, eutrophication, and smog formation by orders of magnitude [27].

Nevertheless, broad regional factors, such as those developed at a state level, may not accurately reflect specific local conditions, leaving some inherent uncertainty in the application of impact characterization factors.

2.3. Economic input–output LCA and hybrid methods

The economic input–output (EIO) LCA method uses a commodity input–output (IO) matrix to trace economic transactions throughout the supply chain for a particular product system. Resource inputs and environmental outputs are then coupled to the economic transactions to construct a LCI. The concept to link environmental burdens to an economic input output matrix was originally proposed by Leontief over 50 years ago [39]. The EIO LCA method was refined and applied relatively recently by Horvath and Hendrickson [40]. Their model utilizes a 1992 commodity IO matrix of the US economy as developed by the US Department of Commerce, which includes 485 industrial sectors. Vectors of resource input coefficients and environmental output coefficients are created for each sector and these coefficients represent resource consumption, emissions, and waste per dollar of industrial output. Specific examples of data sources for computing these coefficients include RCRA (Resource Conservation and Recovery Act) Subtitle C hazardous waste generation, management and shipment biannual report, Toxic Release Inventory Data, and US EPA AP-42 emissions factors.

The attractive feature of EIO LCA is that it has the potential to be more comprehensive than process level LCA. The method accounts for upstream processes and indirect inputs that might not be included in a process level LCA. For example, the steel used to make the stamping press used to stamp the steel for an automotive panel is generally neglected in a process level LCA whereas an EIO LCA would capture this input.

The EIO LCA method suffers from several problems that generally differ from those encountered in the process analysis LCA method [41]. The major limitations of the EIO LCA method relate to the high level of aggregation of industry or commodity classifications both for economic transactions and for resource and environmental coefficients. For example, material production of specific polymers such as PET and ABS are grouped together under plastic materials and resins sector. Coefficients are averaged for the whole sector and will not represent differences between products within a sector. Another limitation results from the fact that monetary value can distort physical flow relations between industries due to price inhomogeneity [41]. For example, the resource and environmental intensity for production of a $50,000 vehicle is not expected to be 2.5 times that of a $20,000 vehicle.

The limited availability of sectoral environmental statistics is a concern particularly for small to medium-size businesses, mobile sources, and non-point sources. Despite these challenges EIO LCA is increasingly being used [40,42,43].

The hybrid EIO process analysis LCA method provides a way to exploit the strengths and overcome deficiencies in each method. One hybrid application utilizes a detailed process level analysis of the manufacturing stage in a process-based LCA and uses the EIO LCA method to evaluate all material and energy inputs into this stage. Although other approaches to simplification of rigorous process level LCA have been proposed [44], hybrid systems utilizing some EIO LCA elements appear most promising.

3. Life cycle energy analysis

Life cycle energy analysis is a subset of LCI that tracks energy flows. LCI of energy use is a valuable tool for identifying the life cycle stages and processes of a product system that consume the greatest energy resources. This metric might be considered the single most significant metric for assessing sustainability for several reasons:

(1) greenhouse gas emissions often correlate strongly with energy use (especially when fossil fuels are used as energy sources);
(2) a wide range of other air pollutant emissions originate from energy production and conversion; and
(3) energy data are relatively more available and with greater accuracy than data for other impact categories.

The life cycle energy profiles of products vary dramatically in magnitude and composition/distribution. Table 7 presents life cycle energy metrics for a variety of product systems.

For many products that require energy to operate, the use phase of the life cycle dominates the energy consumption. This pattern is observed for automobiles, buildings, and appliances. One noteworthy exception is desktop computers where the energy requirements for semiconductor manufacturing are substantially greater than the use phase energy. The relatively short expected service life of computers compared to automobiles and buildings also influences the ratio of use phase energy to total life cycle energy. Although not indicated in Table 7, the end-of-life management stage, in general, is the least energy intensive.

Life cycle energy modeling is useful in exploring strategies to reduce operating energy for products. Tradeoffs can exist if a strategy increases material production energy but reduces the use phase energy requirements. For example, an aluminum body automobile will increase fuel economy through lightweighting but material production energy will increase relative to a steel body vehicle. Adding insulation to a house increases material production energy but the use phase benefits will generally outweigh the difference. Life cycle energy models serve to resolve these tradeoffs.

Table 7
Life cycle energy analysis results for various product systems

Product system (functional unit)	Life cycle energy (GJ)	Life cycle energy/ functional unit	Use phase (%)	Source
Passenger car (120,000 miles, 10 years)	998	8.3 MJ/mi or 100 GJ/year	85	[35]
Residential home (50 years, 228 m^2)	16,000	70 GJ/m^2 or 320 GJ/ year	91	[45]
Energy efficient residential home (50 years, 228 m^2)	6400	28 GJ/m^2 or 128 GJ/ year	74	[45]
Desktop computer (3 years, 3300 h)[a]	16.8	5.1 MJ/h or 5.6 GJ/ year	34	[43]
Mixed use commercial building (75 years, 7300 m^2)	2,300,000	316 GJ/m^2 or 3100 GJ/year	98	[46]
6 oz yogurt packaging (1000 lb yogurt delivered)	0.002	5.23 GJ/1000 lb	38	[47]
32 oz yogurt packaging (1000 lb yogurt delivered)	0.007	3.62 GJ/1000 lb	48	[47]
Household refrigerator (20 ft^3, 10 years)	108	10.8 GJ/year	94	[48]
Office file cabinet (one cabinet, 20 years)	2.4	120 MJ/year	n/a[b]	[49]

[a]Values shown have been recalculated from source to account for electrical grid primary energy efficiency of 0.26.
[b]Energy use during the use phase of the file cabinet life cycle is negligible.

Life cycle energy modeling can also distinguish energy resources used for a product system. In addition to the total energy consumption per functional unit, the renewable energy fraction of total consumption is also an important indicator of sustainability.

3.1. Life cycle energy analysis of energy technologies: net energy ratio

Energy ratios have been used since the 1970s to describe the relative effectiveness of energy technologies in converting input energy into useful output. The initial development of energy metrics was driven by concerns over the viability of fossil fuel substitutes as long-term energy sources. These concerns gave rise to the concept of net energy initially defined as the value of energy to society after energy required for obtaining and concentrating the energy carrier are subtracted [50]. Using this definition, at least eight unique expressions of energy ratio equations have been developed in the literature [51]. These ratios combine one or more energy parameters into a single metric. Key energy parameters include total system energy output, losses of energy within the system, input from supporting energy systems, and energy contained in feedstock resources.

While no one form of this metric may be appropriate in all situations, the original intent of "net energy" as defined by Odum and others should be maintained. This definition calls for an understanding of the relationship between output energy and input energy as it relates to the effectiveness of a given system in providing for energy growth. Therefore, the net energy ratio of an energy system can be defined as the ratio of total energy production (E_{out}) to the sum of total primary non-renewable energy requirements associated with feedstock (E_F) and process operations (E_P). For example, in a biomass electricity generating system this ratio is equal to the electricity generated over the non-renewable primary energy for agricultural production, processing, transport, and construction of the generating facility. In the case of photovoltaics, the denominator would include the primary energy required to manufacture, install and maintain the photovoltaic panels and balance of system components. This is shown mathematically in below.

$$NER = \frac{E_{out}}{E_F + E_P} \tag{2}$$

This definition specifies primary energy as the flow of interest to insure that the full infrastructure system of energy production and delivery, and the associated losses in efficiency, are taken into account.

The resulting metric provides a meaningful assessment of the ability of the system to leverage limited energy resources — an important indicator of sustainability. The net energy ratio for various electricity generating options is indicated in Table 8.

4. Life cycle cost analysis

Economic metrics play a key role in the assessment of sustainability performance. However, traditional accounting systems, those designed to meet fiduciary responsibilities of firms, often fail to provide meaningful metrics for evaluating economic sustainability. An alternative cost analysis approach is required, one that considers the full life cycle of goods and services and accounts for externalities typically ignored in traditional cost accounting systems. Life cycle cost

Table 8
Representative net energy ratio values for electricity generation technologies

Technology/Study	Net energy ratio
Solar – photovoltaic	
BIPV [52]	3.6–5.9
BIPV [53]	5.7
CdS/CdTe [54]	9.5
Hydroelectric	
1296 MW [51]	31
114 MW [55]	24
Wind	
Plains site–Ridge site [51]	47–65
Offshore–On-land [56]	31–46
Inland–Coastal [57]	10–30
Biomass	
Willow [58]	10–13
Hybrid poplar [59]	16
Various crops [60]	15–21
General [55]	7.0
Coal	
Technology range [61]	0.29–0.38
Co-fire biomass [58]	0.34
Natural gas	
Combined cycle [62]	0.40
Combined cycle [53]	0.43
Nuclear	
Pressurized water reactor [51]	0.31

(LCC) analysis is a tool that can be used to study the monetary values for processes and flows associated with a product system. When properly applied, LCC assessment provides economic values for flows identified in a LCI and reports them using a common unit of measure ($), which is often easier for decision makers to consider in contrast to the incommensurable values from an LCA.

There are several approaches to LCC analysis. The most commonly applied method records the purchase (C_p), operating (C_{op}), service and maintenance (C_{sm}), and end-of-life management costs (C_{eol}) for a product system. In this approach the life cycle cost of a product or system is recorded as the sum of the costs in each stage. This relationship is shown as.

$$LCC = C_p + C_{op} + C_{sm} + C_{eol} \tag{3}$$

The simple LCC relationship shown in Equation (3) applies to systems with little temporal difference between or within life cycle stages. For most systems, LCCs occur at different points in time and therefore the time value of money must be considered. Discounting is an economic tool used to compare costs or benefits occurring at different points in time. The goal of discounting is to convert future economic values into present-day monetary terms. The calculation of discounted costs utilizes a discount rate (r) and a period of time (n). The discounted life cycle cost (LCC_{pv}) of a system with a lifetime of n years can be calculated as shown.

$$LCC_{pv} = \sum_{t=0}^{n} \frac{LCC_t}{(1+r)^t} \tag{4}$$

For example, Table 9 provides the LCCs associated with automobile ownership over a 10-year life time. A discount rate of 0% provides the constant dollar LCC of the automobile.

The example shown in Table 9 considers only the transactional costs (also known as private costs) associated with the automobile system. The transactional costs of the product life cycle do not include the external costs (also known as social costs) that are needed for a more comprehensive accounting of costs in the development of sustainability metrics. Examples of social costs associated with automobile ownership include military, air pollution, global warming, safety, congestion, land and roads, parking spaces (unpaid), water pollution, noise, highway litter, police costs, court costs, and disposal [64]. Several researchers and organizations have developed tools and data to support external cost accounting. For example, Ogden has published external costs for a mid-sized automobile as shown in Table 10. The data shown in Tables 9 and 10 suggest that consideration of even a limited set of external costs can increase estimated LCCs for an automobile by 15–19%.

The assessment of external costs of products and services is complicated by the limited data available and a lack of consensus on appropriate valuation of environmental and societal functions that may not be assigned market values. External costs are born by society and are not reflected in transaction cost. Determination of appropriate external cost values generally involves evaluating

Table 9
Life cycle ownership costs for a 2001 family sedan with a 10-year lifetime [63]

Discount rate (real) (%)	Purchase price ($)	Fixed operating cost ($)	Variable operating cost ($)	End-of-life value ($)	Discounted ownership cost ($)
0	20,200	7320	18,800	(1510)	44,800
2	20,200	6660	16,600	(1240)	42,100
4	20,200	6090	14,600	(1020)	39,900
6	20,200	5600	13,000	(843)	38,000
8	20,200	5160	11,700	(700)	36,300

Table 10
External costs for a conventional mid-sized automobile with a 10-year lifetime [65]

Category	Present value of cost[a]($)
Oil supply insecurity (military costs)	2650
Air pollution costs	2640
Greenhouse gas emissions cost	1430
Total external cost	6720

[a]External costs shown here are discounted using a rate of 3%.

Table 11
Example values for external costs of CO_2 emissions related to global warming

Study	Value (1990$/metric ton CO_2)	Discount rate (%)	Source
ExternE: externalities of energy	3.8–126	1–3	[67]
Fankhauser	6.2	0.5[a]	[66]
Ogden	33	3	[65]
Chicago climate exchange	1.7	n/a	[68]

n/a = not applicable.
[a]Range of values studied with upper and lower bounds of 3% and 0%, respectively, and a "best guess" of 0.5%.

the expected environmental and societal damages caused by system outputs. For example, damages attributed to emissions of greenhouse gases include loss of crop yield, damage to property, ecosystem loss, mass migration, and increases in cataclysmic weather events. The cost associated with these damages and/or an individual's willingness to pay to avoid damages depends on the location and population under consideration. Economic estimates of willingness to pay for societal goods (external costs) are typically evaluated using contingent valuation methods (see for example [66]). Results from such studies may provide external cost values that vary by an order of magnitude or more.

A representative range of values for global warming costs associated with CO_2 emissions are shown in Table 11. While most of the data in Table 11 result from the application of the contingent valuation approach or an approach combining contingent valuation with market values, the Chicago Climate Exchange provides an exclusively market-based cost of CO_2. This value represents the current cost to corporations interested in purchasing credits to offset CO_2 emissions. Limited incentives to reduce greenhouse gas emissions lead to a lower than expected market cost for CO_2. Specific study methodology and assumptions may vary, however, one significant source of differences in values is the discount rate applied.

As discussed earlier, the value of money is not constant over time. This holds true for both societal and private costs. While discount rates for private costs are well established and reasonably standard (typically ranging between 3% and 5%), discounting of societal costs is less standardized. Some authors have suggested that future generations will value environmental goods, such as the presence of old growth forests, equivalently to current populations. This would require that societal costs for environmental goods are assigned a discount rate of zero [69]. Values in the range of 1– 3% appear to be most common for societal costs, however, some sources site values of 5% or more [65,67]. One potential solution to this dilemma is the use of sliding scale discount rates that vary over time (also called gamma-discounting), such as those proposed by Weitzman [70]. In the Weitzman sliding scale, the short-term (1–5 years) discount rate (4%) is higher than the intermediate rate (6–25 years, 3%), which is higher than the long-term rate (26 –75 years, 2%).

Case study: Life cycle costs of electricity. The sustainability challenges of the existing fossil-fuel-based electricity generation infrastructure are well documented and include high costs, limited access in developing countries, local air pollution, greenhouse gases, unreliability, and resource depletion. Many renewable alternatives have been proposed as possible solutions to these challenges and the environmental benefits of these alternatives have been documented using LCA. In order to understand the economic implications of this suite of alternatives a LCC assessment is required.

This case study considers LCCs of generated electricity at a utility scale. For this system the traditional stages of purchase, operation, service and maintenance, and end-of-life, become initial capital and construction, fuel, non-fuel operations, and decommissioning. Additionally, the external costs of pollution damage will be considered. All costs are discussed in terms of levelized cost per kWh of electricity generated. The levelized cost represents the net present value of all payments required to cover the cost of the system divided by the total lifetime generation. Values discussed here reflect 20 years of operation assuming operation begins in 1999. Previous research has suggested that substantial increases in evaluation period (from 20 to 30 years) results in only a minimal change in levelized costs (decrease of between 0.2 and 0.3 ¢/kWh) [71].

The cost of electricity generation is typically tracked as the sum of capital costs, fuel costs, and non-fuel operating costs. Capital costs include equipment, materials, labor, land, direct and indirect construction costs, design and engineering, initial loading of consumables (e.g. catalyst) and contingency costs. Fuel costs reflect the price the producer pays for primary fuels used in the production of electricity. Non-fuel facility operating costs include labor, maintenance, administration, and non-fuel operating inputs.

In addition to the private costs of electricity generation, the external costs must be included in a total LCC assessment. The costs of damage caused by pollution is an example of the external costs to society of electricity generation. Costs of damage caused by life cycle pollutant emissions (C_p) from electricity generation systems are calculated as the sum of unit damage costs (u_i) multiplied by emissions mass (e_i) as follows.

$$C_p = \sum u_i e_i \tag{5}$$

As discussed earlier in the chapter, a wide range of values are available for unit damage costs of pollution. Representative values for emissions from utility generation in the mid-western US are shown in Table 12. These values are based on research originally published by Fankhauser [72] and Banzhaf [73], and modified by Lewis et al. [74].

The total LCC is calculated as the linear sum of the four factors discussed above. Values for three renewable and three non-renewable utility scale-generating technologies are shown in Table 13. While the renewable technologies studied generally show lower pollution costs, the nuclear power pollution costs are also relatively low. This can be attributed to the selected methodology, which accounts only for the damage costs of air pollutants and not the potential damages associated with spent nuclear fuels. A more complete accounting of external social costs would incorporate these and other factors, such as the loss of ecosystem function associated with hydropower and the impacts of potential acid mine drainage associated with coal acquisition.

Additionally, wind, biomass, and nuclear technologies generally exhibit higher capital costs than the other systems. In the case of the renewable technologies this is attributed to investor uncertainty regarding long-term technology viability and lower production volumes for core equipment. An exception to this is the direct-fire biomass technology (capital cost of 2.3 ¢/kWh) that utilizes boiler systems similar to those used for over 50 years in coal plants. In the case of nuclear power, the higher capital costs are associated with greater upfront investment in facility, design, verification, equipment, and construction.

Table 12
Unit damage costs for common electric utility emissions [74]

Pollutant	Unit damage cost
Carbon dioxide ($/ton carbon)	30
Carbon monoxide ($/ton)	1
Lead ($/ton)	1965
Methane ($/ton)	172
Nitrogen oxides ($/ton)	218
Nitrogen oxide ($/ton nitrogen)	4498
Particulates ($/ton)	2624
Sulfur oxides ($/ton)	84

Table 13
Life cycle costs of electricity generating technologies (¢/kWh) [51]

Technology	Capital cost	Fuel cost	Non-fuel operating cost	Pollution cost	Total life cycle cost
Hydroelectric, one large-scale installation	1.7	—	0.15	0.1	2.0
Wind, two wind farms	4.7–6.4	—	0.8–1.1	0.01	5.5–7.5
Biomass, three willow conversion technologies	2.3–4.1	1.4–2.2	1.6–2.1	0.10–0.11	5.5–8.4
Coal, average plant	2.1	1.3	0.3	6.0	9.7
Natural gas, combined cycle	1.1	2.0	0.4	1.0	4.5
Nuclear, PWR	3.9	2.6	1.9	0.04	8.4

PWR = Pressurized water reactor.

Ultimately, LCC metrics can assist in evaluating the tradeoffs inherent in technology selection. Some systems, such as coal, exhibit low life cycle transactional costs but place a large external cost burden on society. While others, such as wind, require greater investment in capital, but limit the damage to society caused by air pollution.

5. Life cycle optimization

A critical question regarding the life cycle management of any product system is, "What is its optimal service life?" [75,76] The answer may vary depending on the optimization criteria used, which may include environmental, economic, functional performance, and aesthetic objectives. From an environmental perspective, this is a particularly complex question to resolve for products that consume energy in their use phase. On the other hand, indefinite useful life is generally desired when considering energy and environmental criteria for products that do not require energy inputs in the use phase. In the case of automobiles, household appliances, and computers, there exist multiple tradeoffs between maintaining an existing model and replacing it with one that is more efficient. The efficiency gain from model replacement should exceed the additional resource investments required to produce the new model.

A life cycle optimization (LCO) model was developed recently to evaluate optimal service life from energy, emissions, and cost perspectives [35]. This LCO model is based on a dynamic programming method with inputs derived from LCA. Dynamic programming is a collection of mathematical tools used to analyze sequential decision processes. The planning horizon is divided into multiple stages at which different decisions may be made depending on the state of the system at that time. A given decision will transform the system to a new state with a new corresponding outcome. Dynamic programming seeks the particular sequence of decisions that best satisfies a decision maker's criteria over the complete planning horizon. In a dynamic programming model, the time horizon of the problem is the period of time over which the decisions are made.

Figure 5 provides a schematic example of the LCO model applied to vehicle replacement. The y-axis is the cumulative environmental burden such as NO_x emissions or energy consumption, while the x-axis represents time. The initial vehicle is assumed to be produced at time 0, and a new model vehicle with a different environmental profile is introduced at time T_a and T_b. Decisions to keep or replace vehicles are made at the points marked by black dots. Environmental burdens from materials production and manufacturing are shown as a step function at the time a vehicle is produced. The slope of each line segment represents an energy efficiency or emission factor of a vehicle model. The slopes tend to increase with time, indicating deteriorations of emission controls or energy efficiencies.

Assume that, at time 0, a decision maker tries to minimize the environmental burden of a criterion within the time horizon N based on information the decision maker has regarding the environmental performance of future vehicles. The decision maker seeks a solution of the form "Buy a new vehicle at the start

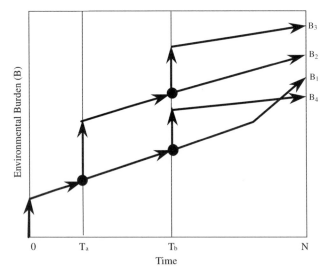

Fig. 5. Schematic example of the life cycle optimization (LCO) model based on four policies. B_1–B_4 represent the final environmental burdens for the four policies [35].

of year 0 and keep it for α years and retire it; then buy a new vehicle at the start of year α and keep it for β years and retire it, etc." As an example, consider four policies depending on the decisions at T_a and T_b.

(1) If the vehicle owner keeps the initial vehicle throughout the time horizon N, the cumulative environmental burden (B) will result in B_1. The slope change between T_b and N represents vehicle deterioration expected for older cars.

(2) If the vehicle owner replaces the initial vehicle with a new vehicle at time T_a and keeps the new vehicle until N, the cumulative environmental burden (B) will result in B_2.

(3) If the vehicle owner replaces the initial vehicle with a new vehicle at time T_a and replaces this second vehicle again at time T_b, the cumulative environmental burden (B) will result in B_3.

(4) If the vehicle owner replaces the initial vehicle at time T_b with a new vehicle and keeps the new vehicle until N, the cumulative environmental burden (B) will result in B_4, which is the minimum possible outcome.

This LCO model was developed and applied to study optimal lifetimes of mid-sized generic cars over a 36-year time horizon (between calendar year 1985 and 2020). Optimal replacement policy was investigated that minimized life cycle energy, emissions, and cost as individual objective functions.

Table 14 gives the optimization results of generic mid-sized model scenarios. The optimal set of lifetimes for the energy/CO_2 objectives in Table 14 can read, for example, "Keep the model year 1985 car for 18 years and retire it at the end of 2002, then buy a model year 2003 car and keep it for another 18 years until 2020 in order to minimize energy/CO_2 emissions when driving a mid-sized passenger car 12,000 mi/yr." For CO, NMHC, and NO_x pollutants with 12,000 miles of annual mileage, automobile lifetimes ranging from 3 to 6 years are

Table 14
Optimal vehicle lifetimes and cumulative burdens for a 36-year time horizon between 1985 and 2020 (12,000 miles of annual driving) [35,63]

Objective minimized	Optimal vehicle lifetimes (years)[a]	Private cost (constant 1985$)	Cumulative environmental burdens				
			Energy (10^3 GJ)	CO_2 (10^5 kg)	CO (10^6 g)	NMHC (10^5 g)	NO_x (10^5 g)
Energy/CO_2	18, 18	77,300	3.34	2.18	4.95	6.18	6.52
CO	3, 3, 4, 6, 6, 7, 7	117,000	3.84	2.46	2.76	4.29	4.54
NMHC	6, 6, 10, 14	94,800	3.53	2.29	2.96	4.07	4.47
NO_x	5, 5, 6, 6, 14	101,000	3.65	2.36	2.86	4.14	4.32
Private explicit ownership cost	17, 19	76,200	6.97	4.54	5.64	9.50	11.0

[a]Replacement intervals for a 36 year time horizon.

optimal for 1980s and early 1990s model years, while optimal lifetimes are expected to be 7–14 years for model year 2000 and beyond. On the other hand, a lifetime of 18 years minimizes cumulative life cycle energy and CO_2 based on driving 12,000 miles annually.

The expected median lifetime of an average car has increased from 12.5 years for model year 1980 to 16.9 years for model year 1990 [77]. Thus, generally, cars are driven for a longer time than is optimal from a regulated emissions perspective, while median automotive lifetimes have been almost ideal from a CO_2 and energy perspective.

The LCO model was also modified to investigate optimal household refrigerator service life. Model runs with a time horizon between 2004 and 2020 show that current owners (2004) should replace typical mid-sized 1994 models and older, which would be an efficient strategy from both cost and energy perspectives [78].

6. Life cycle sustainability indicators

The life cycle framework can also be used to construct a matrix of environmental, social, and economic sustainability indicators for a system. These indicators can be organized by life cycle stage and then categorized into the "triad" of sustainability: economic, social, and environmental. This approach was used to assess the sustainability of the US food system [79]. Table 15 presents the full matrix of sustainability indicators developed by Heller and Keoleian. In many instances, the division of economic, social, and environmental sustainability is somewhat arbitrary since particular indicators often address more than one aspect of sustainability. Also identified in Table 15 are the primary stakeholders involved or influential in each stage of the food system. The indicators evaluated in Table 15 can be both qualitative and quantitative.

This matrix approach can be used to evaluate the sustainability of other product systems. Indicators based on the three dimensions of sustainability developed elsewhere (e.g. Global Reporting Initiative [80]) do not necessarily follow a life cycle framework but can provide useful examples of social indicators.

7. Conclusions

This chapter demonstrated the capabilities of life cycle-based models and metrics for assessing and guiding the sustainability of products and technology. LCA has been applied for over three decades. Increasingly, firms are recognizing the value of life cycle thinking in sustainability metrics and will begin to implement life cycle methods as tools become more accessible. A few final observations regarding these tools and their future development are offered to conclude this chapter.

Table 15
Life cycle sustainability indicators for the food system [79]

Life cycle stage	Stakeholders	Indicators		
		Economic	Social	Environmental
Origin of (genetic) resource — seed production, animal breeding	Farmers Breeders Seed companies	Degree of farmer/operator control of seed production/breeding	Diversity in seed purchasing and seed collecting options Degree of cross-species manipulation	Ratio of naturally pollinated plants to genetically modified/hybrid plants per acre Reproductive ability of plant or animal percent of disease-resistant organisms
Agricultural growing and production	Farm operators Farm workers Agricultural industry Agricultural schools Government Animals	Rates of agricultural land conversion Output/input productivity percent return on investment Cost of entry to business Farmer savings and insurance plans Flexibility in bank loan requirements to foster environmentally sustainable practices Level of government support	Average age of farmers Diversity and structure of industry, size of farms, number of farms per capita Hours of labor/yield and/income Average farm wages vs. other professions Number of legal laborers on farms, ratio of migrant workers to local laborers percent workers with health benefits Number of active agrarian community organizations percent of agricultural schools that offer sustainable agricultural programs, encourage sustainable practices Number of animals/unit, time animals spend outdoors (animal welfare)	Rate of soil loss vs. regeneration Soil microbial activity, balance of nutrients/acre Quantity of chemical inputs/unit of production Air pollutants/unit of production Number of species/acre Water withdrawal vs. recharge rates Number of contaminated or eutrophic bodies of surface water or groundwater percent waste utilized as a resource Veterinary costs Energy input/unit of production Ratio of renewable to non-renewable energy Portion of harvest lost due to pests, diseases

Table 15 (*continued*)

Life cycle stage	Stakeholders	Indicators		
		Economic	Social	Environmental
Food processing, packaging and distribution	Food processors Packaging providers Wholesalers Retailers	Relative profits received by farmer vs. processor vs. retailer Geographic proximity of grower, processor, packager, retailer	Quality of life and worker satisfaction in food processing industry Nutritional value of food product Food safety	Energy requirement for processing, packaging and transportation Waste produced/unit of food percent of waste and byproducts utilized in food processing industry percent of food lost due to spoilage/mishandling
Preparation and consumption	Consumers Food service Nutritionists/ Health pro-fessionals	Portion of consumer disposable income spent on food percent of food dollar spent outside the home	Rates of malnutrition Rates of obesity Health costs from diet related disease/conditions Balance of average diet percent of products with consumer labels Degree of consumer literacy regarding food system consequences, product quality vs. appearance, etc. Time for food preparation	Energy use in preparation, storage, refrigeration Packaging waste/ calories consumed Ratio of local vs. non-local and seasonal vs. non-seasonal consumption
End-of-life	Consumers Waste managers Food recovery & gleaning organizations	Ratio of food wasted to food consumed in the US Dollar spent on food disposal	Ratio of (edible) food wasted vs. donated to food gatherers	Amount of food waste composted vs. sent to landfill/incinerator/ wastewater treatment

1. The life cycle framework is a logical system boundary for defining sustainability metrics to evaluate performance and guide system improvement. Many useful analytical tools including LCA, life cycle energy analysis, LCC analysis, LCO, and life cycle sustainability indicators, have been developed and applied to a variety of systems for measuring aspects of environmental, economic, and social sustainability. LCA provides many important measures for assessing environmental sustainability while LCC analysis can provide a microscale perspective on economic sustainability. Using a life cycle framework to investigate the social dimensions to sustainability can also provide a powerful tool, but this area is much less developed.

2. Data availability and quality remains a major challenge in conducting a LCA. National database initiatives (e.g. US, Japan, Switzerland) and life cycle projects undertaken by industry associations (e.g. APME, IISI, AA) are essential for the development of the life cycle field. In addition to data limitations and time requirements, system boundary issues and truncation are factors that impact the ease and accuracy of conducting a process level LCA. EIO LCA is emerging as an alternative approach that can address some of these challenges. The high level of aggregation of Input/Output tables with respect to products, processes, and technologies limits the accuracy of this approach. Hybrid EIO and process level LCA represents one way to combine the positive attributes of each method.

3. What single metric would best represent environmental sustainability if only one impact category (e.g. greenhouse gas emissions, resource depletion) could be used to assess the environmental sustainability performance of a product system? Life cycle energy consumption might be recommended as a key indicator for serving this general purpose. Declining energy non-renewable energy sources such as petroleum and natural gas is a major sustainability challenge facing society in the 21st century. In addition, greenhouse gas emissions, acidification, and smog formation are important impact categories that often correlate with energy use, particularly fossil fuels. Consequently, life cycle energy analysis might be emphasized if resources are severely limited for environmental sustainability assessment.

4. The net energy ratio is an important sustainability metric derived from life cycle energy analysis of energy carriers including electricity and transportation fuels. This metric indicates the capability of the energy system to leverage non-renewable energy inputs. The net energy ratio is particularly useful in evaluating the sustainability of alternative renewable energy technologies such as photovoltaics, wind, biomass electricity, and biomass transport fuels.

5. LCC assessment indicates how costs are distributed across the supply chain and which stakeholders incur costs and benefits. This can provide valuable insights into the microscale dimensions of economic sustainability. The LCC analysis can be very useful in evaluating the public works projects such as

road infrastructure. For example, design alternatives can be explored that minimize LCCs that include agency costs (construction and maintenance) and social costs (user costs and external costs). Social costs include congestion, lost productivity, vehicle damage, accidents related to poor roads, and traffic during rehabilitation activities [81]. External costs include pollution, which can also be monetized and compared with other LCCs.

6. Life cycle-based social sustainability metrics represent an area in need of development. In addition, methods for evaluating incommensurable environmental, social, and economic sustainability measures and resolving trade-offs when considering alternatives is also an area for investigation. Alignment between social, economic, and environmental sustainability indicators would be the desired outcome in the life cycle design and management of product systems.

7. Developing absolute measures of sustainability may be the most challenging area for research. Most of the metrics for assessing sustainability performance are relative metrics rather than absolute measures. In other words, we can say that less of an impact is better but it is difficult to say what is "sustainable" in the absolute sense. How much carbon dioxide emitted from a product system would be considered sustainable? Even if a sustainable global greenhouse gas emissions target could be established there is no clear method for allocating a global emission threshold to a specific product system. Ecosystems are the foundations of our life support system. Consequently, more research is needed to define life cycle metrics for assessing ecosystem structure, function, and health.

8. Advancements in the field of LCA have been made through professional societies including ISO, SETAC, the Society of Automotive Engineers (SAE), Institute of Electronics and Electrical Engineers (IEEE), International Society for Industrial Ecology (ISIE); academia; and several governmental organizations including the US EPA and UNEP. In academia these tools are being developed in a variety of disciplines including public health, natural resources, environmental science, chemical engineering, mechanical engineering, industrial engineering, and materials science. The field is very interdisciplinary which makes it a rich area for research scholarship.

References

[1] J.T. Houghton, Y. Ding, D.J. Griggs, M. Noguer, P.J.v.d. Linden, X. Dai, K. Maskell, C.A. Johnson (Eds), Contributions of Working Group I to the Third Assessment Report of the Intergovernmental Panel on Climate Change, Climate Change 2001: The Scientific Basis, Cambridge University Press, New York, 2001, p. 881.

[2] C.J. Campbell, J.H. Laherrère, Sci. Am. 278 (1998) 78–84.

[3] UNEP, Persistent Organic Pollutants, [website]. 2005, http://www.chem.unep.ch/pops/.

[4] UNEP, Vital Water Graphics — An Overview of the State of the World's Fresh and Marine Waters, United Nations Environment Programme, Nairobi, Kenya, 2002.

[5] Ecological Society of America, Issues in Ecology, [website]. 2005, http://www.esa.org/science/Issues/.

[6] M.L. Reaka-Kudla, D.E. Wilson, E.O. Wilson, Biodiversity II: Understanding and Protecting our Biological Resources, Joseph Henry Press, Washington, DC, 1997, p. 551.

[7] UNEP, World Population Prospects: The 2002 Revision, 2002, United Nations Environment Programme.

[8] UNDP, Human Development Report, [website]. 2005, http://hdr.undp.org/statistics/data.

[9] G.H. Brundtland (Ed.), Our Common Future, The World Commission on Environment and Development, Oxford University Press, Oxford, 1987.

[10] ISO, 14040 Environmental Management — Life Cycle Assessment — Principles and Framework, International Organization for Standardization, Geneva, 1997.

[11] M.A. Curran, Environmental Life Cycle Assessment, McGraw-Hill, New York, 1996.

[12] US EPA, Life-Cycle Assessment: Inventory Guidelines and Principles, United States Environmental Protection Agency, Washington, DC, 1993.

[13] F. Consoli, D. Allen, I. Boustead, J. Fava, W. Franklin, A.A. Jensen, N.d. Oude, R. Parrish, R. Perriman, D. Postlethwwaite, B. Quay, J. Séguin, B. Vigon (Eds), Guidelines for Life-Cycle Assessment: A Code of Practice, Society of Environmental Toxicology and Chemistry, Pensacola, FL, 1993.

[14] A.d. Beaufort-Langeveld, R. Bretz, R. Hischier, M. Huijbregts, P. Jean, T. Tanner, G.v. Hoof (Eds), Code of Life-Cycle Inventory Practice, SETAC Press, Brussels, Belgium, 2003, p. 160.

[15] D. Allen, F. Consoli, G.A. Davis, J. Fava, J. Warren (Eds), Public Policy Applications of Life Cycle Assessment, Society of Environmental Toxicology and Chemistry, Pensacola, FL, 1997, p. 127.

[16] G.A. Keoleian, K. Kar, J. Clean. Prod. 11 (2003) 61–77.

[17] US EPA, The Use of Life Cycle Assessment in Environmental Labeling, US Environmental Protection Agency, Office of Pollution Prevention and Toxics, Washington, DC, 1993.

[18] ISO, 14041 Environmental Management — Life Cycle Assessment — Goal and Scope Definition and Inventory Analysis, International Organization for Standardization, Geneva, 1998.

[19] ISO, TR 14049 Environmental Management —- Life Cycle Assessment — Examples of Application of ISO 14041 to Goal and Scope Definition and Inventory Analysis, International Organization for Standardization, Geneva, 2000.

[20] T. Ekvall, G. Finnveden, J. Clean. Prod. 9 (2001) 197–208.

[21] G.A. Keoleian, G. Lewis, R.B. Coulon, V.J. Camobreco, H.P. Teulon, Total Life Cycle Conference Proceedings, P-339, SAE International, Graz, Austria, 1998.

[22] J.L. Sullivan, R.L. Williams, S. Yester, E. Cobas-Flores, S.T. Chubbs, S.G. Hentges, S.D. Pomper, Total Life Cycle Conference Proceedings, P-339, SAE International, Graz, Austria, 1998.

[23] E.G. Hertwich, J.K. Hammitt, W.S. Pease, J. Ind. Ecol. 4 (2000) 13–28.

[24] ISO, 14042 Environmental Management — Life Cycle Assessment — Life Cycle Impact Assessment, International Organization for Standardization, Geneva, 2000.

[25] ISO, TR 14047 Environmental Management — Life Cycle Impact Assessment — Examples of Application of ISO 14042, International Organization for Standardization, Geneva, 2003.

[26] H.A.U.d. Haes, G. Finnveden, M. Goedkoop, M. Hauschild, E.G. Hertwich, P. Hofstetter, O. Jolliet, W. Klöpffer, W. Krewitt, E. Lindeijer, R. Müller-Wenk, S.I. Olsen, D. Pennington, J. Potting, B. Steen (Eds), Life-Cycle Impact Assessment: Striving Towards Best Practice, SETAC Press, Brussels, Belgium, 2002.

[27] J.C. Bare, G.A. Norris, D.W. Pennington, T. McKone, J. Ind. Ecol. 6 (2002) 49–78.

[28] G.A. Norris, J. Ind. Ecol. 6 (2002) 79–101.
[29] D.V. Spitzley, D.A. Tolle, J. Ind. Ecol. 8 (2004) 11–21.
[30] E. Lindeijer, J. Clean. Prod. 8 (2000) 273–281.
[31] E. Lindeijer, J. Clean. Prod. 8 (2000) 313–319.
[32] J. Potting, W. Klöpffer, J. Seppälä, G. Norris, M. Goedkoop, in: H.A.U.d. Haes, G. Finnveden, M. Goedkoop, M. Hauschild, E.G. Hertwich, P. Hofstetter, O. Jolliet, W. Klöpffer, W. Krewitt, E. Lindeijer, R. Müller-Wenk, S.I. Olsen, D.W. Pennington, J. Potting, B. Steen (Eds), Life-Cycle Impact Assessment: Striving Towards Best Practice, SETAC Press, Brussels, Belgium, 2002.
[33] W. Krewitt, D.W. Pennington, S.I. Olsen, P. Crettaz, O. Jolliett, in: H.A.U.d. Haes, et al. (Eds), Life-Cycle Impact Assessment: Striving Towards Best Practice, SETAC Press, Brussels, Belgium, 2002.
[34] M. Hauschild, D.W. Pennington, in: H.A.U.d. Haes, G. Finnveden, M. Goedkoop, M. Hauschild, E.G. Hertwich, P. Hofstetter, O. Jolliet, W. Klöpffer, W. Krewitt, E. Lindeijer, R. Müller-Wenk, S.I. Olsen, D.W. Pennington, J. Potting, B. Steen (Eds), Life-Cycle Impact Assessment: Striving Towards Best Practice, SETAC Press, Brussels, Belgium, 2002.
[35] H.C. Kim, G.A. Keoleian, D.E. Grande, J.C. Bean, Environ. Sci. Technol. 37 (2003) 5407–5413.
[36] US EPA, Tool for the Reduction and Assessment of Chemical and Other Environmental Impacts (TRACI), United States Environmental Protection Agency, Washington, D.C., 2002.
[37] H.A.U.d. Haes, E. Lindeijer, in: H.A.U.d. Haes, G. Finnveden, M. Goedkoop, M. Hauschild, E.G. Hertwich, P. Hofstetter, O. Jolliet, W. Klöpffer, W. Krewitt, E. Lindeijer, R. Müller-Wenk, S.I. Olsen, D.W. Pennington, J. Potting, B. Steen (Eds), Life-Cycle Impact Assessment: Striving Towards Best Practice, SETAC Press, Brussels, Belgium, 2002.
[38] J.C. Bare, P. Hofstetter, D.W. Pennington, H.A.U.d. Haes, Int. J. Life Cycle Ass. 5 (2000) 319–326.
[39] W. Leontief, Rev. Econ. Stat. 18 (1936) 105–125.
[40] A. Horvath, C. Hendrickson, J. Infrastruct. Systems. 4 (1998) 111–117.
[41] S. Suh, M. Lenzen, G.J. Treloar, H. Hondo, O. Jolliet, U. Klann, W. Krewitt, Y. Moriguchi, J. Munksgaard, G. Norris, Environ. Sci. Technol. 38 (2004) 657–664.
[42] H.L. McLean, L.B. Lave, Environ. Sci. Technol. 37 (2003) 5445–5452.
[43] E. Williams, Environ. Sci. Technol. 38 (2004) 6166–6174.
[44] T.E. Gradel, Streamlined Life-Cycle Assessment, Prentice-Hall, Upper Saddle River, NJ, 1998.
[45] G.A. Keoleian, S. Blanchard, P. Reppe, J. Ind. Ecol. 4 (2001) 135–156.
[46] C. Scheuer, G.A. Keoleian, P. Reppe, Energy Build 35 (2003) 1049–1064.
[47] G.A. Keoleian, A.W. Phipps, T. Dritz, D. Brachfeld, Packag. Technol. Sci. 17 (2004) 85–103.
[48] Y.A. Horie, Life Cycle Optimization of Household Refrigerator-Freezer Replacement, Master's Thesis, School of Natural Resources and Environment, University of Michigan, Ann Arbor, MI, 2004.
[49] B.A. Dietz, Life Cycle Assessment of Office Furniture Products, Master's Thesis, School of Natural Resources and Environment, University of Michigan, Ann Arbor, MI, 2005.
[50] H.T. Odum, Ambio 2 (1973) 220–227.
[51] D.V. Spitzley, G.A. Keoleian, Life Cycle Environmental and Economic Assessment of Willow Biomass Electricity: A Comparison with Other Renewable and Non-Renewable Sources, Center for Sustainable Systems, University of Michigan, Ann Arbor, MI, 2004.
[52] G.A. Keoleian, G.M. Lewis, Renewable Energy 28 (2003) 271–293.
[53] P.J. Meier, Life-Cycle Assessment of Electricity Generation Systems and Applications for Climate Change Policy Analysis, Ph.D. Thesis, Land Resources, University of Wisconsin, Madison, WI, 2002.

[54] K. Kato, T. Hibino, K. Komoto, S. Ihara, S. Yamamoto, H. Fujihara, Sol. Energy Mater. Sol. Cells 67 (2001) 279–287.

[55] D. Pimentel, M. Herz, M. Glickstein, M. Zimmerman, R. Allen, K. Becker, J. Evans, B. Hussain, R. Sarsfeld, A. Grosfeld, T. Seidel, BioScience 52 (2002) 1111–1120.

[56] L. Schleisner, Renewable Energy 20 (2000) 279–288.

[57] K.R. Voorspools, E.A. Brouwers, W.D. D'haeseleer, Appl. Energy 67 (2000) 307–330.

[58] M.C. Heller, G.A. Keoleian, M.K. Mann, T.A. Volk, Renewable Energy 29 (2004) 1023–1042.

[59] M.K. Mann, P.L. Spath, Life Cycle Assessment of a Biomass Gasification Combined-Cycle System, National Renewable Energy Laboratory (NREL), Golden, CO, 1997.

[60] A. Monti, G. Venturi, Eur. J. Agron. 19 (2003) 35–43.

[61] P.L. Spath, M.K. Mann, D.R. Kerr, Life Cycle Assessment of Coal-Fired Power Production, National Renewable Energy Laboratory, Golden, CO, 1999.

[62] P.L. Spath, M.K. Mann, Life Cycle Assessment of a Natural Gas Combined-Cycle Power Generation System, National Renewable Energy Laboratory (NREL), Golden, CO, 2000.

[63] D.V. Spitzley, D.E. Grande, G.A. Keoleian, H.C. Kim, Transp. Res. Part D 10 (2005) 161–175.

[64] R.C. Porter, Economics at the Wheel — The Costs of Cars and Drivers, Academic Press, Toronto, 1999.

[65] J.M. Ogden, R.H. Williams, E.D. Larson, Energy Policy, 32 (2004) 7–27.

[66] S. Frankhauser, The Energy J. 15 (1994) 157–184.

[67] P. Bickel, S. Schmid, W. Krewitt, R. Friedrich (Eds), External Costs of Transport in ExternE, European Commission, Non Nuclear Energy Programme, Luxembourg, 1997.

[68] Chicago Climate Exchange, CCX Carbon Financial Instruments (CFIs) Monthly Market Summary, December, 2004, http://www.chicagoclimatex.com/.

[69] J. Broome, D. Ulph, Counting the Cost of Global Warming: A Report to the Economic and Social Research Council, White Horse, Cambridge, 1992.

[70] M. Weitzman, Am. Econ. Rev. 91 (2001) 260–271.

[71] R.H. Wiser, Fuel Energy Abstr. 38 (1997) 246.

[72] S. Fankhauser, Energy J. 15 (1994) 157–184.

[73] H.S. Banzhaf, W.H. Desvousges, F.R. Johnson, Resource Energy Econ. 18 (1996) 395–421.

[74] G.M. Lewis, G.A. Keoleian, M.R. Moore, D.L. Mazmanian, M. Navvab, PV-BILD: A Life Cycle Environmental and Economic Assessment Tool for Building-Integrated Photovoltaic Installations, National Science Foundation/Lucent Technologies Industrial Ecology Grant, 1999.

[75] US EPA, Life Cycle Design Guidance Manual: Environmental Requirements and the Product System, United States Environmental Protection Agency, 1993.

[76] W.R. Stahel, The Greening of Industrial Ecosystems, in: B.R. Allenby, D.J. Richards (Eds), The Utilization-Focused Service Economy: Resource Efficiency and Product-Life Extension, National Academy Press, Washington, DC, 1994, pp. 178–190.

[77] US DOE, Transportation Energy Data Book, Edition 22, US DOE, Center for Transportation Analysis, Oak Ridge National Laboratory, 2002.

[78] H.C. Kim, G.A. Keoleian, Y.A. Horie, Energy Policy (in press).

[79] M.C. Heller, G.A. Keoleian, Agric. Sys. 76 (2003) 1007–1041.

[80] GRI, Sustainability Reporting Guidelines, Global Reporting Initiative, Boston, 2002, p. 104.

[81] G.A. Keoleian, A. Kendall, J.E. Dettling, V.M. Smith, R.F. Chandler, M.D. Lepech, V.C. Li, J. Infrastruct. Systems 11 (2005) 51–60.

Sustainability Science and Engineering: Defining principles
Martin A. Abraham (Editor)
© 2006 Elsevier B.V. All rights reserved
DOI 10.1016/S1871-2711(05)01008-1

Chapter 8

Making Safer Chemicals

Ken Geiser

Lowell Center for Sustainable Production, University of Massachusetts Lowell

The physical and chemical sciences have developed largely independent of the environmental and health sciences. The enormous scientific strides of the 19th and 20th centuries in metallurgy, ceramics, inorganic and organic chemistry and polymers developed our capacities to identify, test, characterize, process and synthesize chemicals with a wide range of commercial applications. The result has been a plethora of structural and functional materials from which hundreds of thousands of products have been made. These products have extended our lives, eased our work, secured our homes, created our wealth and enriched our lives.

However, as newer, cheaper and more versatile chemicals have emerged from the laboratories many of them have turned out to be toxic, dangerous or threatening to ecological processes. The hazardous characteristics of these new chemicals were seldom recognized or intended. The organic solvents were not designed to be carcinogenic. The refrigerant gases were not intended to damage the upper atmospheric ozone layer. The synthetic plasticizers were not expected to disrupt hormonal systems. Because knowledge from the emerging fields of toxicology, pharmacology and ecology was not integrated into the physical and chemical sciences that generated innovations in chemistry the hazards of new chemicals has seldom been factored into their design processes.

Instead of trying to create a more comprehensive science for generating highly functional, inexpensive and safe chemicals, the emerging knowledge about the hazards of chemicals was used to construct a vast array of professional guidelines and government public health and environmental regulations. Professional associations wrote voluntary standards and drafted guidance manuals for proper handling of hazardous chemicals. Federal and state agencies were established to regulate toxic chemicals in foods, deadly chemicals used in agriculture, toxic chemicals in drinking water, hazardous chemicals in work-places, dangerous chemicals in products and polluting chemicals in industrial emissions and wastes[1].

The conventional approach for establishing these regulations has focused on the perceived dangers of chemicals and the likelihood of human or environmental exposures. Government agencies employ scientific tests and risk-determining protocols to assess the dangers associated with exposure to chemicals identified by scientific or public concern. Once exposure to a substance is demonstrated to result in unacceptable levels of harmfulness, agency professionals draft regulations to restrict or condition the use of those substances. This represents a *problem-focused* approach to chemicals management.

This approach has positive features. It tends to focus on substances of high public concern. It directs scientific attention to a limited number of potential subjects. It seeks to set guidelines for chemicals use and exposure. However, relying on this problem-based approach alone offers an inefficient procedure for achieving a sustainable chemicals future. We certainly need the resources of science to better our collective understanding of the intricacies of chemical processes and the mysteries of biochemical interactions. However, expanding knowledge about the behavior and effects of chemicals in the environment and within human bodies is only the first step to achieving a future of safer and more effective chemicals.

We need a parallel approach that focuses less on the characterization of problems and more on the development of solutions. Such a *solution-focused* approach would eagerly accept the large accumulation of scientific understanding as a basis for designing chemicals that are safer, cleaner and more environmentally compatible. Solution-seeking approaches to chemical and material development would involve designing chemicals that are not only of high performance and cost effective, but also biologically safe and ecologically sustainable.

1. Seeking safer chemistries

A solution-focused approach to chemicals development and production will not come easily. The transition to safer and more sustainable chemicals requires a significant re-direction of the chemical industry and a reevaluation of its products. Future generations will continue to need chemicals and the industrial transformation of chemicals to meet human needs will continue to require ingenuity and enterprise. However, the types of chemicals and how they are used needs to be reconsidered. Fossil fuels will need to play a much smaller role and

---

[1]For more of this history, see Kenneth Geiser [1]

wastes from production and consumption will need to be managed and recycled in ways that conserve materials and protect the environment. This transition will require a new mission for the industry a mission that promotes human health and environmental quality as seriously as the market promotes economic efficiency and product effectiveness.

The foundations for this development are already being laid. Leading firms in the industry and thoughtful government leaders are exploring new goals and new directions for safer chemicals. Some of the most progressive chemical manufacturing and processing firms have established corporate sustainability policies and many of these firms publish reports on the chemicals they use and the chemicals they avoid. Research on new chemicals, new routes of chemical synthesis, new feedstocks and new chemical services have begun to pay off with cleaner production systems, reduced energy consumption and products that are more easily recycled or biologically disposed. The chemistry and chemical engineering fields have responded with new professional statements, conferences that explore sustainable directions and educational curricula and texts that integrate environmental considerations into conventional education from the primary schools to graduate training [2].

In developing safer and more sustainable chemicals several avenues are emerging. One approach involves assembling a list of all those chemicals characterized by high levels of some type of undesirable hazard or unwanted toxicity and substance-by-substance subtracting undesirable chemicals from the larger list of commonly used chemicals. We can call this *safer chemistry* because it works incrementally to avoid substances that are less safe. A second approach involves reviewing the large body of research and studies in toxicology and pharmacology for guiding principles that are known to lead to toxicity and potential hazards and use these as design criteria for designing chemicals less likely to be dangerous. This is similar to the processes now used in *green chemistry* for designing chemicals that are safer and cleaner. A third approach involves identifying those chemicals commonly employed in natural systems to support life and to study the processes by which organisms make materials and draw from these lessons design criteria for developing chemicals. This could be called *ecological chemistry* because it is based on knowledge gained by studying natural systems.

1.1. Approach one: avoiding dangerous chemicals (safer chemistry)

Safer chemistry means designing substances that avoid chemicals commonly recognized as dangerous. In its most simple form this involves avoiding chemicals that are banned by governments or included on lists of dangerous chemicals. Such lists are not difficult to find. Many government agencies have compiled lists of dangerous chemicals. There are also lists published by professional associations and lists of substances to avoid, often called black lists, drawn up by manufacturing or retail firms.

Some national governments, particularly, in industrialized countries, have banned certain dangerous chemicals. Environmental or health agencies in these governments have used their regulatory powers to phase out or prohibit the manufacture or use of hazardous chemicals, such as various pesticides, organo-metals, or halogenated compounds. In 1991, the Swedish government published a list of eight chemicals and chemical groups that it then proceeded to attempt to sunset. These included methylene chloride, trichloroethylene, lead, organo-tin compounds, chlorinated paraffins, phthalates, nonylphenolethoxylates and brominated flame retardants [3]. The US Environmental Protection Agency has used its authority under the Toxics Substances Control Act to phase out the use of a small number of industrial chemicals and the federal pesticide laws to prohibit the use of a larger number of insecticides and herbicides. The United Nations publishes an *International Registry of Potentially Toxics Chemicals* that lists over 600 substances that have been banned or severely restricted by some national governments. Recently, the Stockholm Convention on the Persistent Organic Pollutants has targeted 12 organic chemicals for international phase out. The European Union has passed a directive prohibiting the use of lead, mercury, cadmium, hexavalent chromium and brominated flame retardants in electronic products.

Some European governments publish lists of dangerous chemicals that, while not prohibited, should be avoided. The Swedish government has developed a hierarchy of lists that range from a short list of substances that are to be phased out of use (see Table 1) to a longer list of substances that should be voluntarily avoided. The Swedish National Chemical Inspectorate, KEMI, publishes an *Observation List* of some 200 chemicals that should be avoided where possible.

Table 1
Partial Swedish list of restricted chemical substances

Arsenic	Asbestos
Benzene	Benzidine
Bis(chloromethyl) ether	Cadmium
Carbon tetrachloride	Chlorofluorocarbons
Chromium	Dichloromethane
Ethylene glycol	Formaldehyde
Lead	Mercury
Nonylphenolethoxylates	Polychlorinated biphenyls
Pentachlorophenol	Phthalates
Tetrachloroethane	Trichloroethylene

Source: Swedish National Chemicals Inspectorate, *List of Restricted Chemical Substances in Sweden*, Solna, Sweden, November, 1995.

The US Environmental Protection Agency (US EPA) lists over 600 substances in its Toxics Release Inventory, which some businesses use as a list of chemicals to avoid. Because the agency's New Chemicals Program must annually review hundreds of new chemicals for possible market entry, the agency has developed a set of procedures for identifying dangerous tendencies. Beginning in 1987 the US EPA began to develop categories of chemicals based on those properties likely to be dangerous. Using the techniques of structure–activity analysis, the agency's first category was acrylates and methacrylates. Today, there are 45 chemical categories and the Chemical Categories List is generally regarded as identifying those substances least likely to be safe.[2]

Governments and international agencies also publish lists of known and suspected human carcinogens, lists of recognized reproductive hazards, lists of neurotoxins, lists of allergens, lists of endocrine disrupting chemicals and lists of substances known to have negative effects on plants, fish or wildlife. Some particularly bioactive substances appear on several of these lists. Table 2 provides samples of these lists.

Some large manufacturing corporations and retail firms also draw up lists of substances to avoid. DaimlerChrysler, Volvo, Canon, Sony and Ben and Jerry's (ice cream) are all firms that maintain lists of chemicals to avoid in their manufacturing processes. Retailers like Boots and the Body Shop in the United Kingdom also use substance avoidance lists in negotiating with suppliers.

Once such avoidance lists have been accepted as guidance, a practicing chemical engineer needs only to conduct a review of existing chemical processes and identify the steps which need to be redesigned in order to eliminate the listed chemicals. Likewise, the synthetic chemist designing a new substance or synthetic process needs only to consult the lists to develop chemistries that avoid the use of the listed chemicals.

This is not unusual. During the 1980s the United Nations negotiated an international treaty to protect the upper ozone layer. This treaty resulted in the Montreal Protocol that required the phase out of a series of ozone-depleting chlorinated and fluorinated compounds. In response industrial chemists and process managers found substitutes that changed the chemistries of refrigeration, foam blowing and degreasing. The state pollution programs in the United States have assisted hundreds of firms in finding safer chemicals to substitute for highly volatile pollutants. For over a decade the Massachusetts Toxics Use Reduction Program has helped manufacturing firms find safer substitutes for some 190 chemicals that are included on the state list of toxic and hazardous chemicals.

---

[2]The most recent Chemical Category for Persistent, Bioaccumulative, and Toxic Substances establishes those properties not desirable in the development of new chemicals.

Table 2
List of Toxic Substances by Effect

Substance	Carcinogen	Reproductive Toxin	Neurotoxin	Mutagen / Teratagen
Acrylonitrile	Probable	Yes		
Arsenic	Known	Yes	Yes	Yes
Asbestos	Known			
Benzene	Known	Yes	Yes	
Benzidine	Known			
Beryllium	Known			
Cadmium	Probable	Yes		Yes
Chromium (+6)	Known	Yes		
Ethylene oxide	Probable	Yes	Yes	Yes
Formaldehyde	Probable	Yes		
Hexane		Yes	Yes	
Lead				
Lead acetate		Yes	Yes	Yes
Methyl Mercury		Yes	Yes	Yes
Nickel	Known	Yes	Yes	
Perchloroethylene	Probable			
Polychlorinated biphenyl	Probable	Yes		Yes
Trichloroethylene	Possible	Yes		
Styrene	Possible	Yes		
Vinyl chloride	Known	Yes		

Source: Curtis D. Klaasesen (Ed.) *Casarett and Doull's Toxicology: The Basic Sciences of Poison*, 5th ed., McGraw-Hill, New York 1996.

Technically, there are many examples of the search for safer chemistries. Substituting aqueous and semi-aqueous (terpines and alchohols) solvents for chlorinated solvents in industrial parts cleaning and degreasing provides a common example. Converting from mineral-based inks to soy-based inks offers an example that swept the newspaper business during the 1980s and 1990s. During this same period many large mills in the pulp and paper industry moved from hazardous chlorine to more benign chlorine dioxide and peroxide for bleaching and delignification. Many conventional hydrocarbon-based paints and coatings have been reformulated into water-based mixtures that have eliminated ingredients such as toluene, methyl ethyl ketone, formaldehyde and various isocyanates[3].

Safer chemistry does not mean safe chemicals. For years, carbon tetrachloride was used as an industrial degreasing agent. During the 1940s evidence of the

---

[3]For further examples see Allen and Rosselot [2].

toxic effects of carbon tetrachloride on the liver and kidneys began to mount and it was gradually replaced by trichloroethylene and perchloroethylene in degreasing. However, by the 1980s these solvents were listed as possible carcinogens. In some applications glycol ethers were substituted for the chlorinated solvents to avoid a suspected carcinogen, but ethylene glycol was found to be a reproductive toxin. Today, ethylene glycol is often replaced by propylene glycol, however, this too may someday need to be replaced.

Safer chemistry provides an effective strategy on a very practical basis, but it is an incremental strategy that promotes minor innovations in a kind of stepwise evolutionary process. Step-by-step safer chemicals replace one another in a long march away from recognized hazards. However, the transition to more sustainable chemistries could be advanced more rapidly by adopting a less incremental and more discontinuous process that seeks environmental compatibility as a direct, self-conscious and normative goal.

1.2. Approach two: designing chemicals based on health and environmental sciences (green chemistry)

The state of chemistry, biology and physics and knowledge about physiology and toxicology has advanced enormously over the past half century. We know far more about what makes chemicals toxic and hazardous and how to make them safer than we once did. For years chemists and chemical engineers have focused their research on questions of functional performance, processing efficiency and cost with little attention to the health or environmental effects of their chemicals. However, the increasing public criticism of chemists and chemistry during the 1980s led some chemists to argue that there is adequate knowledge for designing chemical and chemical processes that pose less risk to human health and the environment. Over this last decade, some in the field began to fashion a more environmentally benign approach to chemistry[4].

The idea of using existing knowledge from the health and environmental sciences to design more environmentally friendly chemicals and chemical processes has opened a rapidly developing new specialty in chemistry often referred to as environmentally benign chemical synthesis, or *green chemistry* [4]. Green chemistry does not focus on incremental substance substitutions; instead, green chemistry focuses on developing alternative chemistries that can be introduced throughout the entire process of chemical manufacturing. Paul Anastas and John Warner, two of the founders of the field of green chemistry, have defined the term green chemistry to mean "the utilization of a set of principles that reduces or eliminates the use or generation of hazardous substances in the design, manufacture and application of chemical products" [6]. In a seminal

---

[4]For an extensive review see Cano-Ruizand McRae [5].

book on green chemistry, Anastas and Warner have drawn up a list of 12 principles that can be used to identify and guide green chemistry initiatives in making more environmentally benign substances. Chapter 1.3 provides substantial background on the 12 principles.

It is important to note that a green chemistry approach, like the previous approach, does not start with a focus on exposure or risk. Green chemistry is directed at the factors that make chemical processes toxic or hazardous. Through a careful consideration of the properties of chemicals that make them toxic or hazardous, it is possible to *design out* those properties and, thereby, reduce or eliminate the hazard. Anastas and a colleague at the US EPA, Tracey Williamson, make this goal clear when they conclude, "(g)reen chemistry seeks to reduce or eliminate the risk associated with chemical activity by reducing or eliminating the hazard side of the risk equation thereby obviating the need for exposure controls and, more importantly, preventing environmental incidents from ever occurring through accident. If a substance poses no significant hazard, then it cannot pose a significant risk ... "[7].

Green chemistry fosters research on alternative feedstocks and intermediaries, environmentally benign solvents, reagents and catalysts, aqueous processing and safer and more readily recyclable chemical products. This involves preferring renewable (bio-based) feedstocks over non-renewable (petroleum-based) sources and seeking starting materials that demonstrate the least hazardous properties (e.g. toxicity, flammability, accident potential, eco-system incompatibility, ozone depleting potential etc.). Employing polysaccharides as feedstocks for polymers is an example of a renewable and non-toxic material for beginning a synthesis pathway. Likewise, glucose can be used as a raw material rather than benzene in the production of hydroquinone, catechol and adipic acid, all of which are important intermediaries in the production of commodity chemicals. Indeed, relatively non-toxic silicon has been suggested as a useful replacement for carbon as a starting base for the synthesis of some organic chemicals [8].

Atom economy is a simple way of describing improvements in yield — getting the maximum amount of chemical product out of each reaction. This involves both selecting the most atom-efficient reactions as well as developing new, more atom-economical ways of carrying out current reactions that often minimize the number of process steps, thereby reducing energy inputs and waste generation.

Additional research focuses on alternative reagents and catalysts. This involves identifying catalysts that function in chemical transformations with minimal environmental harm (e.g. minimizes energy inputs, maximizes yield, minimizes waste outputs, generates the least occupational exposure and the least accident potential). For instance, addition reactions are preferred over subtraction reactions, because they incorporate much of the starting materials and are less likely to produce large amounts of waste. Alternatives to heavy-metal catalysts are sought, because the common metal catalysts are so often extremely toxic. The use of liquid oxidation reactors replaces metal oxide catalysts with pure oxygen and permits lower temperature and pressure reactions with higher selectivity and no metal-contaminated wastes. New catalysis techniques that rely on enzymes, microwaves, ultrasound or visible light obviate the need for harsh chemical catalysts[5].

Organic solvents with significant health and environmental impacts are conventionally used as carriers in chemical reactions. Investigations on alternative solvents have demonstrated the potentially wide applications of aqueous chemistries, ionic liquids, immobilized solvents and supercritical fluids. Water has been shown to be an effective solvent in some chemical reactions such as free-radical bromination. Supercritical fluids which are typically gases (CO_2) liquified under pressure are already commonly used in coffee decafination and hops extraction. Supercritical CO_2 can be used as a replacement for organic solvents in surface preparations, in cleaning and in polymerization reactions and surfactant production. Future work may involve solventless or "neat" reactions such as molten-state reactions, dry grind reactions, plasma-supported reactions, or solid materials-based reactions that use clay or zeolites as carriers[6].

Several firms within the pharmaceutical industry, such as Merck, Pfizer, Bayer and GlaxoSmithKline, have taken leadership positions in promoting green chemistry. Because drug development is so research intensive and the health care industry is so sensitive to health objectives these firms have found competitive benefits in promoting their green chemistry initiatives[7].

The Office of Pollution Prevention and Toxics at the US EPA has several years of experience in trying to promote green chemistry. For instance, the New Chemicals program has developed a computer-based program called the Synthetic Method Assessment for Reduction Technique that helps chemical companies to assess, in advance, the pollution prevention opportunities of new chemistries and a green chemistry expert system that helps companies to identify and design more environmentally benign chemicals.

These green chemistry initiatives have received a substantial boost by the federal government's sponsorship of an annual presidential awards ceremony for the nation's best examples of green chemistry applications. Over the past several years, this awards program has recognized Bayer's environmentally friendly synthesis of biodegradable chelating agents, PPG industry's use of yttrium as a substitute for lead in cationic electro-coatings and Rohm and Hass's design of an environmentally safe marine antifouling coating to replace tributyltin oxides[8].

---

[5]See Centi et al. [9].
[6]See Metzger [10].
[7]For a good review of work at GlaxoSmith Kline see Curzons et al. [11]
[8]For other examples see the Internet site: www.epa.gov/greenchemistry.

Green chemistry goes a long way to promoting avenues for more sustainable chemistry. Avoiding toxic and hazardous substances, optimizing yields, avoiding wastes and minimizing energy consumption generate a broad set of objectives for encouraging innovation and corporate leadership. However, green chemistry stays within the conventional structures of the current chemical industry. An even broader challenge can be envisioned by seeking to follow the paths evolved by nature in the development of chemicals.

1.3. Approach three: modeling chemistry after natural systems (ecological chemistry)

The third approach is based on biology, physiology and ecology and focuses on developing chemicals that are inherently benign because they respect the biological defenses of living organisms and because they are ecologically compatible and degrade naturally under ambient environmental conditions. This involves digestible substances that are safely metabolized and biodegradable and readily compostable materials that fit comfortably into ecological nutrient streams.

While chemists and material scientists have long studied natural systems to discover the structural properties and synthetic processes in the environment, the lessons have focused narrowly on function and efficiency and neglected the health and environmental aspects. Many early polymers were designed as derivatives of natural polymeric compounds and, today, many pharmaceuticals are based on compounds found in living plants and animals. However, the emergence of a specialty of field-based research, currently called *biomimetics* or *biomimicry*, has drawn together a group of natural scientists who study nature to find environmentally compatible processes. These explorations include studies of how organisms make and use materials to compose physiological structures, communication systems, habitats and tools[9].

A starting point for this research is feedstocks. The production of chemicals from biomass is currently receiving renewed attention. This ranges from the production of biopolymers to bio-based cleaners and solvents[10]. Before the synthetic chemicals revolution and the advent of petrochemicals, chemicals were largely derived from plant matter. Still today biomass provides the basis for nearly 6 percent of pigments, 35 percent of surfactants and 40 percent of adhesives. Soy-based inks are the standard for color reproduction in newspapers. In 2000, the US Congress passed a Biomass Research and Development Act to provide new federal support for research on bio-based fuels and chemical

---

[9]Janine M. Benyus [12] has written a non-technical and quite readable review of these investigations.

[10]See R.C. Brown [13].

products. Primarily viewed as a means of expanding markets for agricultural products and reducing export dependence, this initiative, nonetheless, has spurred the search for new uses of bio-based chemicals[11].

However, ecological approaches to chemistry go well beyond agricultural feedstocks. A brief examination of natural processes reveals natural processes for making thousands of different chemicals and material products ranging from pliable polymers, rigid membranes and crack-resistant ceramics to surface coatings, adhesives, gels, lubricants, inks, dyes and disinfectants. There are natural processes for making energy conductors, insulators, data processors, information storage and information translators. Natural processes make proteins, fats, carbohydrates, amino acids and the vitamins and nutrients necessary for life. Considering how organisms make tough, polymeric materials like skin, hair or shells reveals the wide variety of natural chemistries. Indeed, for any synthetic chemical product there is a similar natural product made through relatively benign and renewable processes. Unlike current petrochemical-based processes, natural processes for producing chemicals are often quite sophisticated, occur under ambient conditions, require small amounts of energy, generate high yields and produce limited wastes that are easily consumed by other natural processes.

Ecological chemistry involves studying nature for materials and processes that are safe and ecologically sound. This could begin with a list of those substances that make up the human body[12]. Table 3 lists the essential elements in human anatomy by volume.

Added to this list are those compounds and chemical structures that make up common human foods. People have been exposed to the carbohydrates, starches, proteins, fats, minerals and vitamins that made up the human diet for centuries. In many ways humans and their diets have co-evolved to sustain one another. Indeed, the US Federal Food and Drug Administration maintains a list of food additives that are commonly considered safe. This list contains the names of hundreds of substances that are *generally regarded as safe* (GRAS) by the agency because there is a long safe history of common uses in foods or a strong weight of published scientific evidence suggesting that they are without negative human health effects. Because this list contains a broad array of organic and inorganic chemicals, it provides a useful list of potential chemicals available for industrial production that are to the best of our knowledge compatible with human and biological health.

---

[11]About a dozen federal agencies participate in the federal Biobased Products Coordinating Council. For further information go to the Internet site at www.ars.usda.gov/bbcc and see M. Duncan [15].

[12]A chemical by chemical description in abbreviated, but quite readable, form can be found in John Emsley [14].

Table 3
Essential elements in the human body

Element	Fraction of total body mass (%)
Oxygen	61
Carbon	23
Hydrogen	10
Nitrogen	2.6
Calcium	1.4
Phosphorus	1.1
Sulfur	0.2
Potassium	0.2
Sodium	0.14

Source: John Emsley, *Nature's Building Blocks An A-Z Guide to the Elements*, Oxford University Press, New York, 2001.

The human food source is rich in organics and the processes by which seeds, grains, legumes and other food plants develop could provide a host of lessons on chemical processes that are likely to be compatible with health and well-being. This is not the same as saying that all biomass-produced chemicals are likely to be safe. Ecological chemistry focuses on both the chemical product and its production process. The recent scientific and policy interests in bio-based processing and chemistries based on biomass offer exciting opportunities, but too often careful attention to the production processes is neglected. Studying natural production processes is the key to ecological chemistry and to date this is fairly un-chartered territory.

Ecological chemistry involves taking what biologists, botanists and ecologists study to generate templates and procedures for the development of more environment-friendly chemistries. Starting with processes that are compatible with human cells, some examples of such a list would include [13]:

- design chemicals to be less likely to damage a living organ or cell,
- design chemicals to be less likely to reach a target organ or cell or to be stored there once it has been absorbed into an organism,
- design of the physical properties of a chemical to be less likely to be absorbed into an organism,
- design chemicals that are likely to rapidly degrade in the environment or be converted to nutrients in an organism,
- select production and use processes that occur at normal temperatures and pressures and minimize water and energy use.

Inherently safe chemicals. Biological systems rely on a relatively small number of the chemicals in the periodic table. Healthy, living organisms are quite selective about the chemicals they consume as building blocks and fuel sources. They clearly avoid chemicals likely to damage DNA or interfere with the functioning of RNA. Proteins and enzymes perform a wide range of chemical functions in manufacturing structures, surfaces and barriers and they do so in quite controlled procedures. The self-assembly processes of constructing templated polymer structures is an additive process that generates little or no waste. Where reactions do occur chemicals are "un-zipped" into reactive forms only for the time needed to carry out the reaction and the wastes are typically re-bonded (blocked) to a less-reactive state.

Non-bioavaiable chemicals. If organisms create selectively impermeable membranes to shield vital functions from exogenous chemicals, then chemicals are selected that respect those barriers. Compounds that are composed of large, dissociated, non-lipophilic molecules are less likely to cross the cellular membranes of organisms. Water-soluble chemicals, for example, are less likely to pass through cell membranes than lipid-soluble chemicals. Large molecule polymers are less likely to be absorbed than small molecule polymers. Chemicals that cannot penetrate the membranes of cells or fat molecules are less likely to be stored and accumulate in organic tissue.

Physically benign chemicals. The physical properties of a chemical affect the possibility that it will be easily transported in the environment and be readily absorbed into an organism. For instance, materials in fine powder form are likely to be transported in the air and easily inhaled. The same material in pellet, slurry or solid mass form is less likely to be transported on air currents, dispersed onto food or water supplies, or inhaled into respiratory systems. Materials that easily volatilize are more readily transported and inhaled as vapors or gases, than materials with lower molecule weights. Where a fine dust or volatile state is required, a material could be converted to that state for the minimum time necessary before returning to a less dissipative state.

Biodegradable chemicals. Natural processes display complex processes that assure that structural and barrier materials are durable (resist degradation) for certain time periods, but that at later times are easily biodegraded under natural conditions or readily metabolized inside organisms. Snakeskins that are periodically molted, provide a graphic example. It is often possible to place functional groups into the molecular structure of a chemical so that prolonged sunlight or microorganisms can degrade the substance with relative ease. The biodegradable polymers based on starch and other carbohydrates provide examples and a wealth of lessons.

Ambient condition processing. The ambient conditions at the surface of the planet are quite enough for most natural chemical processes. Chemicals that are produced and used at ambient temperature and pressure are less likely to dissipate, volatilize or leak. Processing chemicals at ambient conditions also reduces the likelihood of exothermic reactions such as explosions and fires. Biosynthesized chemicals, such as carbohydrates and alcohols provide a good example. Such chemicals are also more likely to require less water and energy use over their full life cycle and, thus, generate less environmental pollution or organism exposure in ancillary processes.

Basing chemistry on currently available natural models may appear to limit creativity and innovation. However, this need not be the case. Natural processes are often highly sophisticated and elegant in ways that the conventional synthetic processes avoid because it is so much easier to simply force reactions with intense amounts of heat, pressure and harsh chemistries. Natural processes are also likely to offer superior qualities in terms of efficiency and waste minimization or in terms of waste generation that has readily available uses within ecological systems. Learning how to make chemicals with the cleverness of nature can open up engaging new avenues for creative chemistry.

Ecological chemistry approaches may yield many solutions that are similar to results of green chemistry approaches. However, rather than go through the literature of toxicology and pharmacology to create design criteria, ecological approaches start with how nature makes chemicals and then tries to copy or mimic those procedures. Both green chemistry and ecological chemistry offer truly innovative approaches to making safer chemicals and both provide challenging directions for developing a more sustainable chemical industry.

2. Safer chemicals for the future

Securing a productive, safe and sustainable society is a laudable goal that most would accept. Chemicals and the chemical industry have a large role to play. The planet has limited resources, however, our capacity to manipulate those resources through our increasing knowledge of the physical sciences promises a wealth of products and services. Making the results safe and enduring presents a significant challenge. We can continue to rely on a problem-based approach that addresses the hazards of chemicals only after they have been developed, manufactured and released to the environment. Progress on this approach will be slow and costly.

Instead, we could use our ever-increasing ecological and physiological knowledge to shape and direct our search for inexpensive and productive chemistries. This will require new goals for the discipline of chemistry and new directions for the chemical industry. Significant efforts to redesign educational programs in chemistry, biology, biochemistry and chemical engineering will be needed. Graduates in the chemistry fields of the future will need to be well prepared in the science of chemicals and also well versed in understanding environmental health. While, chemists need not become toxicologists, pharmacologists or ecologists, they should be able to engage in these fields with confidence and appreciation.

We should be expecting a lot from the chemicals of the future. They need to continue to provide low cost and high quality performance. However, they also need to be safe as well as functional and sustainable as well as practical. This paper has attempted to lay out several avenues for making such chemicals. Most of these initiatives are still in early stages. Much more work is needed here, but the successes so far are impressive and promising.

References

[1] K. Geiser, Materials Matter: Towards a Sustainable Materials Policy, MIT Press, Cambridge, MA, 2001.

[2] D.T. Allen, K.S. Rosselot, Pollution Prevention for Chemical Processes, Wiley, New York, 1997.

[3] Swedish National Chemicals Inspectorate (KEMI), Risk Reduction of Chemicals: A Government Commission Report, Solna, Sweden, 1991.

[4] D.T. Allen, D.R. Shonnard, Green Engineering: Environmentally Conscious Design of Chemical Processes, Prentice-Hall, Upper Saddle River, NJ, 2002.

[5] J.A. Cano-Ruiz, G.J. McRae, Environmentally Conscious Chemical Process Design, Ann. Rev. Energ. Env. 23 (1998) 499–536.

[6] P.T. Anastas, J.C. Warner, Green Chemistry: Theory and Practice, Oxford University Press, New York, 1998, pp. 11, 30.

[7] P.T. Anastas, T.C. Williamson, Frontiers in Green Chemistry, in: P.T. Anastas, T.C. Williamson (Eds), Green Chemistry: Frontiers in Benign Chemical Syntheses and Processes, Oxford University Press, New York, 1998, p. 10.

[8] S. Sieburth, Isosteric Replacement of Carbon with Silicon in the Design of Safer Chemicals, in: S.C. DeVito, R.L. Garrett (Eds), Designing Safer Chemicals: Green Chemistry for Pollution Prevention, American Chemical Society, Washington, D.C., 1996, pp. 74–83.

[9] G. Centi, P. Ciabelli, S. Perathoner, P. Russo, Environmental Catalysts: Trends and Outlook, Cata Today 2691 (2002) 1–13.

[10] J. Metzger, Organic Reactions without Organic Solvents and Oils and Fats as Renewable Raw Materials for the Chemical Industry, Chemosphere 43 (2001) 83–87.

[11] A. Curzons, D. Constable, D. Mortimer, V. Cunningham, (A) So You Think Your Process is Green? (B) Using Principles of Sustainability to Determine what is Green: A Corporate Perspective, Green Chem. 3 (2001) 1–6.

[12] J.M. Benyus, Biomimicy: Innovation inspired by Nature, Willam Morrow, New York, 1997.

[13] R.C. Brown, Biorenewable Resources: Engineering New Products from Agriculture, Iowa State Press, Ames, Iowa, 2003.

[14] J. Emsley, Nature's Building Blocks: An A-Z Guide to the Elements, Oxford University Press, New York, 2001.

[15] M. Duncan, U.S. Federal Initiatives to Support Biomass Research and Development, J. Ind. Ecol. 7 (3–4) (2004) 193–201.

Sustainability Science and Engineering: Defining principles
Martin A. Abraham (Editor)
© 2006 Elsevier B.V. All rights reserved
DOI 10.1016/S1871-2711(05)01009-3

Chapter 9

Renewable Feedstocks

L. Moens

National Bioenergy Center, National Renewable Energy Laboratory (NREL), 1617 Cole Boulevard, Golden, CO 80401, USA

1. Introduction

The fifth 'Sandestin Principle' of sustainable engineering relates to the depletion of the resources that have sustained mankind for thousands of years. These resources can be divided into two groups, i.e. *fossil* feedstocks such as petroleum, coal, and natural gas, and *renewable* feedstocks such as trees and other plant materials. The second group is commonly referred to as *biomass* and comprises all plant materials such as wood and its wastes, herbaceous (grasses) and aquatic plants, agricultural crops and their processing residues, and most of the organic portion of municipal wastes. The chemical features of the two groups of feedstocks are very different, but both are based on carbon. Unlike fossil feedstocks that are mostly made up of hydrocarbons, biomass is a much more complex and structurally more diverse material that is highly oxygenated (composed of primarily carbon, hydrogen, and oxygen). Biomass is essentially a heterogeneous material that consists of polymer 'building blocks' such as cellulose (38–50%), hemicellulose (23–32%), and lignin (15–25%). The ratio of these three primary material fractions varies greatly among different plant species, and *lignocellulosic* feedstocks can, therefore, have widely different material properties. In addition, many plants produce significant quantities of oils and extractives (resins) that also have a complex chemical and structural composition. The composition of a biomass feedstock can change significantly after harvesting and processing of the plant material because traditional agricultural practices mostly concentrate on only parts of the original plant or crop materials. An example is the separation of corn kernels that are rich in starch, from the fibrous corn stover that constitutes the remaining part of the plant.

Biomass is formed at the earth's surface through 'photosynthesis', which is the natural process whereby sunlight is used to chemically bind atmospheric carbon dioxide and water from the soil to generate complex plant polymers over short periods of time. Thus, living cells act as chemical reactors that store the solar energy in the form of chemical building blocks. When the biomass is then used as food or as an energy source, it is ultimately returned to the atmosphere in the form of carbon dioxide. This closed 'carbon-cycle' is the crucial issue in the discussion around renewable biomass resources and is driven by the virtually infinite availability of solar energy. The question that arises immediately is whether the growing global demand for carbon-based liquid fuels and materials can be satisfied by renewable biomass carbon without having to rely on fossil carbon resources. The answer lies in the development of new technologies, needed at the supply and demand side, that will allow for a smooth transition from fossil to renewable resources for energy and material production. An important part of such technology development will be the determination through a detailed life-cycle analysis of the energy and carbon (material) balance, as well as the total environmental impact for each new conversion technology or process. The resulting data will then serve as the basis for building so-called *integrated biorefineries*, wherein biomass will be converted into energy and chemicals using a number of efficient pathways. The following sections will address these important issues. However, it should be kept in mind that the 'vision' for biorefineries that will be presented here originates from the perspective of an industrialized society and may not immediately apply to developing countries that have many more challenges to overcome. Nevertheless, it is hoped that it will give the reader a better understanding of the challenges and opportunities offered by biomass feedstocks in the quest for global sustainability.

2. Sustainability of biomass as a carbon resource

The continuous replenishment of biomass carbon through natural photosynthesis implies that the world possesses a virtually unlimited supply of raw materials for the production of energy, chemicals, and materials when biomass-derived carbon is used in a sustainable manner. The Industrial Age, however, witnessed the mining and utilization of the fossil resources to where these have become almost the sole source of energy and organic chemicals in developed nations around the globe [1,2]. There is increasing evidence that the resulting emissions of 'fossil' carbon in the form of carbon dioxide are contributing to global warming and climate changes, and the emission of this 'greenhouse gas' is expected to accelerate as the growing world population will demand more energy and materials. Equally important are the uncertainties about the volume of the remaining petroleum reserves. However, aside from these issues, the bigger picture is that fossil resources are finite, and at some point in the future the industrial world will have to switch over to alternative carbon sources to sustain its energy and materials supplies. And just like the petroleum industry needed many decades of R&D before becoming the mature industry that it is today, it will take a considerable amount of time and effort to create a biomass-based economy [3–6]. An equally important factor in that scenario will be the rapid growth of the global population that may create a competition between land used for the cultivation of crops for bioenergy production and the agricultural land used for food crops [7,8]. It can therefore be expected that other renewable energy technologies such as solar (thermal and photovoltaic), wind, geothermal, and hydro will become important players in the world's energy portfolio.

One absolute fact that must be considered throughout the discussion on sustainability is that biomass-based carbon is the only alternative carbon source available besides fossil carbon sources such as petroleum, coal, or natural gas. It is also the only alternative source of *liquid fuels* and it can, therefore, serve one of the more pressing future energy needs in case access to the fossil resources is compromised for any reason. A major advantage of biomass is that it shares many characteristics with fossil fuels. Biomass-derived fuels (biofuels) exist in solid, liquid, and gaseous form, and thus can be transported and stored, which allows for the generation of heat and power upon demand [9]. The latter will become essential in an energy portfolio with a high dependence on intermittent sources such as wind and solar, and is another reason why biomass is expected to play a major role in future energy scenarios [10–12]. Further support for the hypotheses regarding future demand for renewable energy can be found in a scenario analysis conducted by Shell International Petroleum Company, which showed that after 2020 renewable energy sources will become competitive with fossil fuels [10,13].

An important side effect of biomass growth is that it absorbs CO_2 from the atmosphere, and thus represents a natural route that can mitigate the impact of greenhouse gases. In other words, an economy based on biomass as a source of carbon would alleviate the current concerns about global warming caused by anthropogenic CO_2 that originates mostly from fossil resources. It would stimulate reforestation (restoring existing forest land) or afforestation (establishing forests on new land where it previously did not exist), which would create a sink for atmospheric CO_2 for as long as the biomass accumulates (which is in the order of 100 years). An alternative strategy would be reforestation or afforestation with planned harvesting for the production of a mixture of short- and long-lived products that 'sequester' the CO_2 over variable time periods, as well as energy production that results in the displacement of fossil fuels. Some interesting publications appeared during the late 1990s that focused on the issue of CO_2 sequestration and presented data on the flow of carbon within forests and forest products [14,15]. These studies suggest that the success of such a CO_2 mitigation strategy will strongly depend on a number of variables such as the geographic location, physical environment (climate and soil conditions), economic environment, and the processing efficiency of the harvested forest materials. In response to these challenges, the drive toward 'sustainability' through an economy based on bioenergy and bioproducts has become a very dynamic topic of discussion in numerous reports and international workshops. Several reports and sources on the worldwide web are now available that contain data on global biomass resources in terms of volume and diversity in various geographic regions. They often also include discussions of the socio-economic benefits of a biomass-based economy. Of particular interest are the data and reports available from the international workshops organized under the International Energy Agency Bioenergy Agreement (IEA Bioenergy) [16]. These involve a series of international collaborative research and development projects or 'Tasks' that focus on particular topics within bioenergy development and deployment. The studies conducted under these tasks are directed toward the IEA member countries' collective energy policy objectives of energy security, economic and social development, and environmental protection, and facilitate cooperation in the development of new and improved energy technologies. Another source for international data on biomass resources and management is the 'Food and Agriculture Organization of the United Nations' [17]. It produces an enormous database called FAO Statistical Database ('FAOSTAT') that can be accessed online, and that currently contains over one million time-series records from over 210 countries and territories. The data cover statistics on agriculture, nutrition, fisheries, forestry, food aid, land use, and population. Aside from these continuously updated databases there are also several reviews and discussion groups [11,12,18–21] on current and future scenarios for the use of bioenergy and the production of biomass-derived chemicals and materials. It is generally recognized that the use of biomass as an alternative carbon source is bound to become a significant player in future energy and material supplies.

3. Life-cycle analysis in the use of biomass resources

A major *caveat* in trying to determine the sustainability of biomass as a carbon resource is that the current technologies for biomass processing are still in the early stages of development due to many factors that will be discussed later. A guiding principle in any development of new technology is that a new process must be run with maximum efficiency in order to avoid the generation of emissions that have a negative environmental impact. Ideally, a future biomass-based industry will avoid the use of fossil energy during the cultivation, harvesting, and processing of the biomass in order to achieve a zero-net CO_2 emission balance.

A key engineering tool that is now commonly used for the assessment of biomass systems is an integrated *Life-Cycle Assessment (LCA)* that identifies, evaluates, and helps minimize the environmental impacts of a specific process or competing processes. An LCA uses material and energy balances to quantify the emissions, resource depletion, and energy consumption of all process stages that are applied to convert raw materials into useful products, as well as the disposal of all products and by-products. The results of such a 'cradle-to-grave' study can help identify design improvements that can be implemented to reduce the total environmental impact of the process. A major LCA was recently conducted for biomass power generation relating to the overall CO_2 balance, energy use, and resource consumption for such a process [22]. It incorporated the entire cycle from the production of the plant biomass, to the construction of the power plant, and included emissions from operations at all phases over 37 years of construction, operation, and decommissioning. It even included the starting materials for the production of the fertilizers used in the cultivation of the biomass feedstock, and the petroleum used to produce the diesel fuel that was burned as part of the operations. This study revealed several benefits of using renewable biomass as a raw material for power generation. It was found that without any carbon storage in the soil, approximately 95% of the carbon cycles through the system (a 100% balance would entail a zero-net CO_2 process) (Fig. 1). The largest effect on carbon closure was found to be the degree to which carbon can be sequestered in the soil, and was found to vary between 83% and 200%. The latter would represent a net reduction in the amount of atmospheric CO_2, i.e. a sequestration of CO_2. The energy ratio was found to be

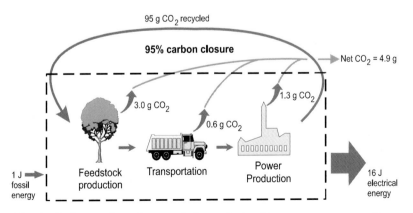

Fig. 1. Carbon-cycle and energy balance during biomass power generation.

16:1, meaning that for every Joule of fossil energy that is consumed by the conversion process, 16 J of 'green' electrical energy are produced. In a worst case scenario the ratio did not drop below 11:1, which means that the biomass-based system will always produce significantly more usable energy than it consumes.

In a subsequent study, power generation was examined for: (a) two fossil-based technologies, i.e. coal-fired power production and natural gas combined-cycle; (b) two biomass technologies, i.e. biomass-fired integrated gasification combined cycle system using a biomass energy crop, and a direct-fired biomass power plant using biomass residue; and (c) a biomass residue/coal co-fired system [23]. For each system, the LCA included the upstream processes necessary for feedstock procurement (mining coal, extracting natural gas, growing dedicated biomass, and collecting residue biomass), transportation, and any construction of equipment and pipelines. As part of this study, CO_2 sequestration was integrated with these systems, and the global warming potential of each was evaluated by looking at the total CO_2, CH_4, and N_2O emissions that make up the greenhouse gas (GHG) emissions produced by each system. The CO_2 sequestration method chosen for the study was the monoethanolamine system whereby the CO_2 is captured, compressed, transported via pipeline and sequestered in underground storage. The overall conclusion of this study was that the production of electricity from biomass instead of fossil fuels with CO_2 sequestration can be a cost-effective solution in helping to reduce the GHG emissions as well as reducing fossil energy consumption from electricity generation. Even with CO_2 sequestration, the amount of GHG emissions per kWh of electricity produced is more for the fossil-based systems than for the biomass power generating systems.

There are some challenges associated with setting boundaries for any LCA in that it raises the question of where to stop tracking the energy and material uses of upstream processes [22,23] (Fig. 2). The system boundaries are determined by the availability of data, and an example of this is the aforementioned LCA of biopower generation. For this particular LCA, data existed on the extraction of natural resources, processing, manufacture, and delivery to the point of use for most process feedstocks, such as diesel fuel and ammonium nitrate fertilizer. Thus, the assessment included nearly all of the major processes necessary to produce electricity from biomass. Examples of operations that were considered to be too far removed from the system, of interest in the study, were the construction of facilities to manufacture transportation equipment, and to manufacture mining equipment. Additionally, because of a complete lack of information, seedling production was not included in the analysis.

That the use of biomass carbon is not always better than using fossil carbon resources for a similar end product was revealed in a recent life-cycle study that compared the production of polyhydroxyalkanoates, a class of microbial

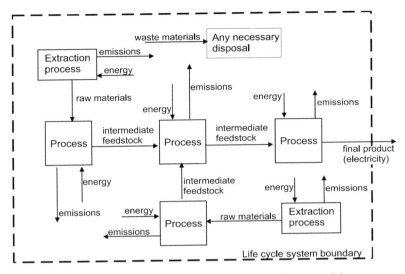

Fig. 2. Life-cycle system boundaries for an LCA model.

polyesters obtained through fermentation of corn sugars, with that of conventional petrochemical polymers [24]. This study showed that the application of LCA methodology could serve as a tool to prove or disprove the environmental benefits of biomass feedstocks for a process on a case-specific basis. More recently, an improvement in the LCA methodology for evaluating biobased polymers was reported that referred energy savings and GHG emission reduction to a unit of agricultural land instead of to a unit of polymer produced [25]. This study concluded that producing biobased polymers scores better in terms of energy savings and GHG emissions than producing bioenergy from energy crops without residue utilization. Thus, the biomaterials create interesting opportunities to reduce the utilization of fossil energy and to contribute to greenhouse gas (GHG) mitigation.

In summary, 'cradle-to-grave' life-cycle assessments represent an important engineering tool that will help in the establishment of *sustainability* in a biomass-based energy and materials economy. Many of these studies are continuing and are undergoing further refinement [26–30]. Furthermore, the diversity in biomass feedstocks necessitates a detailed energy and environmental impact study (i.e. potential for GHG mitigation and soil health) for each crop or residue stream, and such data are now becoming available [31].

4. Biorefinery

During the last two decades, a considerable amount of effort has been directed toward the development of technologies for biopower, biofuels, and

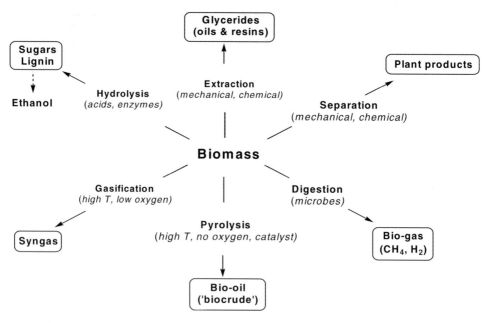

Fig. 3. Known processes for the thermal and biochemical conversion of biomass.

bioproducts as part of a vision toward independence from fossil resources (Fig. 3). Any of the biomass-derived energy products or bioproducts by themselves had to compete from an economically unfavorable position with a very mature industry that uses petroleum, coal, and natural gas as feedstocks, and this situation has hampered the market penetration of biomass applications. A very different scenario would be created if higher-value products could be combined with higher-volume energy production using any combination of conversion technologies, because it would make the production of fuels, power, chemicals, and materials from biomass much more competitive [32]. Such a configuration is now referred to by the general term of *integrated biorefinery*, and the principles of its operation are conceptually similar to those of petroleum refineries in that it would make maximum use of all feedstock material (Fig. 4).

Petroleum is processed along pathways that include a number of thermal conversion platforms to convert high-boiling fractions of crude oil into a wider range of lower-boiling products [33]. The 'cracking' catalysts needed for these processes are well established, and have helped in transforming the petrochemical industry into a mature industry that makes maximum use of the entire crude oil feedstock. The same should apply to future biorefineries, even though the technologies are expected to be more diverse and complex due to the differences in chemical structure and composition of biomass feedstocks. The current vision for integrated biorefineries holds two types of technology platforms, i.e. a *thermochemical* and a *sugar* conversion platform. The thermochemical platform

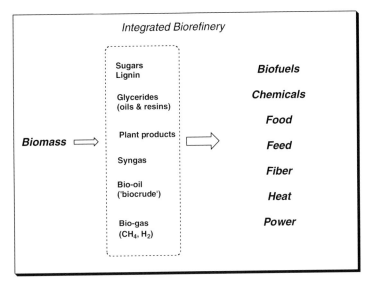

Fig. 4. Integrated biorefinery.

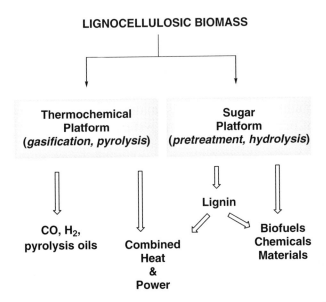

Fig. 5. Technology platforms for the conversion of biomass in an integrated biorefinery.

includes high-temperature processes carried out in the presence of catalysts, while the sugar platform includes biochemical and chemical processes that require milder conditions (Fig. 5).

Ideally, both conversion methods will be flexible enough to deal with a variety of biomass feedstocks, but just as few petroleum refineries use all available

conversion technologies, biorefineries too will use only those technology platforms that are most cost-effective for converting a specific feedstock. An example of such a biorefinery is that of the production of fuel-grade ethanol from
glucose feedstocks through fermentation. In this case, the biochemical route of
the sugar conversion is the only platform used. Other examples of existing
biorefineries are the pulp and paper mill industry, the sucrose-producing sugar
cane and sugar beet mills, the corn and soybean processing industry, the growing biomass power industry based primarily on co-firing of biomass and other
fossil fuels for the production of heat and power, and the more recent specialty
chemical facilities based on fast pyrolysis technology for biomass. In the future,
some, if not all, of these industries can be expected to undergo some kind of
transformation into integrated biorefineries that will be capable of generating a
more diversified product slate. Furthermore, the next generation biorefinery will
include the use of non-traditional sources of biomass, such as agricultural residues (e.g. molasses streams derived from various crops) and energy crops (e.g.
switchgrass), and will require technology that is still under development.

The establishment of next-generation biorefineries will depend on the progress
in science and engineering in five areas where significant technical barriers have
been identified:

Biomass feedstock
Sugar platform (biochemical)
Thermochemical platform
Products (including fuels, chemicals, materials, heat and power)
Integration of process configuration (systems integration)

4.1. Biomass feedstock

The success of the biorefinery is critically dependent on the availability of a
large supply of low-cost, high-quality (ligno)cellulosic biomass that can serve
the sugar and thermochemical conversion platforms. The following sections
describe some of the pressing issues that must be addressed with respect to the
many types of biomass feedstocks that are available from forestry or agricultural sources.

4.1.1. Biomass variability

The characteristics of biomass can vary widely in terms of physical and chemical
composition, size, shape, moisture content, and bulk density. This is especially
important in the case of residues from forestry and agricultural sources such as
the following:

Wood and related residues

- Forest thinnings
- Sawdust
- Wood waste
- Lignins
- Black liquor
- Municipal solid waste (lignocellulosics)

Agricultural residues

- Corn stover
- Wheat straw
- Rice hulls
- Sugar cane bagasse
- Sugar cane and sugar beet molasses
- Soy hulls
- Soy molasses
- Cheese whey
- Hulls from peanuts and other nuts
- Animal wastes

Energy crops

- Switchgrass
- Hybrid poplar
- Willow

In addition, the chemical composition of a biomass feedstock itself varies as a function of many other factors such as plant genetics, growth environment, and the harvesting and storage methods used. Controlling the factors that cause the biomass variability is difficult and poses a real challenge for biorefineries that require a steady supply of feedstocks with a consistent quality.

The measurement of biomass composition through standard wet-chemical analysis method is time-consuming and very costly, and is thus unsuitable for use in continuous biorefinery operations. However, new, rapid, and inexpensive methods have been developed and have already been proven to be excellent tools to monitor the chemical composition of corn stover and corn stover-derived samples [34]. The development of these new techniques has been driven by an ongoing effort to commercialize the conversion of stover into fuels and chemicals. It involves the use of near-infrared (NIR) spectroscopy combined with multivariate analysis, and is capable of delivering a large amount of data in a very short time at a minimal cost per sample. Because of these characteristics, this technique can be applied online in a process-control setting.

4.1.2. Engineering systems

The development of large biorefineries will require the development of a reliable and cost-competitive feedstock infrastructure that combines several engineering systems to create total feedstock supply chains. The future systems will be able to deal with the biomass variability as well as the various needs imposed by the different feedstocks. Collection today requires removal of the residue in a second pass through the field after harvest. Furthermore, current biomass harvest and collection methods do not have the ability to selectively and with minimal impact harvest the desired components of the biomass. In response to this challenge it would be desirable to develop a single-pass harvester that is capable of handling multiple components of the crop, which can selectively harvest those portions of the plant that are optimally suited for various downstream processors, while leaving behind components that best meet the needs of good soil management. The elimination of a second harvesting step will significantly reduce soil erosion.

Once all the biomass has been collected, its storage must be engineered such that decay is prevented that could otherwise degrade the biomass to a point where it becomes less suitable for further processing [35]. This is especially important in the case of high-moisture biomass that is highly susceptible to spoilage, rotting, spontaneous combustion, and odor problems.

There is little doubt that the optimization of the advanced engineering systems will benefit greatly from the development of better sensors and controls, as well as rapid analysis methods such as the NIR method mentioned earlier, that will allow the continuous monitoring of the quality of the biomass feedstock in real time.

4.1.3. Soil requirements

There is a lack of information and predictive tools to predict effects of the envisioned residue removal in the context of soil erosion, soil health, and productivity. This can have longer-term consequences on the sustainability. A recent life-cycle model was developed that addressed the impacts of corn stover collection on soil health under a 'no-till' scenario [26]. This model incorporates results from individual models for soil carbon dynamics, soil erosion, agronomics of stover collection and transport, and bioconversion of the stover to fuel ethanol.

4.1.4. New crop development

In the long term, large-scale replacement of petroleum calls for the introduction of a new generation of energy crops that can dramatically increase the potential supply of energy from biomass, without sacrificing the important role that agriculture plays in meeting our society's need for food and fiber. This will be particularly true for energy crops that can be grown in areas where poor soil conditions exist. From a socio-economic perspective, the cultivation of such new crops will provide farmers with more options for sustaining their livelihood.

4.2. Sugar platform

Lignocellulosic biomass is a heterogeneous composite material consisting of interlinked hemicellulose, cellulose, and lignin polymers. Cellulose, a crystalline glucose polymer, and hemicellulose, a non-crystalline polymer of hexoses (D-glucose, D-galactose, and D-mannose) and pentoses (D-xylose, L-arabinose, and uronic acids) together make up the carbohydrate portion that constitutes about two-thirds of biomass on a dry weight basis. Lignin, which is a phenolic polymer composed of substituted phenylpropane units (i.e. C_9-units), makes up for most of the remainder of the biomass besides minor components such as protein, oils, waxes, and other extractives. The purpose of the sugar conversion platform is to separate the different fractions into mixed sugar streams that can subsequently be processed into a wide product slate under relatively mild conditions. The traditional pulping process to produce paper is essentially a separation of the lignin portion from the cellulosic and hemicellulosic fibers under rather harsh conditions; e.g. through the addition of sulfur-containing reagents under strong alkaline conditions (*inter alia*). The mixture of spent reagents ends up in the lignin stream ('black liquor'). The valuable reagents can then be recovered through combustion of the black liquor whereby the lignin serves as the fuel. However, in a future biorefinery configuration, the goal would be to obtain a multitude of sugar and lignin streams (besides the desired cellulose) from which a larger number of higher-value products can be generated.

A more advanced version of a cellulose-based sugar conversion platform is being developed for future ethanol biorefineries. The envisioned commercial use of (ligno)cellulose feedstocks for ethanol production will greatly expand the supply of sugars (esp. glucose) for the fermentation process that currently uses only cornstarch as a source of glucose [32]. The pretreatment process involves the use of relatively mild methods that disrupt the cellulose–lignin matrix, followed by an enzymatic saccharification step wherein the cellulosic polymers are hydrolyzed to simple sugars. Several types of such pretreatment methods are currently being studied [36]. Ideally, one could envision the fermentation of the hemicellulosic pentoses to ethanol, and the chemical or biochemical conversion of lignin into higher-value bioproducts. Integration of these conceptual processes could lead to a biorefinery that would maximize the use of the biomass feedstock. Within this concept, the process configuration could also incorporate a thermochemical platform based on a gasification or pyrolysis process that would lead to gaseous products, as well as Fischer–Tropsch products from an additional synthesis step. The combination of the sugar and thermochemical platforms is shown in Fig. 6 that represents a conceptual flow diagram of a next-generation ethanol biorefinery. In the sugar platform, the mixed sugar stream, which contains hexoses and pentoses, may be subjected to various fermentation processes to produce fuel-grade ethanol as well as a number of other higher-value bioproducts depending on the choice of microorganism. Aside from the efficiency in product formation, the biorefinery should be configured such that it becomes self-sufficient in terms of energy requirements. Combined heat and power may be generated through combustion of any low-value residues originating from the various process units. This energy would feed the various biorefinery units, and any excess power could be delivered to the grid.

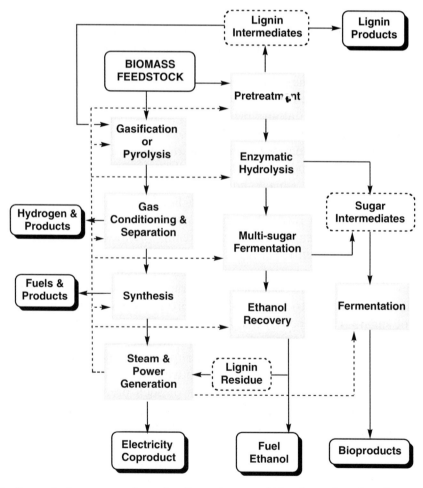

Fig. 6. Conceptual process configuration for an integrated biorefinery for ethanol production.

Much R&D work remains to be done to make this integrated biorefinery a reality because several technical barriers still exist. With respect to the sugar platform, the pretreatment and the enzymatic saccharification represent two primary technical barriers that must be conquered for emerging lignocellulosic biorefineries. As mentioned earlier, the pretreatment process is essentially a step needed to open up the polymer matrices in the biomass, and to render it more susceptible to the subsequent hydrolytic action of cellulase enzymes. However, the optimization of the pretreatment step is hampered by the lack of sufficient insight into the root causes of the recalcitrance of biomass, and more fundamental studies of the biomass structures within various feedstocks as well as their interaction with (bio)chemical agents are urgently needed. In addition, the engineering of the resulting process(es) will only be successful if better analytical tools are developed that can measure sugar yields in real-time at all stages of the process. The key here is to minimize any losses in sugar yields to potential side reactions, such as, the formation of furan-type derivatives of the sugars that are known to have an inhibitory effect on subsequent microbial fermentation of the sugar solutions. The saccharification step (i.e. enzymatic hydrolysis) requires inexpensive cellulase enzymes that can hydrolyze the cellulose fibers [37]. Enhancements in this area may come from the development of novel thermostable enzymes that promote higher process rates at higher temperatures. There is also a need for more fundamental insight into the enzymatic decrystallization process wherein the crystalline structure of the cellulosic fraction is deconvoluted [38], as well as insight into the self-inhibitory effect between the enzymes and the sugar end product [39].

Sugar solutions resulting from the pretreatment/saccharification process are impure as they contain a mixture of sugars, and are contaminated with a variety of non-sugar components. The latter can include acetic acid formed during the hydrolysis of the hemicellulosic fraction, lignin-derived phenolics solubilized during the pretreatment process, inorganic acids and bases introduced during pretreatment, and sugar degradation products. These impurities can have an inhibitory effect on the biocatalysts (enzymes or whole cells) or they can poison chemical catalysts in the downstream process(es) [40,41]. In addition, there is a need for enzymes or chemical catalysts that can efficiently convert the pentoses that originate from the hemicellulosic fraction, as well as chemical and biochemical catalysts that can convert mixed sugar streams [42]. In the shorter term, this will require the development of improved methods for separating sugars prior to conversion, including the removal of inhibitory agents, or for separating products from sugar mixtures after conversion. This will facilitate the deployment of existing microbial, enzymatic, and chemical catalysts, but in the longer term the processing of mixed sugar streams will require new and more robust catalysts that can bypass the need for intensive separation or purification steps. Future developments are expected in new concepts, e.g. consolidated bioprocessing, in which the hydrolysis and conversion of carbohydrates to products is achieved using a single organism, or in the development of new enabling technologies that offer new possibilities for leapfrog improvements in yield and cost.

The lignin by-product generated during the processing of lignocellulosic biomass creates new opportunities as a potential source for higher-value products [43]. This material is often isolated as a moisture-rich material that is notoriously resistant to chemical attack. From the perspective of engineering an integrated process, it remains to be seen if its fuel value will remain more important than its potential value as a source of new products for the chemicals market. An alternative approach would be to gasify lignin to generate synthesis gas (a mixture of carbon dioxide and hydrogen) from which high-energy liquids can be formed.

4.3. Thermochemical platform

The thermochemical platform involves the use of elevated temperatures under *pyrolysis* or *gasification* conditions to convert biomass or biomass-derived biorefinery residues to intermediates that may be used directly as raw fuels or products, or that may be further refined to produce fuels and products that are interchangeable with existing commercial commodities [44]. Intermediate products from these thermochemical processes include syngas, pyrolysis oils, hydrothermal oils, and gases rich in hydrogen or methane. These intermediate products can be used directly for heat and electric power generation, or may be upgraded by various processing technologies to products such as gasoline, diesel, alcohols, olefins, oxo chemicals, synthetic natural gas, and high-purity hydrogen. In contrast with the sugar conversion platform, the thermochemical processes are much more destructive in that the biomass material is broken down into much smaller molecules at elevated temperatures, and thus the sugars and lignins lose most, if not all, of their structural complexity and properties. However, the temperature and the concentration of oxygen strongly determine the types of products that are formed. *Pyrolysis* is defined by a thermal conversion in the absence of any oxygen at *ca.* 400–650°C, and generates liquid oils that often contain significant amounts of carbohydrates and phenolics in ratios that depend on the composition of the biomass feedstock [45]. Gasification is conducted at *ca.* 650–900°C with limiting amounts of air or oxygen, and results in the formation of gaseous products (CO, CO_2, H_2, and CH_4) [44]. The third type is *Hydrothermal Upgrading* (HTU) and consists of treating the biomass during 5–20 min with water under subcritical conditions (300–350°C and 10–28 MPa) to maintain a liquid phase [46]. This type of process produces heavy organic oils that consist of a complex mixture of sugar degradation products such as low molecular weight aldehydes and acids. Because of the use of water as a reaction medium, this technology lends itself well to a variety of wet biomass feedstocks, but the resulting oils have high reactivity, and must be stabilized.

Unlike the biological conversion pathways, thermochemical conversion provides an efficient approach for producing fuels and products from all the fractions of the biomass feedstock, including the lignin. The gasification of biomass to syngas is an especially powerful technology in that the gas can serve as a feedstock for a wide range of liquid fuels and chemicals through technologies that are already well known to the chemical and petrochemical industries. One of the challenges that is currently being addressed is the cleanup of raw biomass-derived syngas that is contaminated with tars (i.e. condensable organic compounds heavier than benzene). Biomass gasification is important for providing a source of fuel for electricity and heat generation, and can make use of the energy contained in process residues that are not suitable for conversion into primary products. Thus, residues can be used for combined heat and power production, which maximizes the efficiency of the biorefinery. Pyrolysis oils, which are often referred to as 'biocrude', are liquids that can serve as intermediates to produce fuels, products or power, and constitute a transportable energy carrier.

There are some specific technical barriers that must be addressed in order to achieve high efficiency in the thermochemical platforms. First, it will require a supply of uniform biomass feedstocks and reliable feed preparation, storage, and handling systems. Feed systems in particular create an important engineering challenge in that their reliability relies heavily on a narrow range of physical properties such as particle size, moisture content, etc., that is specific for any feedstock. Other areas of investigation that have been identified include densification of the feedstock, and also the removal of undesirable chemical components, such as, alkali species, halogens, sulfur, and nitrogen that can have a negative impact on the gasification catalysts. With regard to pyrolysis oils, much work remains to be done in the areas of handling, processing, and upgrading of these chemically complex liquids. There is a need for more data regarding the instability of these oils due to the presence of reactive components, phase separation, acidity, and environmental safety and health issues associated with long-term exposure. Consequently, new specifications and American Society for Testing and Materials (ASTM) standards must be developed for pyrolysis oils. Similar to the biochemical conversion processes, there is also a need for online analytical tools and sensors that can provide real-time information about the thermochemical process such that high plant efficiency and performance can be maintained.

4.4. Products

By analogy with existing petroleum-based industries, a number of primary biomass products have been identified that can serve three major market sectors.

The largest market is that of *transportation fuels*, wherein biofuels can play a prominent role because of the diversity in chemical composition that can be obtained. These include alcohols such as ethanol and its blends, methanol, biodiesel (methyl and ethyl esters of fatty acids), Fischer–Tropsch liquids (derived from syngas), oxygenates (especially ethers), methane and hydrogen. Biobased *chemicals and materials* may be able to penetrate the existing market by direct replacement of existing commodity chemicals, or they may become new commodity chemicals for existing or new applications. In the latter case, *biobased materials such as polymers may display improved performance and properties that give them a very competitive edge in the marketplace.* An example is the production of polylactic acid (PLA) that was commercialized recently as a new commodity polymer for the production of fibers. PLA is derived from sugars through fermentation, and is an example of a biochemical platform process. Depending on the source of the biomass feedstock and the pretreatment processes used by the conversion platforms, a number of other biomass components such as oils, plant extractives, and polymeric carbohydrates such as starch, cellulose, and hemicellulose can be isolated and converted directly into higher-value products that currently already have large-volume markets. The third sector is that of the *utility market* that may benefit from heat and power provided by the biorefineries. This scenario will only make sense in those cases where the biorefinery produces an excess amount of energy above and beyond what it required for feeding its internal processes. That excess amount of energy could then be sold to the grid.

Although there already exist large-scale commercial ethanol production facilities that use fermentation of starch to produce fuel-grade ethanol, there are a number of barriers that must be overcome in order to establish integrated biorefineries for the production of fuels and products from lignocellulosic biomass. In the arena of fermentation technology, there is a need for improvement in yields, rates of the bioreactions, degrees of conversion, and product selectivities, while the economic viability will depend on the development of new and robust microbes that can convert all five biomass sugars mentioned earlier. The robustness that is being pursued relates to higher thermal stability, efficiency in higher salt concentrations, and resistance to inhibitors. An engineering challenge will be to continually remove products and inhibitors from the fermentation broth through advanced bioreactor designs. Combined with improved process control techniques, these new biocatalysts may then become applicable to the production of commodity chemicals and materials. There is also a strong interest in bioreactions that can be carried out under more extreme temperature, pressure and pH environments, and also in gas/liquid biphase systems and non-aqueous phase media.

Aside from the use of biocatalysts such as enzymes or whole cells to effect conversion of biomass fractions, there is a real need for chemical catalysts that can function in aqueous media and that are compatible with the highly oxygenated nature of most of the biomass fractions. Sugars and lignins have solubility characteristics that are unlike those of traditional chemicals known to the current chemical industry, and new heterogeneous or homogeneous catalysts must be developed that can interact with these new substrates.

The following are areas where current catalyst systems must be improved or where innovation is needed:

- Catalysts and processes for hydrogenation of sugars, lignins, and oils
- Catalysts for oxidation under milder conditions and with higher product specificity
- Acid catalysts for dehydration reaction (sugars)
- Catalysts for selective bond cleavage in sugars and lignins
- Catalysts for selective hydrogenolysis reactions (i.e. reductive bond cleavage using hydrogen and a catalyst)
- Better catalytic systems to effect Fischer–Tropsch technology that can be applied to the more complex biomass-derived syngas streams.

It is clear that fundamental studies are needed to create these new or improved catalyst systems, but equally important are the engineering issues that will have to be addressed to make large-scale operations feasible. Improved separation technologies must be developed in order to deal with the isolation of liquid biobased products such as alcohols and polyols, and ionic species such as carboxylic acids and amino acids. The higher degree of chemical functionality and thermal instability of many biobased products often precludes the use of traditional distillation methods that are commonplace in the chemical industry.

4.5. Integrated biorefineries

The efficiency of an integrated biorefinery will depend on how well its process configuration can generate a combination of fuels, chemicals and materials, and combined heat and power from various biomass feedstocks. The engineering challenges associated with each of the units (i.e. the biomass feedstocks, the sugar and thermochemical platforms, and the downstream products) will have to be integrated in order to maintain a technically and economically viable biorefinery. Furthermore, the complexity of the future biorefineries can increase significantly as more enabling technologies become available that can lead to a wider slate of products.

The vision for future biorefineries is already becoming reality to some degree in that the R&D activity in this arena is picking up momentum. Part of the reason is that biomass represents a resource that holds much promise from a business point of view. The perception that biomass derivatives and bioenergy are not competitive in terms of quality and market potential is indeed starting to change, and several major chemical manufacturers are starting to invest in biomass-related technologies. An excellent example is the recent development by Cargill Dow LLC of its NatureWorks^{TM} process for the production of PLA through fermentation of glucose derived from cornstarch [47]. This polymeric material serves as a building block for Cargill Dow's Ingeo^{TM} fiber. The PLA produced here is a new, versatile, bulk polymer family that is now enjoying many applications in consumer products, e.g. food containers and packing materials, and fibers for use in various textile products. Key to its commercial success is that the PLA resin offers performance in the fiber and packaging markets that is on par with existing materials that are derived from petrochemicals. Cargill Dow's new facility in Blair, Nebraska, has a full capacity for lactic acid production of 400 million pounds per year. The company has announced plans to expand into technologies that make use of sugars derived from lignocellulosic feedstocks such as corn stover, wheat, rice straw, and bagasse. These feedstocks are abundant, are often cheaper than corn, and could provide farmers with a secondary source of income. It would also provide lignin that can be used as a fuel for combustion or gasification to produce process steam for the conversion platforms. In other words, the entire envisioned process configuration becomes one of an integrated biorefinery. Additional energy requirements will be met by tapping into wind energy in order to replace the use of electricity that is currently generated from fossil and nuclear fuels. Thus, the company strategy is to use local biomass and renewable energy sources to sustain its commercial operations.

More recently, DuPont announced a collaboration with the National Renewable Energy Laboratory in a new project that aims at the development of an integrated biorefinery to produce both fuels and value-added chemicals from not only cornstarch but also from the fibrous material in the stalks, husks, and leaves of the corn plant [48]. One of the value-added chemicals that is targeted is 1,3-propanediol (PDO), which is the key building block for DuPont^{TM} Sorona®, the company's newest polymer platform that can be used in textiles, carpets, and packaging materials. It is projected that the use of biomass as a feedstock will offer more benefits compared to the production of PDO from petrochemical pathways.

A considerable amount of effort has been dedicated to the production of succinic acid from corn-derived glucose, and represents another potentially useful process for a sugar platform in an integrated biorefinery [49]. This high-value dibasic acid is a precursor for numerous chemicals, e.g. 1,4-butanediol, tetrahydrofuran, γ-butyrolactone, N-methyl-pyrrolidinone, and 2-pyrrolidinone. The fermentation was demonstrated in a 500 L fermenter, and the entire exploratory study made use of expertise present in four US national laboratories. The succinic acid is formed by fermentation using a new

Escherichia coli strain (ATCC 202021), which overproduces succinic acid under anaerobic conditions. After the fermentation, the succinic acid is recovered through a two-stage desalting and water-splitting electrodialysis. The purified succinic acid is then catalytically reduced to the final products. Using an economic analysis, it was shown that the production of 1,4-butanediol along this integrated process can compete favorably with currently used chemical processes.

At the time of this writing, several other major industrial companies are working on integrated biorefinery projects, but further details cannot be discussed yet. Nevertheless, it is clear that the leaders in the chemical industry are starting to show serious interest in the use of biomass resources for future operations.

5. Conclusion

As we face a combination of rapid growth of the global population, changing climate patterns that are possibly caused by GHG emissions, and a rapid depletion of fossil energy reserves, future generations will have no other choice but to tap into alternative carbon sources for energy, chemicals, and materials. This transcends any discussion of whether biomass is a sustainable resource or not, because it will depend greatly on how fast and efficiently we will use up the biomass that is available to us. The 'rate of depletion' of the biomass will depend on social and cultural behaviors around the world, as well as the development of new and improved technologies, and their application within the process configuration of an integrated biorefinery. While the former is hard to predict, the development of better technologies is more likely to be successful because it can be done using the technical tools developed over the last century. The goal of this chapter was to highlight the key elements of a rational technical approach embodied in the biorefinery vision that will allow us to make better use of the biomass resources that are currently available around the globe, with the hope that it will create sustainability for future generations.

References

[1] J. Swinnen, E. Tollens, Bioresource Technol. 36 (1991) 277.
[2] R.C. Kuhad, A. Singh, Crit. Rev. Biotechnol. 13 (1993) 151.
[3] E.S. Lipinsky, Science 212 (1981) 1465.
[4] R.M. Busche, Biotechnol. Prog. 1 (1985) 165.
[5] S.A. Leeper, G.F. Andrews, Appl. Biochem. Biotechnol. 28/29 (1991) 499.
[6] I.S. Goldstein, For. Prod. J. 31 (1981) 63.
[7] C. Okkerse, H. van Bekkum, Green Chem. 1 (1999) 107.

[8] M. Hoogwijk, A. Faaij, R. van den Broek, G. Berndes, D. Gielen, W. Turkenburg, Biomass Bioenergy 25 (2003) 119.

[9] P.A.M. Claasen, J.B. van Lier, A.M. Lopez Contreras, E.W.J. van Niel, L. Sijtsma, A.J.M. Stams, S.S. de Vries, R.A. Weusthuis, Appl. Microbiol. Biotechnol. 52 (1999) 741.

[10] D.O. Hall, J.I. Scrase, Biomass Bioenergy 15 (1998) 357.

[11] D.O. Hall, J.I. House, Sol. Energy Mater. Sol. Cells 38 (1995) 521.

[12] M. Parikka, Biomass Bioenergy 27 (2004) 613.

[13] The Evolution of the World's Energy System, Shell International Ltd. Group External Affairs, Shell Centre, London, 1996.

[14] G. Marland, B. Schlamadinger, Biomass Bioenergy 13 (1997) 389.

[15] K.E. Skog, G.A. Nicholson, For. Prod. J. 48 (1998) 75.

[16] http://www.ieabioenergy.com

[17] http://www.fao.org, and for statistical database http://www.fao.org/waicent/portal/statistics_en.asp

[18] M. Eissen, J.O. Metzger, E. Schmidt, U. Schneidewind, Angew. Chem. Int. Ed. 41 (2002) 415.

[19] H. Danner, R. Braun, Chem. Soc. Rev. 28 (1999) 395.

[20] J.J. Bozell (Ed.), Chemicals and Materials from Renewable Resources, ACS Symposium Series No. 784, American Chemical Society, Washington, DC, 2001.

[21] Center for Renewable Energy and Sustainable Technology (CREST) http://www.crest.org

[22] M.K. Mann, P.L. Spath, Life Cycle Assessment of a Biomass Gasification Combined-Cycle System, NREL Report No. TP-430-23076, 1997.

[23] P.L. Spath, M.K. Mann, Biomass Power and Conventional Fossil Systems with and without CO_2 Sequestration — Comparing the Energy Balance, Greenhouse Gas Emissions and Economics, NREL Report No. TP-510-32575, 2004.

[24] T.U. Gerngross, Ch. 2, p. 10, in: J.J. Bozell, (Ed.), Chemicals and Materials from Renewable Resources, ACS Symposium Series No. 784, American Chemical Society, Washington, DC, 2001.

[25] V. Dornburg, I. Lewandowski, M. Patel, J. Ind. Ecol. 7 (2004) 93.

[26] J. Sheehan, A. Aden, K. Paustian, K. Killian, J. Bremmer, M. Walsh, R. Nelson, J. Ind. Ecol. 7 (2004) 117.

[27] L.R. Lynd, M.Q. Wang, J. Ind. Ecol. 7 (2004) 17.

[28] B. Cunningham, N. Battersby, W. Wehrmeyer, C. Fothergill, J. Ind. Ecol. 7 (2004) 179.

[29] M.C. McManus, G.P. Hammond, C.R. Burrows, J. Ind. Ecol. 7 (2004) 163.

[30] E. Gasafi, L. Meyer, L. Schebek, J. Ind. Ecol. 7 (2004) 75.

[31] S. Kim, B.E. Dale, J. Ind. Ecol. 7 (2004) 147.

[32] An excellent resource for biorefinery development is the multi-year task plan that can be downloaded from http://www.eere.energy.gov/biomass/

[33] J.H. Gary, G.E. Handwerk, Petroleum Refining—Technology and Economics, 3rd Ed., Marcel Dekker, Inc., New York, 1994.

[34] B.R. Hames, S.R. Thomas, A.D. Sluiter, C.J. Roth, D.W. Templeton, Appl. Biochem. Biotechnol. 5 (2003) 105–108.

[35] R. Jirjis, Biomass Bioenergy 28 (2005) 193–201.

[36] N. Mosier, C. Wyman, B. Dale, R. Elander, Y.Y. Lee, M. Holtzapple, M. Ladisch, Bioresource. Technol. 96 (2005) 673–686.

[37] S.R. Decker, W.S. Adney, E. Jennings, T.B. Vinzant, M.E. Himmel, Appl. Biochem. Biotechnol. 689 (2003) 105–108.

[38] C.W. Skopec, M.E. Himmel, J.F. Matthews, J.W. Brady, Protein Eng. 16 (2003) 1005.

[39] M. Gryta, A.W. Morawski, M. Tomaszewska, Catal. Today 56 (2000) 159.

[40] B.R. Hames, T.K. Hayward, N.J. Nagle, A.D. Sluiter, US Patent No. 6,737,258, 2004.

[41] C. Luo, D.L. Brink, H.W. Blanch, Biomass Bioenergy 22 (2002) 125.

[42] A. Mohagheghi, N. Dowe, D. Schell, Y.C. Chou, C. Eddy, M. Zhang, Biotechnol. Lett. 26 (2004) 321.

[43] Examples of pathways and references for converting lignins into products can be found at http://www.eere.energy.gov/biomass/lignin_derived.html.

[44] P.L. Spath, D.C. Dayton, Preliminary Screening — Technical and Economic Assessment of Synthesis Gas to Fuels and Chemicals with Emphasis on the Potential for Biomass-Derived Syngas, NREL Report No. TP-510-34929, 2003.

[45] A.V. Bridgewater (Ed.), Pyrolysis and Gasification of Biomass and Waste, CPL Press, UK, 2003.

[46] Z. Srokol, A.-G. Bouche, A. van Estrik, R.C.J. Strik, T. Maschmeyer, J.A. Peters, Carbohydr. Res. 339 (2004) 1717.

[47] http://www.cargilldow.com/corporate/

[48] News release by DuPont Corp., October 6, 2003, http://www1.dupont.com.

[49] N. Nghiem, B.H. Davison, M.I. Donnelly, S.-P. Tsai, J.G. Frye, Ch. 13, p. 160, in: J.J. Bozell, (Ed.), Chemicals and Materials from Renewable Resources, ACS Symposium Series No. 784, American Chemical Society, Washington, DC, 2001.

Sustainability Science and Engineering: Defining principles   201
Martin A. Abraham (Editor)
© 2006 Published by Elsevier B.V.
DOI 10.1016/S1871-2711(05)01010-X

Chapter 10

When is Waste not a Waste?

J.B. Zimmerman[a,b], P.T. Anastas[c]

[a]National Center for Environmental Research, Office of Research Development, United States Environmental Protection Agency, 1200 Pennsylvania Avenue, NW (8722F), Washington, DC 20460, USA
[b]Department of Civil Engineering, University of Virginia, Thornton Hall D219, 351 McCormick Road, PO Box 400742, Charlottesville, VA 22904-4742
[c]Green Chemistry Institute, American Chemical Society, 1155 Sixteenth Street, NW, Washington, DC 20036, USA

1. Introduction

"It is better to prevent waste than to treat or clean up waste after it is formed" is Principle 1 of Green Chemistry and Principle 2 of Green Engineering [1,2]. Regardless of its nature, waste consumes resources, time, effort, and money first in its creation and then in its handling and management, with hazardous waste requiring even greater investments for monitoring and control. As has been stated on numerous occasions, creating, handling, storing, and disposing of waste do not add value to the product or the enterprise. In addition, the traditional mechanisms for managing wastes often move it from one media, such as wastewater, to another, such as the landfill [3]. In processes of production, therefore, waste is always undesirable in all its forms.

Ideally, molecules, products, processes, and systems would be designed not to create waste. In other words, across all scales, inputs are designed to be a part of the desired output eliminating the concept of waste. The concept of "zero waste" is not forgetful of the laws of thermodynamics but rather it is a goal of perfection. This model has been described at the molecular scale as "atom economy" [4] and can be extended a cross design scales as the "material economy [2]." While, perfection may not be attainable, it provides a guiding direction to designers, which results in product, processes, and systems that are more productive and also more sustainable. Whether the waste is material, energy space, time or the derivative of all of these, money, there are design strategies that can and are being implemented in Green Chemistry and Engineering to eliminate the concept of waste.

While it could be observed that efficiency has always been a fundamental aspect of good engineering design, the current state of our process and product infrastructure can call into question how well this design rule has been historically or systematically applied. One could argue that during the 20th century, it was far more of an engineering focus to design strategies for dealing with waste in a socially acceptable manner, such as treatment or disposal, rather than to focus on innovative, disruptive technologies based on efficiencies. It became far more important due to regulations, laws, and invested capital, to make existing inefficient and unsustainable processes continue through the use of elegant and expensive technological bandages than it was to engage in fundamentally efficient and sustainable design. The result of this skewed focus is an extensive engineering portfolio of ways to monitor, control, and remediate waste. Green Engineering simply refocuses efforts on sustainable, efficient design. This means that waste is avoided in the first place and the concept of waste is eliminated wherever possible.

So, when is a waste not a waste? When it has a productive purpose. While there may be current barriers, including scientific, technical, or economic, to inherent zero-waste design, it is important to note that the concept of waste is human. In other words, there is nothing inherent about materials, energy, space, or time that makes it waste except that a defined use for it has not been imagined or implemented. Therefore, if the creation of the waste cannot be avoided under given conditions or circumstances, designers and engineers should consider alternative mechanisms to effectively exploit these resources for value-added, beneficial use as a feedstock by capturing it within the process, the organization, or beyond. This turns a cost and liability into a savings and benefit. It is important to consider that the materials and energy that were utilized and are now "waste" have embedded entropy and complexity representing an investment in cost and resources. This indicates that the recovery of waste as a feedstock represents both potential environmental and economic benefits, as the following case studies will illustrate.

2. Eliminating the concept of waste

Waste can be eliminated through the design of disruptive technologies that move toward sustainable products, processes, or systems that inherently reduce unnecessary inputs or the hazard associated with those inputs while still attaining the desired outcome. Another strategy to eliminate the concept of waste is to design molecules, products, processes, and systems, to incorporate all of the inputs. These strategies suggest that the designer or engineer design with intention such that all of the resources used are desirable, value-added investments. This approach has both environmental and economic benefits in terms of reducing the life-cycle impacts eliminating the costs and impacts of resources, and the associated costs of procurement and disposal for these inputs that are no longer necessary. The following are several example products, processes, and systems where this notion has been successfully applied including:

- solventless processes or benign solvents,
- self-separation,
- process intensification,
- material deposition, and
- coatings.

2.1. Solventless processes or benign solvents

A solvent is a liquid that has the ability to dissolve, suspend, or extract other materials, without chemical change to the material or the solvent. Solvents are used throughout the economy to process, apply, clean, and separate materials. Given the versatility and broad applicability, solvents are used across a wide variety of sectors including paints and coatings, pharmaceuticals, cleaners and cleaning, printing, and adhesives. The majority of solvents used in industry are classified as volatile organic compounds (VOCs) indicating that they have a high vapor pressure and low water solubility. VOCs are typical groundwater contaminants and are involved in the formation of ground level ozone as well as depletion of the ozone layer. They also contribute to the greenhouse effect in that methane and photochemical oxidants produced from the use of VOCs are both greenhouse gases.

The environmental impacts of these chemicals become an even larger issue when one considers that solvents are inputs to a product, process, or system, that are not intended to be in the final product. In fact, there are examples in which any solvent remaining in the final outcome is considered a hazardous contaminant as in the case of vegetable oil extraction. Designing products, processes, and systems that do not include solvents can lead to clean, efficient, and cost-effective science and engineering with: improved safety and security; reduced cost associated with solvent procurement and disposal; and enhanced reactivities and sometimes selectivities without dilution [5]. As such, there has been increasing research and development in solventless processes, products, and systems as well as the use of more benign solvents such as supercritical carbon dioxide (sCO_2) and water as highlighted in the following examples.

Conventional production of aromatic polycarbonates involves interfacial polycondensation between phosgene ($COCl_2$) and bisphenol-A (BPA). This $COCl_2$ process has several drawbacks such as environmental and safety problems involved in using $COCl_2$, an extremely toxic chemical intermediate with acute (short-term) inhalation exposure, as a reagent. This process results in the formation of chlorine salts of a stoichiometric amount, and uses copious amounts of methylene chloride as a solvent with estimations of 10 times the weight of final products [6,7]. For these reasons, polycarbonate producers are increasingly using a solvent-free, melt-phase polymerization technology [8], based on diphenyl carbonate and BPA, instead of the traditional interfacial route, based on $COCl_2$ and BPA. Many other solvent-free processes are sure to emerge as scientists and engineers respond to customer demands for "dry" processes [9]. Indeed, recent research efforts have succeeded in achieving important chemical transformations using solvent-free processes including efficient catalytic enantioselective syntheses of unsaturated amines [10] and for Wittig reactions [11].

The replacement of conventional solvents with sCO_2 has been successfully demonstrated in a variety of applications including dry cleaning [12], photoresists for silicone wafers [13], seed oil extraction, and other applications in the food sector including coffee decaffeination [14], polymer foaming [15], and in analytical applications for separation and chromatography [16]. Supercritical water has also found success as a substitute for conventional organic solvents in a range of applications including oxidation of sewage sludge [17], oxidation of polymers for recycling [18], as well as the degreasing and cleaning of machine parts. In addition, researchers have demonstrated that fundamental chemical synthesis such as Grignard-type reactions can occur in water rather than organic solvents [19]. In each one of these examples, the use of a hazardous solvent has been reduced or eliminated thereby reducing or eliminating the generation of the associated hazardous and general waste. This would suggest that these designers are contributing to the elimination of the concept of waste by reducing the amount of inputs that are not intended to add value to the final outcome representing both environmental and economic benefits.

2.2. Self-separation

Product separation and purification is one of the single largest consumers of energy and material in many manufacturing processes. Much of the downstream processing in manufacturing is associated with separation and purification including distillation, membrane techniques, evaporation, extraction, chromatography, etc. These processes are highly energy and resource intensive. For example, the total energy consumption of the distillation columns in operation in the United States is more than three times as high as the total energy consumption of Switzerland on an annual basis [20]. This energy and resource investment leads to significant environmental impacts that, similar to solvent systems, are not value-adding to the final product. To avoid investing large amounts of energy and resources to drive the system toward separation, up-front design can allow products to self-separate using intrinsic physical/ chemical properties such as solubility and volatility. This can include reactions where the product precipitates out of solution since it is insoluble in the reaction media.

One of the most interesting developments in the area of self-separation is a recyclable catalyst that precipitates at the end of the reaction [21] as shown in Fig. 1. The catalyst is soluble in one of the reagents and remains soluble when the other reagent is added. As the reaction goes on, and the product builds up, the catalyst precipitates from the mixture as oil. This oil–liquid clathrate — remains to be an active catalyst, as the reagents are able to penetrate into it. When all the reagents are converted into products, the oily catalyst turns into a sticky solid, which can be easily separated and recycled. Using a self-separating, recyclable catalyst in solvent-free conditions can lead to environmental and economic benefits by eliminating the solvent used in the reaction and to separate the products at end of life as well as avoiding costs associated with procuring or recovering the catalyst for the next reaction.

There are numerous additional examples where up-front design can lead to improved ease and accuracy of separation again reducing the associated waste. Several examples of designs that may aid in separation include threaded fasteners, heat-activated reversible fasteners, and fasteners made out of the same materials being joined. These types of designs facilitate the separation of materials that are in a relatively pure form without residues or contamination from

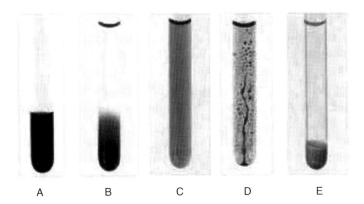

A   B   C   D   E

Fig. 1. Photographs of the catalytic hydrosilylation of a carbonyl by a ketone complex. (A) Ketone complex before adding triethylsilane ($HSiEt_3$). (B) $HSiEt_3$ added, liquid not yet mixed. (C) Mixed and homogeneous. (D) Liquid clathrate formed. Reaction nearing completion. (E) End of reaction. Catalyst has precipitated. Taken from Dioumaev [21].

other materials. Unlike these designs, rivets, welds, or chemical bonds, join materials in such a way that they cannot be separated without significant energy and/or resources, and even then, the materials are likely contaminated or damaged.

Consideration of materials separation up-front will result in easier and more accurate separation of materials for recovery, reuse, or recycling and can lead to environmental and economic benefits. For example, the materials being separated are now available and usable for more beneficial end of life strategies such that this material is not considered waste. This can be achieved without significant investments of energy and resources. Also, the demand for procuring virgin materials is reduced subsequently reducing the environmental and economic impacts associated with acquiring new materials.

2.3. Process intensification

Process Intensification (PI) involves making fundamental changes to processing technologies to yield improved product quality, throughput, and energy efficiency. PI is a highly innovative concept in the design of a process plant. The aim of intensification is to optimize capital, energy, environmental, and safety benefits by radical reductions in the physical size of process plants or a reduction in number of reactors necessary to achieve the desired outcome. Examples of process intensification include microreactors, reactive and catalytic distillation, and disk and plate reactors.

The small dimensions and energy efficiencies of microreactors make it possible to synthesize products on a continuous basis in increasingly larger amounts. The reactors, heat exchangers, mixers, pumps, and valves can be fabricated in configurations measured in micrometers. It was found that devices with these small dimensions are more efficient for mass and heat transfer, resulting in greater selectivity and higher product yield [22]. One such example is the Sub-Watt Power Generation system developed by Pacific Northwest National Laboratory. This device integrates a catalytic combustor, methanol reformer, vaporizers, and a heat exchanger into one micro-system (less than $20\,mm^3$ in volume) to produce a hydrogen-rich stream for fuel cells [23,24].

Reactive and catalytic distillation are hybrid combinations of separation and reaction for chemical synthesis that offers improved efficiencies (e.g. reduced energy requirements, lower solvent use, reduced equipment investment, and greater selectivity) [25]. Methyl acetate production for acetic acid and methanol, invented and practiced by Eastman Chemical Company, is probably the best-known example of reactive distillation in the literature [26–28]. This example of reactive distillation shifted the process from the conventional technology based on 11 major units to a single hybrid unit (Fig. 2). In this case, the switch to reactive distillation lowered the investment as well as the energy demand by

80% as compared to the traditional technology [25]. In addition, two solvents were eliminated from the production process for methyl acetate by moving to reactive distillation. These benefits are obtained not only by overcoming limitations due to chemical reaction equilibrium but also by avoiding or overcoming difficult separation [25]. Although the data are not in public domain, the quantity of materials used in this reactive distillation process is presumably substantially smaller than compared to the conventional method.

The spinning disk reactor (SDR), a small continuous reactor, was established as an alternative to traditional stirred tank processing technology and offers distinct advantages with respect to mixing characteristics, heat transfer, and residence time distribution. This leads to increases in time, space, energy, and mass efficiency resulting in waste minimization and environmental, and often, economic benefits. Spinning disk reactors have successfully been demonstrated for increased efficiency and waste minimization in the following applications: polymers, fine chemicals, food industries, and nanoparticle production. For example, SmithKlineBeecham investigated the use of SDRs for one of its processes and reported that it displayed distinct advantages over batch processing techniques than several commercially relevant processes for the manufacture of pharmaceuticals. In comparison to presently used batch processes, the reaction utilizing an SDR resulted in a reaction with 99.9% reduced reaction time, 99%

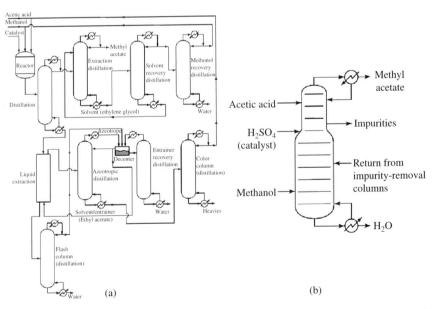

Fig. 2. Schematic of (a) conventional and (b) reactive distillation technologies for the production of high-purity methyl acetate. Taken from Siirola [28].

reduced inventory, and 93% reduced impurity level for the same product yield [29].

2.4. Material deposition

Traditional parts and product manufacture has been mainly a process of material removal. That is, parts and products are conventionally designed from a quantity of material that undergoes etching, machining, or cutting to yield the desired result in terms of size and shape. Necessarily, this approach begins with excess material that is known from the outset to ultimately become waste. In other words, these processes consciously and knowingly build waste into the design. In contrast, there are numerous examples of how Green Engineering approaches can instead design processes that utilize virtually all material inputs by "building" the desired product or part rather than relying on the removal of excess materials.

Through material deposition or "building" strategies, materials requiring multiple conventional processes can be formed in a single step, eliminating the need for dies and molds, capital equipment, and space associated with each additional process [30]. This results in reduced waste and subsequent environmental and economic impacts. There are several areas of research that have successfully demonstrated these approaches to designing and creating the desired part or product such as direct metal deposition.

Direct metal deposition (DMD), a recently developed additive process, has the potential to reduce the environmental impacts associated with mold making and metalworking. DMD is a type of rapid tooling process that makes parts and molds from metal powder that is melted by a laser by building up layers of material until the desired product or part is created. This process allows for the production or reconfiguration of parts, molds, and dies that are made out of the actual end material, such as tool steel or aluminum, eliminating the need to produce scale models and casts. DMD yields molds, parts, and products with higher quality and better accuracy in less time and at lower cost than conventional manufacturing processes [31]. The environmental benefits of this technology include reduced waste associated with casting and scale models, with parts or products outside of specifications that need to be scraped, and with precision working and finishing processes necessary after traditional fabrication processes.

2.5. Coatings (dipping, spraying)

Coatings are ubiquitous in the current portfolio of materials that we use in products ranging from automobiles and appliances to machine parts. Historically, the application of these coatings developed as a wasteful process that resulted in significant amounts of materials utilized in the process that were not transferred to the final product. Different coating technologies all have performance attributes, such as type of finish or evenness of coating, that make them well suited for certain applications. Historically, air atomized spray systems were widely used in coating operations, but they are starting to be replaced because of their relatively low transfer efficiency. During conventional spray painting, some of the paint is deposited on the surface being painted, while much of it, in the form of overspray, is sprayed into the air. The overspray is collected, concentrated into paint sludge, and either disposed of in a landfill or incinerated. This process presents an inherently wasteful approach to the application of coatings.

An alternative coating application technology that embraces the Green Engineering approach is one that has an increased transfer efficiency such as applying coatings by dipping. In dipping operations, the parts to be coated are immersed completely in the coating material and then allowed to drain. The material that is not successfully transferred to the final product is returned to the dipping tank for continued use in the coating process. Dipping is commonly used for primer systems, where a topcoat applied by another method covers imperfections. This would suggest that dipping may not be appropriate for all coating applications; however, from a material conservation and transfer efficiency perspective, dipping should be a preferred option whenever feasible.

In either spraying or dipping for coating applications, water-based rather solvent-based coatings are environmentally preferable from the perspective of reduced VOC emissions and subsequent impacts on air quality and worker exposure. Whatever type of paint and application method that is chosen, the best environmental solutions may be to redesign the product to eliminate unnecessary coating. For example, if the coating is being applied for strictly aesthetics reasons, such as color, it may be possible to mix the pigment directly into the material prior to machining, molding, or blowing it into the desired final product. This eliminates the environmental and economic costs associated with the life cycle of the entire coatings process.

3. Utilizing waste as a feedstock

While the goal of zero waste is one to constantly strive for, there will always be limitations to eliminate the creation of waste in the design of all molecules, products, processes, and systems. It is important therefore to consider that the waste generated has value in terms of invested resources. Recovering the waste, preferably in an immediately usable form, and utilizing it as a feedstock for another molecule, product, process, or system provides a mechanism to prevent the handling and disposal of the waste. This not only prevents the need to handle, manage, and dispose of the waste, it offsets the need to use virgin resources in the next molecule, product, process, or system in arriving at the desired result.

The overriding reason that it is undesirable to generate waste is that it is always desirable to use resources for their highest value application. Therefore, any resource that results as an output of a process or system should be evaluated for its highest value application. While a first level analysis would suggest that any output from a process or system would have its highest value as a commercial product, its next highest value within a closed loop, followed by recycling the resource into a separate product, process, or system, and lastly disposed of a waste, other factors may need to be considered such as economics, technical feasibility, and logistics.

An important consideration in deciding what recycle/reuse options to pursue is the state of purity of the waste. If the substance is a small component of a complex mixture, the amount of time and energy that would need to be invested in separating out the material could be substantial. In this case, it is possible that the high net value-added option could involve an application for the waste material that, while doesn't maximize its performance potential, it minimizes the purification expenses that would otherwise be incurred.

The design decisions on how best to utilize waste at the highest value include factors relevant to a particular process, a particular company, and a particular locality. The final conclusions will be specific to the context and constraints of the situation where the decision is made. However, the most important point is that the factors to evaluate how to best utilize waste material need to include a consideration of impacts not merely for near term and the immediate circumstances but also for the consequences over time.

Table 1 shows several example products, processes, and systems successfully designed to utilize waste as a feedstock .

3.1. Recovery of a waste as a feedstock, in-process

Reclaiming and reusing a waste in-process is highly desirable because it eliminates the need to transport the material. By cycling the waste/feedstock material in a closed loop, you can eliminate the transportation step that is necessary for off-site recycle/reuse options. In addition, in-process recycling also results in a reduction in the potential for losses from handling and shipping.

3.1.1. Nitrous oxide (N₂O) from adipic acid production as a feedstock in phenol production

Nitrous oxide (N_2O) is a greenhouse gas that contributes to global climate change and has a global warming potential of 310 times higher than that of carbon dioxide [32]. The main industrial sources of N_2O in the United States are

Table 1
Examples of utilizing waste as a feedstock across design scales

Design scale	Product, process, or system
In-process	Nitrous oxide (N2O) from adipic acid production as a feedstock in phenol production Acetic acid recovery from ibuprofen production as a source of acetic anhydride in ibuprofen production
Within an organization	Material from end of life carpet as a feedstock for next generation of carpet Heat from braking as a feedstock for battery charging
Beyond organizational boundaries	Crustacean shell waste as a feedstock for chitin and chitosan-based products Gypsum to wall board and sulfuric acid from flue gas desulfurization Waste cooking oil as a feedstock for biodiesel

adipic acid and nitric acid production [33] with the production of adipic acid accounting for approximately 13% of the annual US release in atmospheric nitrous oxide [34]. Adipic acid is mainly used in the manufacture of the nylon fiber, nylon 6,6, but is also used in the production of plasticizers, lubricants, insecticides, and dyes.

The traditional production process of adipic acid begins with benzene as a feedstock (see Fig. 3A) and after several reaction steps yields adipic acid and N_2O. In order to limit the emission of N_2O, a heavily regulated chemical and potentially value-adding feedstock, Solutia, Inc. developed a one-step hydroxylation of benzene to phenol (see Fig. 3B) with the nitrous oxide waste from adipic acid production using a zeolite catalyst. The phenol generated from this process can then be sold as a product on the market or further reduced to cyclohexanol or cyclohexanone for producing additional adipic acid. Either would result in capturing the N_2O from adipic acid production thereby minimizing or eliminating the emission of waste nitrous oxide. This has the further benefit of providing a cost-effective means to produce phenol as an end product or an intermediate in the production of additional adipic acid. This example demonstrates the concept that utilizing a process waste as a feedstock can lead to both environmental and economic benefits.

3.1.2. Acetic acid recovery from ibuprofen production as a source of acetic anhydride in ibuprofen production

The traditional industrial synthesis of ibuprofen was developed and patented by the Boots Company of England in the 1960s [35]. This synthesis is a six-step process and results in large quantities of waste chemical by-products.

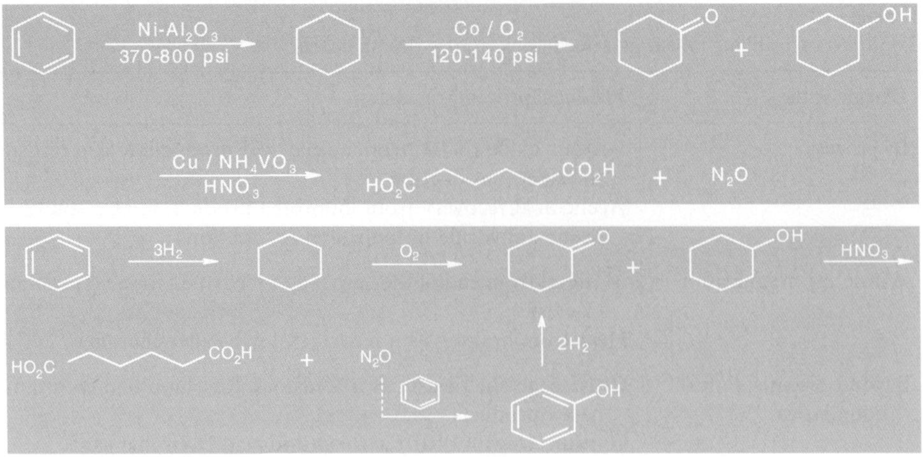

Fig. 3. The chemical synthesis of adipic acid (A) by the conventional method and (B) by the method developed by Solutia, Inc. to capture the waste nitrous oxide and utilize it as a feedstock in the production of phenol and/or adipic acid.

In 1991, BHC company filed patents [36,37] for a new greener industrial synthesis of ibuprofen that is only three steps (Fig. 4). In Step 1, acetic anhydride is added as a reactant with acetic acid produced as a by-product. The resulting acetic acid can be captured and reacted to form acetic anhydride that can then be used as the reactant feedstock in Step 1.

The recovery step coupled with BHC's green synthesis results in 99% atom economy for this reaction up from 40% atom economy in the traditional ibuprofen production process [38]. That is, 99% of the atoms that are utilized in the green synthesis are included in the desired final product(s). By capturing the waste stream and utilizing it to replace virgin feedstock, the environmental and economic impacts associated with waste handling and disposal is eliminated. In addition, there is an added benefit in that virgin acetic anhydride no longer needs to be purchased and stored for the production of ibuprofen.

3.2. Recovery of waste as a feedstock, within an organization

Recovering of waste within an organization leads to a known supply stream with known properties for an organization. This known quantity can then be used as a feedstock in the organization's own designs eliminating issues associated with disrupted supply streams and lack of control over meeting specifications. This strategy also represents a direct savings to the organization by eliminating the disposal and handling costs associated with the waste stream as well as the procurement of virgin feedstock.

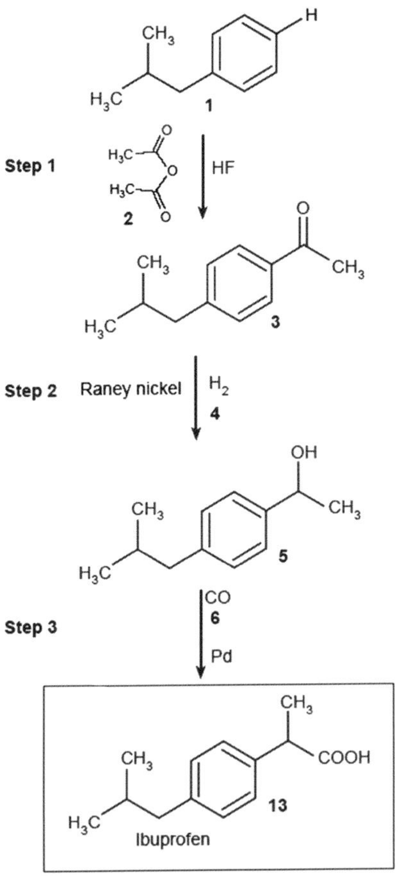

Fig. 4. The BHC Company green synthesis of ibuprofen.

3.2.1. Material from end of life carpet as a feedstock for next generation of carpet

Every year, 4 billion pounds of carpet are discarded in the United States [39], of which only about 1% is recycled [40]. For commercial facilities, government agencies, and institutions such as schools and hospitals, disposing of large amounts of used carpet can be a major cost issue. Carpet also poses waste management problems because it consumes enormous amounts of landfill space due to its bulk, 2% of solid waste by volume, according to the US Environmental Protection Agency [41]. In addition, the incineration of discarded carpet, especially products with polyvinyl chloride (PVC) backing, can release toxic chemicals, including dioxin, into the air. PVC often contains phthalate-based softening agents, which are recognized as reproductive toxins that may also contribute to indoor pollution.

In order to address these issues in a way that adds value, several carpet manufacturers have developed methods to recover carpet at end of life and utilize the material as the feedstock to produce the next generation of carpet. For example, Shaw Inc. developed the EcoWorx system, a recyclable carpet tile product [42]. A typical carpet tile comprises two main elements, the face and the backing. In carpet tile, the backing represents a more significant investment in materials, engineering, and performance than backings related to typical broadloom carpeting. By replacing the traditional PVC-based backing with a metallocene polyolefin, the EcoWorx system makes it possible to recycle both the face and the backing components into the next generation of face and backing components, respectively, for future generations of EcoWorx carpet tile. The EcoWorx system also utilizes Shaw's EcoSolution Q nylon 6 premium branded fiber system, which is designed to use recycled nylon 6, and currently embodies 25% post-industrial recycled content in its makeup. The EcoSolution Q nylon 6 branded fiber system can be recovered and used as a feedstock through a reciprocal recovery agreement with Honeywell's Arnprior depolymerization facility without sacrificing performance or quality or increasing cost. This allows Shaw's carpet tile products to return to manufacturing, with backing and fiber from end-of-life tiles to be made into more backing and fiber (see Fig. 5).

It is now in Shaw's best interest to receive carpet back at end of life, as that is the feedstock for it's next generation products. To help facilitate this, each EcoWorx tile is back printed with a toll free number to contact for disposition of the material for recycling. A value recovery system is in place to handle the flow of material based on projected return rates. Not only does this reduce or eliminate Shaw's EcoWorx product from becoming waste, it is also Shaw's findings that reclaimed EcoWorx product components are projected to be less costly to process than virgin components. This afterlife value end-of-life removes EcoWorx tiles from the category of waste, placing them instead in a category of raw material feedstocks.

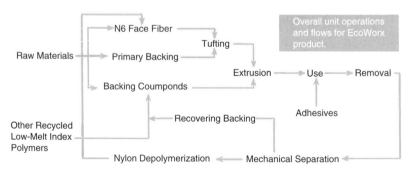

Fig. 5. Overall unit operations and flows for Shaw Inc.'s EcoWorx product. Taken from Segars et al. [42].

3.2.2. Heat from braking as a feedstock for battery charging

One of the most important differences between a hybrid electric vehicle (HEV) and a conventional vehicle is HEV's ability to reclaim a portion of the energy otherwise lost to braking through a process called regenerative braking. In a conventional vehicle, brakes slow and stop a car by converting kinetic energy into heat energy through braking. The energy generated from braking is then lost to the air as waste heat. In other words, fuel is burned to make heat to put kinetic energy into the car, then that energy is given off as waste heat.

The goal of regenerative braking is to recover some portion of that heat and store it in the form of battery capacity so it is available to convert back into kinetic energy to power the car (see Fig. 6). When the driver slows or stops an HEV, it does not create friction to slow down, instead the electric motor is reversed turning it into an electric generator. Any time a hybrid-electric vehicle slows down, lifting the accelerator or application of the "brake" causes the system to use the vehicle's momentum to generate electricity. It is estimated that regenerative braking can eventually be developed to recover about half of the energy lost as waste heat from braking [43].

Regenerative braking, by capturing waste heat and converting it to usable energy, leads to a reduction in fuel consumption and the associated air emissions. While the fuel is not directly captured as a waste in the combustion process and used as a feedstock, a waste product is recovered from the process and offsets additional fuel required to power the vehicle. This strategy has the potential for reductions in direct and indirect environmental impacts associated with exploring and drilling for crude oil, transporting the crude oil with potential for accidental releases, distilling and separating the crude oil components, combusting gasoline in the engine producing air emissions including green house gases, etc., as well as those associated with using a non-renewable resource.

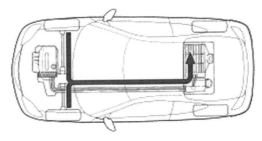

Fig. 6. The path of energy flow from the wheels to the battery in a commercial HEV. Taken from [44].

3.3. Recovery of waste as a feedstock, beyond organizational boundaries

The recovery of waste as feedstock beyond organizational boundaries represents a potential additional revenue stream for the organization. While there may be significant costs associated with preparing and purifying the waste for use as feedstock, this once liability now represents a potential benefit. This strategy presents an opportunity for positive corporate image and strong community relations that may not be directly quantifiable in terms of revenue but may contribute to the bottom line in intangible ways. There is also the potential for significant benefits in terms of externalities that the organization may not capture directly but benefit society in terms of improved air and water quality, reduced toxic emissions, and improved human health.

3.3.1. Crustacean shell waste as feedstock for chitin and chitosan-based products

Chitin is a high-molecular weight linear polymer that is highly insoluble and has low chemical reactivity. Similar to cellulose, chitin functions as a structural polysaccharide and is most abundant in crustaceans, insects, and fungi. Chitosan is the N-deacetylated derivative of chitin, though this N-deacetylation is almost never complete. As most polymers are synthetic materials, their biocompatibility and biodegradability are much more limited than those of natural polymers such as cellulose, chitin, chitosan, and their derivatives. In this respect, chitin and chitosan are recommended as suitable resource materials, because the natural polymers have desirable properties including biocompatibility, biodegradability, non-toxicity, and high-adsorption capacity.

The production of chitin and chitosan is currently based on crab and shrimp shells discarded by the canning industries in Oregon, Washington, Virginia, and Japan and by various finishing fleets in the Antarctic. The production of chitosan obtained from crustacean shells, a food industry waste, has the potential to be economically feasible [45].

For every pound of picked crabmeat, there are approximately six pounds of shell waste generated, and about 20% of each shell is chitin [46]. To produce 1 kg of 70% deacetylated chitosan from crustacean shells, approximately 6.3 kg of hydrochloric acid and 1.8 kg of sodium hydroxide are required in addition to nitrogen, process water (0.5 tons), and cooling water (0.9 tons) [46]. While this indicates that there may be environmental impacts associated with recovering chitin, and subsequently chitosan, from this waste stream, there are numerous applications in which chitin and chitosan outperforms competitive products while remaining biodegradable and non-toxic.

The use of chitin and chitosan in a wide variety of sectors with significant success has been reported in the literature. For medical applications, it has been suggested that chitin and chitin derivatives may be used to inhibit fibroplasia in wound healing or can be utilized as absorbable sutures and wound-dressing materials [47]. In water treatment and remediation applications, chitosan has been demonstrated to be an effective chelating agent for metals including mercury, cadmium, nickel, and chromium [48]. Chitin and chitosan have been successfully used to sorb dyes from effluent streams from fabric and paper coloring operations [49]. Chitin fibers are also being explored to improve wet paper strength and for use in fabrics and textiles. All of these applications represent the use of "waste" that is naturally complex with low entropy. By recovering this investment as a value-adding feedstock, a significant organic waste concern can be reduced or eliminated with environmental and economic benefits.

3.3.2. Gypsum to wall board and sulfuric acid from flue gas desulfurization

The burning of pulverized coal in electric power plants produces sulfur dioxide (SO_2) gas emissions. The 1990 Clean Air Act and its subsequent amendments mandated the reduction of power plant SO_2 emissions. The best demonstrated available technology for reducing SO_2 emissions is wet scrubber flue gas desulfurization (FGD) systems. The sludge that results from these systems can be treated with several processes including forced oxidation that will eventually result in the formation of useful, value-added products such as gypsum and sulfuric acid.

Dilute sulfuric acid generated from these systems can be concentrated to economically obtain the desired concentration and purity. Gypsum is also a by-product with potential economic value. There is significant interest in using calcium sulfate FGD scrubber material in wallboard construction and in Portland cement production (as a gypsum source) [50] due to the highly consistent and predictable performance characteristics of synthetic gypsum versus natural gypsum [51]. In fact, wallboard production represents the single largest market for FGD scrubber materials [52] with over 1.6 million tons of FGD gypsum being used in wallboard in 1998 [53]. This reduces the need for mining natural gypsum and once again demonstrates that the recovery of a waste stream for a feedstock prevents not only the environmental impacts associated with that waste stream but also the environmental impacts associated with acquiring or manufacturing the virgin feedstock that the waste stream is replacing.

3.3.3. Waste cooking oil as a feedstock for biodiesel

Exploring new energy resources, such as biodiesel, has received increased attention in recent years. Biodiesel, derived from vegetable oil or animal fats, is recommended for use as a substitute for petroleum-based diesel mainly because biodiesel is a renewable, domestic resource with an environmentally friendly emission profile and is readily biodegradable [54]. Biodiesel, in and of itself, represents a potential fuel that can advance the goal of sustainability with reductions in criteria of air pollutants, such as carbon monoxide, nitrogen oxides, and particulate matter, when compared to conventional petroleum-based diesel fuel in the use phase.

Compared to petroleum-based diesel, the high cost of biodiesel is a major barrier to commercialization. It costs approximately 50% more than petroleum-based diesel depending on the feedstock oil and it is reported that approximately 70–95% of the total biodiesel production costs arises from the cost of raw material; that is, vegetable oil or animal fats [54]. This presents an even more promising strategy for moving toward sustainability: the production of biodiesel fuel from a waste product such as used cooking oil.

The production processes for biodiesel from waste oil and fats are well known and include three basic routes to biodiesel production: base-catalyzed transesterification of the oil, direct acid-catalyzed transesterification of the oil, and conversion of the oil to its fatty acids and then to biodiesel. Most of the biodiesel produced today is done with the base-catalyzed reaction for several reasons: low temperature and pressure yields high conversion (98%) with minimal side reactions and reaction time, direct conversion to biodiesel with no intermediate compounds, and no exotic materials of construction are needed. The chemical reaction for base-catalyzed biodiesel production is depicted in Fig. 7. One hundred pounds of fat or oil (such as soybean oil) are reacted with 10 pounds of a short chain alcohol in the presence of a catalyst to produce 10 pounds of glycerin and 100 pounds of biodiesel [54]. This process is another example of other waste recovery strategies for feedstocks including recovered methanol and purifying waste glycerin into a value-added product for the pharmaceutical sector.

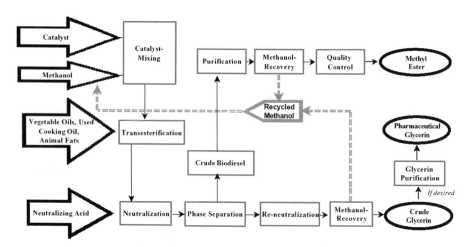

Fig. 7. The base-catalyzed production of biodiesel from virgin or recovered oil with several additional waste recovery for feedstock examples. Taken from [55].

The use of waste cooking oil can greatly reduce the cost of biodiesel because waste oil is available at a relatively low cost. This will encourage the use of biodiesel, potentially increasing its marketshare, and thereby potentially increasing the associated benefits of biodiesel. In this case, the benefits of recovering waste cooking oil for conversion to biodiesel represent a win–win situation. There are economic benefits in using a waste to produce a value-added product. In addition, there are environmental benefits both from offsetting the use of virgin crop for the conversion to fuel. In addition, recovering used cooking oil also prevents it from entering the waste handling and management system. Furthermore, the use of biodiesel has a more preferable emissions profile than petroleum-based diesel leading to better environmental protection as well as a reduction in the health affects associated with exposure to conventional diesel exhaust.

4. Conclusions

The advances that are being made in Green Engineering to avoid waste through thoughtful up-front design and innovative utilization and integration of materials and energy flows are models for what needs to happen on a more general scale in the future. While each of the examples cited are creative solutions to specific problems, the approaches illustrated need to be incorporated as an essential part of the engineers design protocols. When engineering designs are fundamentally judged on their degree of elegance by the inherent efficiency and sustainability embedded rather than how they remedy the problems of their own creation, Green Engineering will be engineering with a fundamental understanding of the environmental, economic, and social systems within which the design will function.

References

[1] P. Anastas, J. Warner, Green Chemistry: Theory and Practice, Oxford University Press, London, 1998.
[2] P. Anastas, J. Zimmerman, Environ. Sci. Technol. 37 (2003) 94A–101A.
[3] G. Hunt, Overview of Waste Reduction Techniques Leading to Pollution Prevention, in: H. Freeman (Ed.), Industrial Pollution Prevention Handbook, Vol. 9, McGraw-Hill Inc., New York, NY, 1995.
[4] B. Trost, Science 254 (1991) 1471–1477.
[5] M. Kidwai, Dry Media Reactions, Proceedings of ECSOC-5, The Fifth International Electronic Conference on Synthetic Organic Chemistry ECSOC-5, September 1–30, 2001.
[6] W.B. Kim, U.A. Joshi, J.S. Lee, Ind. Eng. Chem. Res. 43 (2004) 1897–1914.
[7] K. Komiya, S. Fukuoka, M. Aminaka, K. Hasegawa, H. Hachiya, H. Okamoto, T. Watanabe, H. Yoneda, I. Fukawa, T. Dozona, in: P.T. Anastas, T.C. Williamson (Eds),

Green Chemistry: Designing Chemistry for the Environment, Vol. 20, American Chemical Society, Washington, DC, 1996.

[8] D.W. Fox, US Patent 3,153,008, 1964.

[9] J.M. DeSimone, Science 297 (5582) (2002) 799–803.

[10] S.J. Dolman, E.S. Satterly, A.H. Hoveyda, R.R. Schrock, J. Am. Chem. Soc. 124 (2002) 6991.

[11] V.P. Balema, J.W. Wiench, M. Pruski, V.K. Pecharsky, J. Am. Chem. Soc. 124 (2002) 6244.

[12] J.M. DeSimone, T. Romack, J.B. McClain, J. DeYoung, R.B. Lienhart, K. Huggins, US Patent 6,248,136, June 19, 2001.

[13] L.B. Rothman, R.J. Robey, M.K. Ali, D.J. Mount, Supercritical Fluid Processes for Semiconductor Device Fabrication, Proceedings IEEE International Symposium on Semiconductor Manufacturing Conference, Boston, MA, April 30–May 2, 2002, pp. 372–375.

[14] H. Peker, M.P. Srinivasan, J.M. Smith, B.J. McCoy, AIChE J. 38 (5) (1992) 761–770.

[15] S.K. Goel, E.J. Beckman, Polym. Eng. Sci. 34 (14) (1994) 1137–1147.

[16] D.L. Hjeresen, Crit. Rev. Anal. Chem. 28 (2) (1998) 9–12.

[17] A. Shanableh, E.F. Gloyna, Water Sci. Technol. 23 (1–3) (1991) 389–398.

[18] C. Fromonteil, Ph. Bardelle, F. Cansell, Ind. Eng. Chem. Res. 39 (4) (2000) 922–925.

[19] C.J. Li, ACS Sym. Ser. 767 (2000) 74–86.

[20] Annual Energy Review 2000, US Department of Energy, Energy Information Administration, 2000.

[21] V.K. Dioumaev, R.M. Bullock, Nature 424 (2000) 530–532.

[22] R.C. Costello, Chem. Eng. 111 (4) (2004) 27–31.

[23] C. Tsouris, J.V. Porcelli, Chem. Eng. Prog. 99 (11) (2003) 50–55.

[24] C. Tsouris, J.V. Porcelli, Chem. Eng. Prog. 99 (7) (2003) 15.

[25] M.F. Malone, R.S. Huss, M.F. Doherty, Envir. Sci. Technol. 37 (23) (2003) 5325–5329.

[26] V.H. Agreda, L.R. Partin, US Patent 4,435,595, 1984.

[27] V.H. Agreda, L.R. Partin, W.H. Heise, Chem. Eng. Prog. 86 (2) (1990) 40–46.

[28] J.J. Sirola, Foundations of Computer Aided Process Design, in: L.T. Biegler, M.F. Doherty, (Eds), AIChE Symposium Series 304, AIChE, New York, 1995.

[29] P. Oxley, Ind. Eng. Chem Res. 39 (2000) 2175–2182.

[30] G.K. Lewis, E. Schlienger, Mater. Design 21 (4) (2000) 417–423.

[31] S. Krar, A. Gill, Adv. Manuf. Mag. 5 (2002), www.advancedmanufacturing/may02/exploringant.htm.

[32] US Greenhouse Gas Inventory Program, Office of Atmospheric Programs, US Environmental Protection Agency, Greenhouse Gases and Global Warming Potential Values, 2002.

[33] R. Lempert, P. Norling, C. Pernin, S. Resetar, S. Mahnovski, Next Generation Environmental Technologies: Benefits and Barriers, MR-1682-OSTP (2003) A107.

[34] EPA, US Adipic Acid and Nitric Acid N20 Emissions 1990–2020: Inventories, Projections and Opportunities for Reductions, December 2001, http://www.epa.gov/ghginfo/pdfs/adipic.pdf.

[35] U.S. Patent 3,385,886.

[36] U.S. Patent 4,981,995.

[37] U.S. Patent 5,068,448.

[38] M.C. Cann, M.E. Connelly, The BHC Company Synthesis of Ibuprofen in: Real World Cases in Green Chemistry, American Chemical Society, Washington, DC, 2000, pp. 19–24.

[39] B.K. Fishbein, Carpet Take-Back: EPR American Style, Environ. Qual. Manag. 10 (1) pp. 25–36 (Autumn 2000), http://www.informinc.org/carpettakeback.php.

[40] US Environmental Protection Agency, Characterization of Municipal Solid Waste in the United States: 1998 Update, 146 (July 1999).

[41] US Environmental Protection Agency, Carpet, Product Stewardship Fact Sheet, April 30, 2001, http://www.epa.gov/epr/products/carpet.html.

[42] J.W. Segars, S.L. Bradfield, J.J. Wright, M.J. Realff, Environ. Sci. Technol. 37 (23) (2003) 5269–5277.

[43] J. Gordon, Motor Age 9 (2003).

[44] http://www.insightcentral.net/enregenerativebraking.html.

[45] M.N.V.R. Kumar, Bull. Mater. Sci. 22 (1999) 905.

[46] M. Leffler, Maryland Marine Notes 15 (1997) 2.

[47] F.-L. Mi, Y.-B. Wu, S.-S. Shyu, J.-Y. Schoung, Y.-B. Huang, Y.-H. Tsai, J.-Y. Hao, J. Biomed. Mater. Res. 59 (3) (2002) 438–449.

[48] C.L. Lasko, B.M. Pesic, D.J. Oliver, J. Appl. Polym. Sci. 48 (9) (1993) 1565–1570.

[49] B. Smith, T. Koonce, S. Hudson, Am. Dyest. Rep. 82 (10) (1993) 18–36, 66.

[50] R. Taha, D. Saylak, The Use of Flue Gas Desulfurization Gypsum in Civil Engineering Applications, in: Proceedings of Utilization of Waste Materials in Civil Engineering Construction, American Society of Civil Engineers, New York, NY, September 1992.

[51] M. Holigan, Project House 2001–2002, Segment 7006, www.michaelholigan.com.

[52] US Department of Energy, Inventory of Utility Power Plants in the United States. US Government Printing Office, Report No. 061-003-00934-4, Washington, DC, 1995.

[53] ACAA, 1998 Coal Combustion Product (CCP) Production and Use table, www.acaa-usa.org/whatsnew/ccpcharts.html.

[54] Y. Zhang, M.A. Dube, D.D. McLean, M. Kates, Bioresource Technol. 90 (2003) 229–240.

[55] www.biodiesel.org/pdf_files/prod_quality.pdf.

Sustainability Science and Engineering: Defining principles
Martin A. Abraham (Editor)
© 2006 Published by Elsevier B.V.
DOI 10.1016/S1871-2711(05)01011-1

Chapter 11

Socially Constructed Reality and Its Impact on Technically Trained Professionals

Peter Melhus

Department of Urban and Regional Planning, San José State University, One Washington Square, San Jose, CA 95192, USA

Contemporary engineers, especially those working on public benefit projects, work in an environment far different from their predecessors. The contemporary engineer often works in an environment where many people have opinions on and can influence whether an engineered solution is the correct one to pursue, even if its technical attributes are beyond question. These opinions are based not only on the technical expertise of the commentator, but also on his life experiences and his experience within the group established to address the task/problem at hand. Planning and engineering in many societies, especially in the developed world, include far more public participation than in the past and the participatory trend continues. Stakeholder engagement, addressed elsewhere in this volume, is one example of this higher level of public participation.

Today's engineers need to be aware of the trend of increasing public participation in their work and be prepared to work within it. They need to be aware that politics can and will influence the implementation choice among the potential technical solutions to an issue/problem. Accordingly, they need to be aware of the social sciences as they are trained in the physical sciences. They need to be aware of the social construction of the reality of the people with whom they are engaging as they develop technically preferable and, equally important, socially acceptable solutions to publicly related tasks/problems.

The purpose of this chapter is to discuss the styles of planning most often in use today, the social learning paradigm and its influence on the collaborative style of planning and the engineering profession, and illustrate the effectiveness of the participatory paradigm in the context of two examples. One example led to a fundamental change in the direction of California energy policy in the late 1980s. The other addresses a contemporary engineering/planning effort underway in mid-2005.

1. Contemporary planning styles

Using the dimensions of both the interdependence and diversity of people, interests and perspectives of those involved in and affected by strategic planning efforts, Innes and Booher [1] describe four styles of planning — rational/technical, pork barrel, ideological vision and collaborative.

Rational/technical planning involves a so-called "value-free" technical analysis, usually conducted by professional planners/engineers, to arrive at a solution to a societal issue. Ideally, this solution is then implemented by the authorized organization(s) and the issue is resolved. The extent to which the rational/technical planning process considers either the interdependence of the people affected by the decision or the diversity of their views is most likely determined by the ideology and experience of the professional planner/engineer conducting the analysis; hence the "so-called" phrase preceding "value-free technical analysis" above. Most human beings inherently bring their values and biases to any task tackled. Logic, logistics and human nature suggest that it would be a rare case when the interdependence and diversity of the people affected by a decision are adequately represented, especially if adequacy of representation is evaluated from the perspective of the people whose views are supposed to be considered. For example, the environmental community would probably not feel adequately represented if the technical analyst was an engineer with a reputation for being environmentally insensitive.

Pork barrel planning has as its basis the consolidation of political influence, usually by a current political office holder. Resources are distributed by the political office holder to the individuals and organizations interested in development/planning in a manner that obligates the recipients to continue their loyalty to the politician. Planning for the benefit of the community is more of a fringe objective to the political objective in pork barrel planning. The interdependence of pork barrel planning is vertical in nature, that is, a "mutual need" relationship is established between the recipient of the resources and political office holder distributing the resources. This contrasts with the desired consideration of the interdependence of the people affected by the planning/engineering decisions.

Individuals and organizations focused on a specific social issue, for example, the environment, homelessness, and so on, generally use the *ideological vision* style of planning. Tactics employed in this style typically involve interaction between the group or its representatives and the legislative or judicial branches of government, either directly through lawsuits and proposed legislation or indirectly through the development of larger, grassroots organizations to influence policymakers and legislators. The ideological vision is inherently focused and not diversified — the movement itself evolves from a group of people with similar views on an issue. The interdependence of the participants in this planning mode is high since the group itself is united due to a common cause. However, the interdependence of varying perspectives is usually non-existent. It is usually a case of "singing to the choir".

Collaborative planning involves discourse among the stakeholders of the issue to be addressed. It attempts to bring the views of all major stakeholders to the table in the process of developing recommendations and coming to decisions. Collaborative planning is inherently diversified because one of its tenets is to include all major stakeholders. "Stakeholder" can be defined as anybody (both individuals and organizations) who "have an interest or share in any undertaking" (*Webster's Ninth New Collegiate Dictionary*). A more pragmatic definition might be anybody (individuals and/or organizations) capable of preventing the implementation of the ultimate decision(s)/recommendation(s) emanating from the process. An integral part of the process of collaborative planning is the development, or perhaps more appropriately, the shared understanding, of the interdependence of the interests of the people involved in the process. Collaborative planning is an offshoot of social learning theory, an understanding of which is becoming more essential for contemporary engineers.

2. Social learning

The social learning paradigm and its contemporary multi-stakeholder collaboration offshoot offer the professional engineer the opportunity to help frame a problem to be solved and shape the operative paradigm of the problem-solving effort.

The social learning approach is focused on non-coerced, task-oriented action by a group of individuals. Two kinds of theory are involved in such group-related action — theories of reality and practice. The theory of reality is related to the lifelong education and experience of an individual that shapes her view of the world and her understanding of the problem/task the group faces. The theory of practice is related to the norms of behavior of individuals in specific roles within the task-oriented group. Each of these theories influences the other [2].

Social learning is a cumulative process that lasts for the duration of the existence of the group. It takes place primarily through face-to-face dialogue. Objectives and sub-objectives tend to emerge during the course of the group interaction. However, "double-loop" learning, that is, learning which results in adjustments in previously accepted norms of engagement and/or additions to the lifelong experience/learning of individuals — in other words, changes in the individuals' reality and practice — may need to occur before significant changes in objectives are accepted.

In his 1987 work, Friedmann cites several examples of the application of the social planning paradigm in the public domain and criticizes the participatory efforts for their acceptance of the existing power relations in the society of focus. Among the work cited and criticized by Friedmann are his own work, *The Good Society* [3], the work of David Korten [4], Korten and Alfonso [5], Rain Umbrella, Inc., and citizen participation efforts. *The Good Society* relates to small, task-oriented, temporary action groups addressing certain social problems, for example, the role of women, governance, etc. The Korten work relates to citizen involvement in regionally based rural economic development in the developing world. The work of Rain Umbrella, Inc. relates to a citizen-based visioning process in Portland, Oregon [6].

Based on the social learning paradigm, multi-stakeholder collaboration can overcome the shortfall of the acceptance of the existing power structure as illustrated in the examples that follow later in this essay. Multi-stakeholder collaboration can also result in the creation of social, intellectual and political capital, as will also be illustrated in the examples. Indeed, social, intellectual and political capital are necessary for a multi-stakeholder, collaborative organization to reach consensus and/or develop agreement [7]. Social capital will be discussed in further detail subsequently in this paper but for present purposes it suffices to say social capital is what enables individuals to work together.

Intellectual capital creates a framework for the participants to move toward consensus. Intellectual capital is created by jointly developing technical or scientific information to the point where all the participants in the collaboration accept it. Intellectual capital is enhanced by ordinary conversation among the participants.

Political capital is the collective ability of the participants in a collaborative process to implement their agreed upon changes. In successful, broad-based, multi-stakeholder consensus building efforts, political capital will be synergistic since the entire group, rather than a subset of it, will endorse the consensus reached by the participant organizations. While these three forms of capital often reinforce one another in their development, social capital is the fundamental building block without which the other forms are less likely to evolve.

2.1. Social capital theory

In recent years the concept of social capital has been gathering interest and acceptance, particularly among social scientists and policymakers. In *Making Democracy Work: Civic Traditions in Modern Italy*, Robert Putnam [8] defines social capital as the "features of social organization, such as trust, norms, and networks that can improve the efficiency of society by facilitating coordinated actions" (p. 167). Putnam further suggests that in the complex societies of today's modern world, trust evolves at the societal level from the continuing use of the norm of generalized reciprocity and social networks of civic engagement.

Generalized or diffuse reciprocity is one of two types of reciprocity norms that Putnam describes; the other being balanced or specific. Balanced or specific reciprocity refers to the approximately simultaneous exchange of items of similar value, while generalized or diffuse reciprocity refers to the continuing long-term relationship or exchange that may or may not be balanced in the short-term. While two friends alternating driving to the same place over a 2-day period illustrates the concept of balanced reciprocity, a long-term friendship embodies generalized reciprocity. For example, a person might do something for a friend today with the understanding that sometime in the future the favor may be returned, if needed. Putnam points out the highly productive potential of generalized reciprocity in communities worldwide.

Networks of civic or social engagement or "horizontal associations," which represent the interaction of people that generally have equal societal status and/or power, encourage robust norms of reciprocity. The more often people interact, and more varied the context, the more communication occurs and the more likely peoples' mutual expectations will be understood. These social networks also facilitate the exponential spreading of societal trust as people tend to trust those who are trusted by others that they trust. In addition to spreading the breadth of trust in this manner, the depth of the trusting relationships is increased over time as an implicit sanction is inculcated, that is, as the frequency of interaction increases and the value of the relationship increases, the potential risk/cost increases for an individual not to meet the expectations or betray the trust of those within his network.

James Coleman [9] suggests that social capital be defined by its function and that it is embodied in interpersonal relationships. The various forms of social capital have two things in common according to Coleman: they consist of a social structure by which individual interests can be achieved and they facilitate certain actions within the social structure, either by individuals or corporations. Coleman suggests that three forms of social capital exist:

1. obligations and expectations that depend on the trustworthiness of the social environment, for example, each individual on an engineering team will meet his/her deadline while delivering quality work;
2. information flow capability of the social structure, for example, an individual can rely on another to be kept informed on an issue if he knows that person stays current on it, rather than having to stay current on it himself; and

3. norms accompanied by sanctions, for example, an engineering firm's
   behavioral norm that expects certain behaviors from the firm's members.

Coleman's three forms of social capital are consistent with Putnam's features
of social capital — trust, networks and norms of engagement, including reciprocity, respectively.

Effective norms of engagement depend in part on a "closure" mechanism that
leads to the effective monitoring and guiding of behavior, essentially a
mechanism providing the potential for feedback within the network. Coleman
illustrates this concept by suggesting if individual A has a relationship with
individual B, individual B has a relationship with individual C and individual
C has a relationship with individual D; closure is achieved when individuals D
and A also have a relationship. According to Coleman (p. S108), "closure
creates trustworthiness in a social structure".

Finally, Coleman suggests that social capital is less tangible than human
capital, defined essentially as individually achieved education and developed
skills, since social capital is created and exists within the relationships between
individuals or entities, while human capital is created by changes within individuals to bring about heightened skills and capabilities. In addition to the
tangibility issue suggested by Coleman, the nature of social capital makes it
more difficult to create (it inherently depends upon more than one individual)
and easier to lose (betray trust or act outside the norms of engagement and you
jeopardize the social capital developed).

2.2. Examples of multi-stakeholder collaboration in action

The examples that follow are real and occurred in the San Francisco Bay Area
in California. The first example tells the story of collaborative planning
involving Pacific Gas & Electric Company (PG&E) and the Natural Resources
Defense Council (NRDC), and focuses on California's energy policy. The second example tells the ongoing and evolving story of Sustainable Silicon Valley,
and focuses on the creation and implementation of a regional environmental
management system.

While the following examples would likely be considered public benefits
projects, the principles described would work equally as well in private engineering decisions, within a single firm for example. Rational/technical analysis
can lead to different results to the same problem emanating from different
departments within the same firm or different firms supposedly working
together to resolve the problem. A lack of inter-departmental or inter-firm
social and intellectual capital can lead to an unhealthy level of competition that
could be avoided if these capital currencies existed. Internally focused politics

can lead to the inter-organizational equivalent of pork barrel decision-making, especially as manifested by advancement opportunities within the organization.

2.2.1. Collaboration in action: energy policy in California's regulated energy utilities

Historically, the planning style of choice for the environmental community, when dealing with energy utilities, has been ideologically — rather than collaboratively — based. As such, the environmental community would engage with the legislative or judicial branches of government through lawsuits, proposed legislation, and/or the development of grassroots organizations to influence policymakers and legislators. Ideologically based planning styles also prevailed historically in the utility community in its dealings with environmentalists. These realities and practices [2] inherently led to adversarial relations between energy-focused environmental organizations and PG&E, the energy utility serving most of the northern two-thirds of California. Similar adversarial relations also often developed between the individuals representing the organizations. These adversarial relations are not surprising when one considers the normal historical venue for interaction between these people — usually a courtroom or its ancillaries.

Historical trends began to change in 1989 when PG&E and the NRDC entered into negotiations with one another and other major stakeholders with the objective of developing a policy blueprint for California's energy use in the 1990s. In hindsight, this process was clearly a collaborative one rather than a traditional negotiation for these entities and the process exemplified the conditions of social learning as delineated by Friedmann. The process led to emancipatory knowledge, that is, knowledge beyond self-fulfilling rationalization [1], and double-loop learning [2] for many of the participants. It also led to the development of social, intellectual and political capital as will be illustrated subsequently.

Over the course of a year, PG&E and the other investor-owned utilities in California (Southern California Edison, Southern California Gas and San Diego Gas & Electric), the NRDC and other environmental organizations which focus on energy-related issues (for example, Environmental Defense known as the Environmental Defense Fund at the time), the California Public Utilities Commission (CPUC) and other regulatory agencies (for example, the California Energy Commission, etc.) and constituency-based groups representing customers (for example, the Association of California Water Agencies, Toward Utility Rate Normalization, etc.) met to discuss and ultimately agree on state policy on energy efficiency for the regulated energy utilities. Initially, the atmosphere was tense due to the almost exclusively adversarial long-term relationships between the utilities, environmentalists, and regulators (and often customers).

Each stakeholder group came to the table not only with a position, but also with a description of its interests [10]. The willingness of the parties to bring to the table their interests as well as their positions was very significant. By expanding the focus from "this is what we want" (our position) to "this is why we want what we want" (our interests), the negotiators were able to find common ground that met the interests of all the stakeholders without necessarily meeting their positions.

The environmentalists wanted renewed emphasis on energy efficiency and renewable electric generation technologies in order to reduce the environmental impact of electricity generation. Their engineers knew that energy efficiency was cost effective both at the consumer and societal levels.

Customer groups wanted lower energy costs for their constituents.

The utilities were focused on their fiduciary responsibility to their shareholders. They did not want to invest vast amounts of money in activities that provided no return. Since the utilities' profits were de-coupled from their sales, the utilities were "indifferent" to increasing/decreasing sales whether due to energy efficiency or otherwise. They knew that "safe" investments with a reasonable rate of return were the result of prudent investments in electric generation, transmission and distribution facilities, but not in energy efficiency and cost-ineffective (relative to traditional generation) renewable electric generation technologies. Clearly, this regulatory institution provided a perverse incentive to utilities in an era when California society was increasing its focus and awareness of the environmental and other benefits of energy conservation and efficiency.

The regulators, that is, those with the majority of the power in this "society", wanted a long-term solution to this energy-related societal dilemma. They did not want a repeat of the fuel oil-related electric energy shortages of the mid-1970s. They also recognized and accepted the connection between fossil fuel-powered electric generation and air pollution, a critical issue in California politics.

As the stakeholders became aware of each party's interests as well as its positions, their discourse led to the evolution of alternative mechanisms that might be able to achieve all the parties' interests simultaneously. The ultimate product, released in 1990, was called the *Report of the Statewide Collaborative Process: an Energy Efficiency Blueprint for California* [11].

The "blueprint" advocated expanded utility investment in energy efficiency and provided shareholder incentive mechanisms (financial returns) to encourage significantly more reliance on energy efficiency rather than on the construction of new power plants.

In the case of PG&E, a "shared savings" mechanism was proposed. Under this mechanism, the life-cycle savings attributable to the installation of energy-efficient technologies at customer sites would be shared by PG&E's shareholders and its customers. (Energy efficiency or "demand-side management" was less costly than constructing new power plants or "supply side" options in California at the time. The life-cycle saving to be achieved was the difference in life-cycle cost of the supply side options that were avoided and the demand-side options that were deployed, discounted for the time value of money). Eighty-five percent of these life-cycle savings were returned to customers in the form of energy rates lower than they would have been had power plants been constructed. Fifteen percent of the savings went to PG&E's shareholders.

Clearly, this collaborative process led to a desirable outcome from the perspective of all of the stakeholders. But it was an outcome that none of the stakeholders individually had brought to the table. Each team of negotiators and engineers had its own solution (its position). But by working collaboratively, they came up with a better solution that met the interests of all parties.

The utilities were able to encourage their customers to use energy-efficient technologies while simultaneously increasing shareholder earnings, a desirable outcome. By their willingness to share their traditional, synoptically based power with the participants in this collaborative process, the regulatory agencies found a way out of their societal dilemma, also a desirable outcome. The customer groups could offer their constituents lower energy costs, provided they took advantage of the energy-efficiency technologies and the subsidies that were made available. And, the environmental community saw an increase in the reliance on energy efficiency over supply side options and the resultant environmental benefits.

Between 1990 and 1996, PG&E's energy-efficiency programs produced savings of almost 11 billion kWh of electricity and 480 million therms of natural gas. Even with PG&E's relatively clean fuel mix for electric generation (compared to the industry averages nationwide), these results were comparable to taking about 775,000 vehicles off California highways [12].

To PG&E, the financial benefits were significant enough — in the range of $4–$30 million per year or approximately 1–10 cents per share — to get the wholehearted endorsement of its Board of Directors. But PG&E's benefits were not limited to these short-term earnings increases that continued until later in the decade when the electric utility industry in California was restructured (or deregulated). There were also second and third generation benefits. Social, political and intellectual capital had been generated between PG&E and the environmental community, especially NRDC.

The traditional relationship between PG&E and the NRDC was fundamentally changed. Their realities and practices had changed [2]. In 1991 the two organizations, along with the Atlantic Richfield Co. (ARCO), filed joint testimony before the California Energy Commission on a recommended greenhouse gas emissions policy for California.

In 1992 the two organizations spearheaded the development of the Recycled Paper Coalition, a voluntary organization comprised of organizations using large amounts of paper that commit to increase their usage of recycled paper products in order to close the recycling loop. The organization currently has more than 250 members nationwide.

Benefits of national significance also began to accrue to PG&E. Recognizing PG&E's energy efficiency and renewable electric generation programs, the first Bush Administration awarded PG&E a gold medal in the President's Environmental & Conservation Challenge Awards program in 1991. PG&E was the only industrial company in the nation to receive a gold medal.

The first President Bush also invited PG&E's CEO to join the President's Commission on Environmental Quality (PCEQ). PG&E, the only utility in the nation invited to join the PCEQ, was an integral player along with the executive director of the NRDC. Among the recommendations of the PCEQ to President Bush was to establish a presidential commission with a broader mission, one which would provide recommendations on sustainable economic development. Reflecting the high-public approval rating of President Bush at the time of the recommendations, most PCEQ members expected that this commission on sustainable economic development would be established in a second Bush Administration.

However, when Bill Clinton won the 1992 presidential election, the PCEQ recommendations were passed on to his Administration. PG&E's CEO was invited to join President Clinton's Council on Sustainable Development (PCSD). As with the PCEQ, PG&E was an integral player on the PCSD along with the executive director of the NRDC; and again, PG&E was the only utility represented on this national council.

PG&E had become a nationally recognized environmental leader in the energy utility industry. As such, PG&E was invited to participate in policy-level discussions related to the energy industry, its regulations and regulators. Along with representatives of other stakeholders in the energy sector, the environmental community and government, PG&E was able to influence the shaping of the regulatory policies under which it would operate in the future.

Trust, as well as norms of engagement and reciprocity [8], had been established between PG&E and the NRDC in the late 1980s and blossomed in the 1990s as the frequency of interaction between the organizations increased. With this increased frequency of interaction and the high value attributed by both organizations on their future interactions, cooperation between the organizations evolved consistent with Robert Axelrod's theories on the evolution of cooperation [13]. The NRDC would call upon PG&E when it needed an entry, heretofore unattainable, into certain portions of the business world. Similarly, PG&E would call upon the NRDC when it needed to establish relations with an organization in the environmental community that had been unapproachable.

In addition, the relationship between PG&E and the environmental community spread well beyond the NRDC. Solid working relationships continue in place today between PG&E and the Sierra Club, Environmental Defense, The Nature Conservancy, the Trust for Public Land and others.

To date, the culmination of the financial benefits to PG&E for the relationship that began in the collaborative effort in 1989 was the environmental community's, and especially NRDC's, key role in supporting the 1996 legislation to restructure California's electric utility industry (AB1890). Of most concern to PG&E and the other investor-owned utilities in California was the opportunity to recover the costs already incurred for the construction of power plants and related facilities and included in the utilities' rate base. Notwithstanding the fact that the costs that were included in the utilities' rate base had already been through the regulatory agencies' prudency review process, there were those that advocated no utility recovery of these costs — that the utility shareholders should bear them. While the exact value of this opportunity to recover these costs is arguable, the range of benefits to PG&E and its shareholders was $1–$4 billion.

But, like many stories this too was to get muddier. The electricity crisis of the late 1990s, in conjunction with the partial "deregulation" of the electricity market (the wholesale market was deregulated but not the retail market), forced PG&E into bankruptcy. In mid-2004, PG&E emerged from bankruptcy and once again seems to be on solid financial footing.

Another issue to consider is whether these benefits would have been forthcoming even if it were not for the collaborative planning process undertaken in 1989 and the subsequent joint activities, which increased the social, political, and intellectual capital of both organizations. The question is really academic, since an ex post facto analysis would be impossible. Perhaps a more fundamental question, and one that could be answered, could be posed to the current CEO of PG&E and Executive Director of the NRDC — would they make the same decisions, if they were able to relive the 1989 situation? Without having asked the question, I feel confident that both would answer emphatically in the affirmative, if for no other reason than for risk management.

2.2.2. Collaboration in action: a regional environmental management system in Silicon Valley

The Sustainable Silicon Valley initiative (SSV) began as an idea discussed in a white paper drafted by one of the members of the Board of Directors of the Silicon Valley Environmental Partnership (SVEP), a public benefit organization in Silicon Valley that focuses largely on the development and dissemination of data related to the condition of the environment in Silicon Valley (generally defined geographically as Santa Clara County, California in the San Francisco Bay Area). In essence, SSV's idea was to develop and implement a regional environmental management system (EMS).

While some of the most environmentally responsible corporations are implementing EMS, it is not a widely adopted practice. An EMS is similar to other management systems widely used in organizations, such as financial management systems, human resource management systems and others. Management systems usually entail a planned approach to managing issues/resources and striving for improvement. The more comprehensive systems encompass organizational policy, planning, implementation and operation; checking and corrective action; and management review. Management systems have three common components: an assessment of the current conditions; targets and timelines for the desired conditions; and specific action steps that will achieve the targets within the timeframes identified. Similarly, a comprehensive EMS is a planned approach to manage an organization's resource use and the environmental consequences of its activities (environmental pressures) while improving environmental performance.

The California Environmental Protection Agency (Cal/EPA)[1] embraced the SSV idea and assumed a pivotal leadership position. Recognizing the need to include the business community as a key stakeholder if actions were to be taken to achieve the targets inherent in an EMS, Cal/EPA approached the Silicon Valley Manufacturing Group (SVMG), an organization representing almost 200 members that works cooperatively with government officials to address major public policy issues affecting the economic health and quality of life in Silicon Valley. Cal/EPA presented a more detailed proposal to SVMG to create an environmental sustainability-based EMS for Silicon Valley. The SVEP was brought back in to participate in the development of EMS and to act as the principal source of, and repository for, project data. An SSV Development Committee, consisting of representatives from these three sectors (business, government and the environmental community) and other interested parties, was formed to more fully develop this proposal and introduce it to a wider audience.

The initial meetings of SSV included representatives from some of the most easily recognized corporations in the Silicon Valley high-tech industry; several locally based environmental groups; academics from San Jose and San Francisco state universities; and local, regional and state government. The participants included policy experts, attorneys and engineers. Some of the participants were there to protect their organizations' positions and/or interests. Others were there to help meet their organizations' position of goading industry

---

[1]The California Environmental Protection Agency (Cal/EPA) is a cabinet-level office reporting to the governor with a mission to restore, protect and enhance the environment, to ensure public health, environmental quality and economic vitality. Cal/EPA comprises six Boards, Departments and Offices: the Air Resources Board, Department of Pesticide Regulation, Department of Toxic Substances Control, Integrated Waste Management Board, Office of Environmental Health Hazard Assessment and State Water Resources Control Board.

into being more environmentally responsible. Most had an idea of what should be done, although there was little overlap in these ideas. Many also had an idea what should not be done; for example, business people knew that their businesses should not be hamstrung.

Rather than focus on *how* things should be done, the participants initially engaged in social learning by discussing *what* should be done in the broadest sense, the thought being that they can decide how to do it after they decided what they wanted to achieve. After a series of meetings, the group, which involved into the "Organizing Committee" for the SSV project, came to consensus on what they wanted to do:

- Ensure that Silicon Valley's air, land and water are maintained and/or improved in order to maintain human health and ecological stability, and ensure that human use of natural resources and systems remains within sustainable limits.
- Identify the trends in environmental changes in Silicon Valley, the current state of environment, the impacts that represent risks to human and environmental welfare, and the driving forces and pressures producing those risks.
- Identify joint and collaborative actions that can be taken to reduce the pressures and monitor the effectiveness of these responses.
- Determine the efficacy of a regional EMS.

As meetings continued with the same general group of participants, social and intellectual capital began to evolve. The people comprising the SSV Organizing Committee began to trust one another and accept the norms of accepted behaviors and expectations for their engagement. They began also to learn from one another. And, they reached consensus on the process to be used to determine how they should do what they wanted to do:

- Develop a broad partnership of stakeholders representing business, environmental groups, government, private citizens and others in Silicon Valley to create an environmental and resource sustainability management system for the region.
- Undertake collaborative projects involving businesses, government agencies, non-government organizations and private citizens to significantly reduce the environmental or resource impact in selected areas.

The SSV Organizing Committee agreed that the EMS should encompass business and public sector processes and the products, services and wastes they produce as well as the ecological efficiency of the material, human and other resources used in those processes and products. Significantly, they agreed that the EMS would not focus on the discrete operations of specific organizations, but rather on the environmental risks and impacts resulting from the operations of these organizations, grouped into broad categories.

With the initial stages of planning behind them, the SSV Organizing Committee was now ready to begin the work of: (a) measuring the current state and the trends in environmental changes in Silicon Valley; (b) determining which environmental pressures to focus on; and (c) agreeing on the appropriate goals and timelines to include in the EMS.

The Organizing Committee started with what was already available — the information developed by the SVEP for its *1999 Silicon Valley Environmental Index*, a report listing 23 environmental indicators for the Valley. This list was expanded by the participants to the 40 environmental and resource impacts (also called "pressures" or "aspects") identified on Table 1.

Recognizing its earlier commitment to be inclusive (a broad partnership of stakeholders representing business, environmental groups, government, private citizens and others in Silicon Valley) the organizing committee arranged for an email letter, signed by the secretary of Cal/EPA, the CEO of SVMG and the chair of SVEP, to be sent to 450 people in the Valley inviting them (a) to forward the letter to others they thought would be interested and (b) to participate in a two-phase web-based public input process.

In the first phase [14], participants were invited to review the list of significant environmental and resource impacts posted on a website and add to it any other impacts. With the input received, the organizing committee reviewed, revised and repackaged the environmental and resource impacts as indicated in Table 2.

In the second phase (February 2002) participants were invited to prioritize the environmental pressures (Table 2), either positive or negative, that may exert pressure on the natural surroundings, living conditions and economic well-being of Silicon Valley. A short descriptive paragraph was included for each pressure. Participants were asked to rank each pressure according to significance of its impact using the following choices:

- *Do More.* I see this pressure as a significant threat or a significant opportunity to address a threat. New policies and actions should be developed by governments, businesses and others, and significant additional resources should be allocated.
  (a) Threat: Current policies and actions are insufficient to address this pressure.
  (b) Opportunity: New policies and actions are needed to take advantage of this opportunity.
- *Watch.* I think this pressure poses a potential problem, or current understanding is inadequate. This could become serious, so governments and businesses should apply sufficient resources to monitor it carefully.

Table 1
Significant environmental and resource impacts

Impact	Impact
Air quality	*Water quality*
Air pollution — ozone generating releases	Watershed health — controlled and accidental releases to surface waters
Air pollution — particulate emissions	Controlled and accidental releases to groundwater
Air pollution — indoor	Biological contamination of drinking water
Controlled toxic chemical releases	Chemical contamination of drinking water
Accidental toxic chemical releases	Urban run-off to rivers and streams
Global warming gas releases	Controlled toxic chemical releases to water treatment plants
Ozone depleting gas releases	Unregulated toxic chemical releases to water treatment plants
	Particulate discharge to water bodies
Hazardous materials	
Hazardous waste generation	*Discarded materials*
Hazardous waste recycling or incineration	Discarded material /product disposal – business
Household hazardous waste collection	Discarded product disposal — consumer
Hazardous waste land disposal — controlled	Discarded product — reuse/remanufacture
Hazardous waste — unregulated disposal	Discarded product recycling
Superfund sites — uncontrolled releases	Use of recycled raw materials
	Pesticide and fertilizer runoff
	Species and habitats
Resource use	Loss of indigenous species and communities
Energy use	Non-native species introduction
Water use	Tidal marsh ecosystem services reduction
Water recycling	Terrestrial habitat destruction or fragmentation
Urban land use	Riparian habitat destruction or fragmentation
Urban brownfield reuse	
Land and open space	
Loss of undeveloped space	
Loss of unique natural features	
Loss of agricultural land	

- *Maintain.* I believe this pressure is safely under control. Recent trends have been positive and are expected to continue in the future. Current government and business policies are addressing the pressure, and a strategy of continual improvement within current resources should be pursued.
- *Don't Know.* I don't have enough information to form an opinion.

Table 2
Significant environmental pressures

Impact	Impact
Materials and resources	*Materials management*
Use of energy from non-renewable sources	Material waste—business
Use of energy from renewable sources	Material waste—consumer
Use of fresh water	Construction material waste
Use of recycled water	Organic material waste
Use of non-renewable raw materials	
Use of renewable/sustainably harvested raw materials	*Releases to water*
	Discharges of toxic chemicals to surface and groundwater
Releases to air	
Ozone generating releases (ground level)	Run-off to surface waters
	Non-permitted discharges of toxic chemicals to sewage systems
Ozone-depleting releases (upper atmosphere)	
Particulate emissions	*Hazardous materials*
Acid rain generating releases	Production, use, storage and transportation
Discharges of toxic chemicals to the air	Disposal
Greenhouse gas emissions	Accidental releases
Indoor air pollution	
	Species and habitat
The built environment	Habitat development or fragmentation
Urban sprawl	Introduction of non-native species
Availability and cost of housing close to work	
Transit system availability	*Governance*
Use of urban brownfields for redevelopment	Regulations
Urban development of flood plains	Coordination of improvement efforts
	Public education
	Land use decisions
	Regional planning

Armed with this public input, a subcommittee of the SSV Organizing Committee was asked to prioritize the 35 environmental pressures (Table 2) by considering the following criteria:

- Environmental concern as determined by probability, intensity and duration:
  (a) Probability – the likelihood of occurrence of impact from the pressure;
  (b) Intensity – the magnitude of impact on human health or environmental health; and
  (c) Duration – the time for impact of the pressure to dissipate.

- Effectiveness of existing institutional controls; and
- Public concern.

   After several meetings and much discourse, the subcommittee reached consensus on the six highest priority environmental pressures:

- Use of energy from non-renewable sources
- Use of fresh water
- Urban sprawl
- Habitat development and fragmentation
- Use of non-renewable raw materials
- Discharges of toxic chemicals to the air.

   Recognizing the reality of SSV's very limited resources, the Organizing Committee agreed to address all six of these environmental pressures but address them sequentially rather than concurrently. Energy use was selected as the first issue of focus.

   The SSV Energy Subcommittee, later renamed as the CO_2 Subcommittee, met through the remainder of 2002 and into 2003. By engaging regularly and consistently, this ideologically diverse group of individuals representing a similarly diverse group of organizations was able to agree on a target and timeline for addressing the first of the environmental pressures in the SSV regional EMS. The target and timeline was publicly announced in April 2003. The participants in SSV would strive to reduce CO_2 emissions in the Valley by 20 percent by 2010, using 1990 as a base year.

   CO_2 emissions were selected as the metric rather than Btu's or other units of energy because, from an environmental perspective, both the amount and environmental impact of energy use are important. The carbon content of fossil fuels can be used as a proxy for the relative environmental impacts of their combustion. While some emissions occur naturally when a fuel is burned and can be cleaned or "scrubbed" (e.g., nitrogen oxide and oxides of sulfur), the carbon in these fuels cannot, and is released, most often as CO_2. In addition to serving as a proxy for the relative cleanliness of different fossil fuels, the carbon compounds released when they are burned (most significantly CO_2) have been implicated as major contributors to global climate change. Therefore, CO_2 emissions are representative of both energy efficiency (the amount of energy used per unit output) and energy effectiveness (the relative carbon intensity of fuels or the amount of renewable energy used).

   Over the course of the next 12 months, the CO_2 Subcommittee continued to meet and expand its social and intellectual capital by working on the reporting protocol for the CO_2 emissions reduction target and recruiting organizations to pledge to help SSV meet its target. In March 2004, at a press conference in San

Jose City Hall, SSV announced the initial pledging organizations, the names of which follow:

- Hewlett-Packard
- LifeScan
- Oracle
- City of San Jose
- Lockheed Martin Missiles and Space Company
- Calpine
- Santa Clara Valley Water District
- Akeena Solar
- NASA Ames Research Center
- Pacific Gas and Electric Company
- Alza Pharmaceuticals

The SSV CO_2 Subcommittee continues its efforts to recruit additional organizations to help meet the CO_2 emissions reduction target. A September 2004 press conference is expected to announce a doubling in the number of pledging organizations, followed by another press conference in 4–6 months announcing another doubling. Organizations being approached include large and small businesses; colleges and universities; and municipalities.

Over the next several years, the CO_2 Subcommittee will compile annual reports from pledging organizations to measure progress toward the regional goal, and report the results and highlights of energy efficiency and CO_2 emissions reductions at public events and on websites.

The SSV Organizing Committee continues its efforts to develop the regional EMS through a multi-stakeholder collaborative process and is now focusing on the second of the six highest priority environmental pressures — fresh water. The SSV publicly announced its water initiative at a workshop in April 2004. The workshop, which focused on water quality, quantity and reliability issues, was attended by about 100 participants from industry, government, non-government organizations and the public. Participants provided input relating to a vision; goals, objectives, and targets; strategies; and metrics and challenges.

The SSV participants in the water initiative are currently identifying goals and strategies to ensure the sustainable use of fresh water in the Valley, and will be identifying specific practices and monitoring key indicators of progress toward the goal of water sustainability.

Starting in 2006, SSV will begin addressing the next environmental pressure from the top six, either urban sprawl, habitat development and fragmentation, use of non-renewable raw materials, or discharges of toxic chemicals to the air. The specific pressure to address next will again be chosen by group consensus.

The SSV has refined and rearticulated its mission: "to collaborate with organizations and the community to achieve environmental sustainability in Silicon Valley" [15]. SSV is using existing social, economic, business, and regulatory structures to achieve its mission. It is making decisions based on consensus, building relationships among all participants, and acting in ways that are collaborative and supportive, rather than adversarial. The SSV has now progressed through planning, data gathering and priority setting stages, and is moving into implementation.

3. Lessons learned and applications for contemporary engineers

Perhaps without realizing it, the participants in the SSV project and the PG&E/NRDC effort have put into practice the theories of social learning. In these efforts they have shed the paradigm of old, that of rational/technical planning and engineering, and embraced multi-stakeholder collaboration. They have developed social, intellectual and political capital and have achieved results that likely would not have been achieved without recognizing the social and political aspects of group problem-solving.

Tradition and the theory of reality would have led the engineers and negotiators involved in these efforts to rely upon their traditional methods of negotiation. With the traditional goals agreed upon and the means to achieve the desired end known, Karen Christensen would likely characterize the role of the leaders in these cases as programmers, rule-setters, regulators, analysts and administrators. Yet these leaders were willing to go beyond tradition. They were willing to revisit the agreed-upon goals and redefine the goals. They were willing to go beyond the known technologies for achieving the goals and into heretofore-unknown mechanisms. The role of these "daring" people would likely be characterized as charismatic leader and problem-finder [16]. These leaders seized the opportunity to frame the problem to be solved and shape the operative paradigm of the problem-solving effort. The results warranted the risk. Contemporary engineers need to be prepared and properly skilled to take similar risks.

References

[1] J. Innes, D.E. Booher, Planning Institutions in the Network Society: Theory for Collaborative Planning, Paper prepared for presentation at the Colloquium "Revival of Strategic Spatial Planning" sponsored by the Royal Netherlands Academy of Arts and Science, Amsterdam February 25–26, 1999.

[2] J. Friedmann, Planning in the Public Domain: From Knowledge to Action, Princeton University Press, Princeton, NJ, 1987.

[3] J. Friedmann, The Good Society: A Personal Account of its Struggle with the World of Social Planning and Dialectical Inquiry into the Roots of Radical Practice, MIT Press, Cambridge, MA, 1979.

[4] D.C. Korten, Community Organizations and Rural Development: A Learning Process Approach, Public Admin. Rev. 40 (5) (1980) 480–511.

[5] D.C. Korten, F.B. Alfonso (Eds), Bureaucracy and the Poor: Closing the Gap, Asian Institute of Management, Manila, 1981.

[6] The Editors of RAIN, Knowing Home: Studies for a Possible Portland, RAIN, Portland, 1981.

[7] J. Gruber, Coordinating Growth Management through Consensus Building: Incentives and the Generation of Social, Intellectual and Political Capital; Institute of Urban and Regional Development Working Paper, 1994.

[8] R.D. Putnam, Making Democracy Work: Civic Traditions in Modern Italy, Princeton University Press, Princeton, NJ, 1993.

[9] S.C. James, Social Capital in the Creation of Human Capital, Am. J. Sociol. 94 (Suppl.) S95 (26) (1988).

[10] R. Fisher, W. Ury, Getting to Yes: Negotiating Agreement Without Giving In, Houghton Mifflin Company, Boston, 1981.

[11] Report of the Statewide Collaborative Process: An Energy Efficiency Blueprint for California, 1990.

[12] Pacific Gas and Electric Company's 1996 Environmental Report: Environmental Quality.

[13] R. Axelrod, The Evolution of Cooperation, Basic Books, Inc, 1984.

[14] Sustainable Silicon Valley meeting minutes, New York, October 15, 2001.

[15] Sustainable Silicon Valley SSV Fundraising Proposal, June 2004.

[16] K. Christensen, Coping with Uncertainty in Planning, J. Am. Plann. Assoc. 51 (1) (Winter 1985) 63–73.

Sustainability Science and Engineering: Defining principles
Martin A. Abraham (Editor)
© 2006 Elsevier B.V. All rights reserved
DOI 10.1016/S1871-2711(05)01012-3

Chapter 12

Be Creative: Develop Engineering Solutions Beyond Current or Dominant Technologies, and Improve, Innovate, and Invent Technologies to Achieve Sustainability

Heather M. Cothron

Engineering and Infrastructure Business Unit, Science Applications International Corporation, 151 Lafayette Dr., Oak Ridge, TN 37831.

1. Introduction

The wide-spread application of sustainable engineering principles is still in its infancy. Although there are many examples of sustainable design, construction, and other applications, there are still many issues to resolve before these principles become embedded in the typical business approach and culture.

In this chapter we will explore some innovative and creative ways sustainable engineering technologies have been applied. Each of the case studies herein has incorporated, to some degree, the holistic principles that will eventually lead to true "cradle-to-cradle" sustainability. These examples of sustainable engineering provide a wide range of applications, from technology intensive solutions to more traditional approaches used in construction engineering. This chapter differs from others within this book in that it incorporates a number of individual examples, conducted and described by individual authors, to provide a broader understanding of the role of creativity in green engineering.

The diverse set of case studies presented includes an internet-based evaluation, approval and control mechanism for chemical usage and waste generation at a US Department of Energy national laboratory; the life cycle assessment of a brewery in Estonia using a computer model to optimize sustainability and efficiency; and the conversion of large quantity process waste streams in India (gypsum and iron oxide) into valuable commercial products. Each of these case studies presents a new approach to old "waste" issues, whether addressing the potential value of a traditional solid or hazardous waste or the inefficiency in processes that leads to wasted resources. As the case studies illustrate, the savings, both economic and environmental, can be substantial especially if adapted to large-scale operations.

The challenges of sustainable building design and construction are also highlighted in two case studies. The Department of Energy has estimated that the cost of heating, cooling, and lighting in residential and commercial buildings could be reduced by 80% using only currently available technologies. When coupled with emerging technologies, the potential for energy and cost savings is a compelling incentive. However, there are still many hurdles to tackle for the wholesale application of sustainable building practices. The development and testing of alternate building products made from recycled materials is explored in one case study. In addition, participation in a unique competition to design and construct a house powered only by solar energy is reported.

All of the case studies in this chapter represent a small effort within a much larger emerging trend. Engineering solutions beyond current technologies are being developed and transferred into the commercial and industrial sector more rapidly than ever before. However, there is still substantial need to further embrace the ideals represented by sustainable engineering and apply this thinking to both the developed and developing world community. Sustainability can no longer remain an ideal for the future; it must be transformed into ideas for the present, ideas such as those presented in these case studies.

2. Case study i: web-based chemical purchasing system—facilities operations go lean and green[1]

Heather M. Cothron

Engineering and Infrastructure Business Unit, Science Applications International Corporation, 151 Lafayette Dr., Oak Ridge, TN 37831, U S A.

Oak Ridge National Laboratory (ORNL) has an inventory of thousands of chemicals, ranging from exotic reagents for specific research to common household cleaners. A wide variety of these chemicals are used by the Facilities and

[1]The submitted manuscript has been authored by a contractor of the U.S. Government under contract No. DE-AC05-00OR22725. Accordingly, the U.S. Government retains a nonexclusive, royalty-free license to publish or reproduce the published form of this contribution, or allow others to do so, for U.S. Government purposes.

Operations (F&O) Directorate, which is responsible for maintenance, upgrades, utilities, and services for all infrastructure, buildings, and grounds at ORNL. As part of an initiative to "green" the laboratory, F&O management has implemented a web-based system to control the purchase and use of chemicals and to reduce the number and volume of chemicals that generate hazardous waste. Chemicals are evaluated prior to purchase so hazardous waste generating chemicals can be identified and "green" substitutes identified.

2.1. Way we were...

Prior to this initiative, operations personnel used many chemicals that were flammable, corrosive, or solvent containing. Operations planners purchased chemicals with little evaluation or waste planning. In addition, each shop area operated independently, procuring and storing chemicals based on the shop manager's criteria. As a result, the use of many products generated hazardous waste.

The chemical ordering system was developed to limit the generation of hazardous waste by requiring pre-purchase disposal path evaluation of chemicals and to provide easy access to essential environment, safety, and health (ES&H) information. Figure 1 illustrates the type of chemicals used by F&O before the chemical purchasing system was developed.

2.2. System design and development

The chemical ordering system was *envisioned* as a holistic approach to chemical management. The system is designed to accept a request for a chemical approval

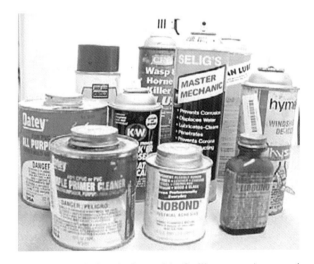

Fig. 1. Typical chemicals used in facility operations work.

or a purchasing order for approved chemicals from laboratory desktop computers and to automatically direct the request to the appropriate personnel. The electronic chemical approval request forms are automatically forwarded by e-mail to specified evaluators for review, signature, and database entry. ES&H experts evaluate the requests as they are generated. For consistency, the evaluators use a list of possible evaluation choices for waste disposal path, safety and health issues, handling , use, and storage. Approved chemicals are added to the database and can then be purchased online. Each chemical's ES&H information and approved use is stored in the database. The system is user friendly, and contains extensive prompts for the required input from the requester for both approval and procurement.

2.3. System capabilities

The online chemical procurement system allows the user to:

- Browse a web-based, color-coded list of approved chemicals and easily assess the approved "green" chemicals (hazardous waste-generating chemicals are color coded red).
- Order approved chemicals online through an internal bench stock location.
- Request approval of a chemical that is not on the approved list.
- Request re-approval of a chemical for which approval is about to expire or has expired. (Chemical approval length varies depending on hazards and waste associated with the chemical.)
- Receive automatic e-mail notification of chemical approval expirations that will occur within 30 days.
- View an online list of chemicals (updated daily) for which approval will expire within 30 days.

2.4. Implementation

Initial implementation plans targeted large quantity solvent processes/usage and janitorial chemicals as the first groups of products to evaluate and populate the database. Engineering flow diagrams of paint shop and fabrication processes were developed and modified to significantly reduce waste generation. For example, the paint shop researched and purchased computer-based sign making equipment to completely replace the labor- intensive process using large quantities of solvent-based products. This change essentially eliminated the hazardous waste associated with this operation.

System operation was planned to be as user friendly as possible by using icons and information prompts with embedded explanations. To request approval, chemical users complete a short electronic form. The form is then forwarded to evaluators, who assess the ES&H aspects of the requested use of the chemical and code the request with the appropriate information. Evaluator, requester, and manager signatures are obtained, and the evaluation is entered into the database or a substitution is suggested. The managerial level of signatures depends on the hazards associated with the chemical; division directors approve the use of, and accept accountability for, potential hazardous waste generation.

Following evaluation, approval and database entry, the chemical is then available for purchase. To purchase an approved chemical, an electronic procurement form must be filled out from a desktop computer. When the order is placed, the electronic form is automatically sent to procurement clerks. The clerks fill the orders from bench stock or through the laboratory-wide procurement system. Orders are typically delivered to the requester within one week or less, although rush orders can be filled in a day.

At the present time, signatures must be obtained on paper, and hard copies are retained for backup documentation. Upgrades and improvements that are currently in progress include complete automation of the system with electronic approvals and backup, substitution of all hazardous waste-generating janitorial chemicals with "green" products, and establishment of direct desktop links between the facilities operations chemical system and the laboratory-wide procurement system.

2.5. ES&H evaluation

The F&O chemical procurement database contains essential ES&H information specific for each chemical use request. Because a portion of the work involves hazards, such as radioactivity, that are site-specific, many of the evaluations are tailored to conditions and disposal options encountered only at ORNL. Providing specialized information, such as site-specific requirements for radiological waste generation, assists with efficient planning and execution of project work and multiple skill job coordination. In some cases, the work process has been reevaluated and the radiological waste stream completely eliminated.

The ES&H data can be accessed and printed by clicking on the name of the product in the approved list of chemicals. The information provided includes:

- Approved use of the product, the requester's name, and names of approving managers.
- Waste path information, including classification (hazardous or non-hazardous) and US Environmental Protection Agency (EPA) waste codes, collection, packaging, and staging requirements, and disposal or recycling options.
- Health and safety information, including industrial health evaluation, chemical-specific Occupational Safety and Health Administration training requirements, and safety support information.

- Chemical handling and use limitations, precautions, and requirements.
- Chemical storage requirements and incompatibilities with other chemicals.
- Direct links to the ORNL Material Safety Data Sheet system, the ORNL Hazardous Material Inventory System, and desktop training on the use of the system.

2.6. Benefits

Many of the benefits of the web-based system can be directly measured. Thus, there are specific criteria, such as cost savings, that have been used as indicators of success. Other benefits, such as health and safety aspects of using less hazardous materials are more difficult to quantify. The primary advantages of the chemical purchasing system include decreased generation of hazardous waste and associated transportation and disposal liability and cost; increased accountability for the generation of hazardous waste; one-stop shopping for identifying, selecting, and procuring environmentally preferable chemicals; decreased inventories of hazardous chemicals in individual shop areas; reduction in the variety of hazardous chemicals used by operations personnel; and elimination or extreme reductions in the most common solvent-based chemicals, such as paints, paint strippers, and lubricants.

2.7. Summary and conclusions

The web-based chemical ordering system developed by F&O provides one-stop desktop shopping for environmentally preferable materials. The web-based system allows users to easily access ES&H information and to request approval for and order chemicals online. Primary benefits include management accountability for waste planning and disposal, decreased generation of hazardous waste, and a projected annual savings of ~$40,000. Other benefits include the potential among workers for reduced chemical exposure, increased ES&H awareness, and improved accountability for chemicals. The overall percentage of hazardous waste–generating chemicals in use by facilities operations continues to drop as more and more "green" substitutes are identified and process re-engineering continues.

The success of the desktop system has generated interest in demonstrations by other organizations at ORNL. Pollution prevention proposals have been prepared to expand the web-based procurement system to some of the laboratory research organizations as a pilot project. ORNL is taking extensive measures to be a flagship of the environmental stewardship within the DOE system. The web-based chemical procurement system is just one initiative in the rapidly growing area of sustainable engineering, science, and research being conducted at ORNL.

3. Case study ii: life cycle assessment as a green engineering tool for internal decision making [1]

Siret Talve

CyclePlan Ltd., Leina str. 14A, 11612 Tallinn, Estonia

This case study outlines the assessment process used in the design and construction of a new brewery based on the decision to relocate operations and upgrade an existing old brewery. The old brewery had been in the same location for decades, but finally the decision was made to relocate outside the city and construct a completely new plant.

In the design stage of the new brewery, it was considered useful to study the old unit with respect to the sustainability and effectiveness of different internal partprocesses. In addition, the consumption of auxiliary process materials was revisited to address environmental aspects. The project was designed to assess the potential environmental effects of the different stages in the beer life cycle. The life cycle assessment (LCA) method was selected as the best tool for such a study.

The model system for the whole life cycle consists of distinct processes such as the cultivation of barley, malt production, production of other raw materials, production of auxiliary materials, transport of raw materials, beer production, transport of residues and beer, and transport of auxiliary materials.

A computer modeling tool called KCL-ECO (http://www.kcl.fi/eco/in-dexn.html) was used for all LCA calculations. KCL, a pulp and paper research company, originally designed this tool for performing LCA on wood and forest products. (KCL is owned by the Finnish pulp, paper and board industries and their task is to provide new information and research results to the owners.). In this investigation, the KCL-ECO was used for the first time to study the life cycle of a food product. It appeared to be easily adaptable to beer production and was useful for explaining and graphically presenting the results. This flexibility of analysis was due to a feature allowing the whole scheme of the life cycle, along with numerical data flows, to be depicted on the same figure.

For impact assessment, the KCL-ECO program has software that includes Ecoindicator 95 and DAIA (Decision Analysis Impact Assessment) 1998 methods. Because DAIA 1998 is considered to be a more suitable impact assessment method for Nordic conditions [2] it was chosen and used. In the environmental impact assessment climate change, acidification, eutrophication, oxygen depletion, and summer smog were included.

The environmental information and knowledge to conduct the assessment were spread between different specialists at various locations. Thus, it was assumed that required data were available, but in reality, some important data were missing. As LCA is a complex task that needs active inputs in the data collection phase from all staff levels in the production unit at the enterprise, the general environmental awareness of employees was increased. Also the links between the different production operations and resource conservation and waste generation were established and clarified. This demonstrated to all staff the need for communication and sharing of data on environmental issues.

The primary issue encountered while conducting this study was the varying data quality from enterprises other than the brewery that must be included in the life cycle of the beer. It was especially difficult to obtain data from the central Europe and US companies, even when requested by the brewery (i.e. client). To ensure a thorough LCA for any product, cooperation between different enterprises within the life cycle of the product should be obtained.

An important task of this study was to determine the resource consumption within the brewery, based on a consistent unit of production, defined as 505 multipacks of bottled beer (10 hl of beer) in the shop. Because it was possible to monitor the electrical consumption at all key-points within the production system, this aspect was investigated in fine detail (see Fig. 2). Likewise, all the waste flows of the brewery were mapped (for example see Fig. 3). Finding ways to diminish waste generation and to reduce electricity use in the brewery will provide both environmental and economic benefits. The collected information serves as the basis for such improvements.

The modeled LCA results are presented in Table 1 It is important to note that the highest projected environmental impact scores originate outside of the brewery—from agriculture. This prediction underlines the need for cooperation and exchange of information along the product chain in order to improve environmental performance of the end product.

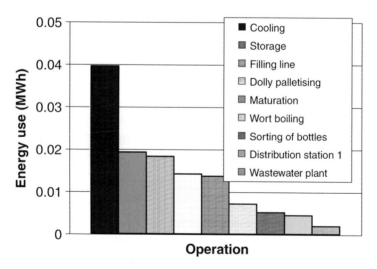

Fig. 2. Electricity consumption in the brewery, MWh per functional unit.

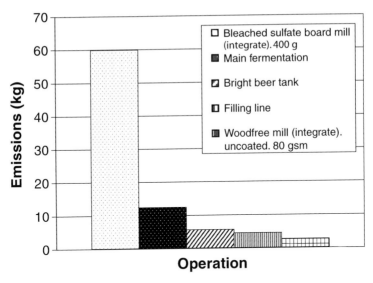

Fig. 3. Emission sources of biogenic CO_2 from the investigated brewery system, kg per functional unit.

Table 1
Results of LCA-weighting (DAIA 98, scores of different impacts)[a]

	Agriculture	Production of beer in 6-pack	Production of raw material	Production of auxiliary material	Transport of auxiliary material	Transport of beer, residues, raw materials	Sum
Climate change	4.864	10.913	1.860	8.357	0.212	4.362	30.569
Acidification	4.172	6.456	2.341	7.909	1.399	22.410	44.686
Eutrophication	485.077	9.070	0.468	6.739	0.232	5.559	507.145
Summer smog	2.238	1.662	1.492	3.489	0.749	17.901	27.53
Oxygen depletion (COD)	0.016	2.832	0.003	10.528	0	0	13.3781
Sum	496.365	30.933	6.165	37.020	2.592	50.232	

[a]Reported values are $\times 10^{-5}$.

The suggestion to seek life cycle inventory product quality data from suppliers of auxiliary materials was made. These data would make it easy to update the LCA modeled results and then choose between various offers from suppliers. Traditionally, the transport of materials has been the suppliers' responsibility; the brewer has never before assessed the environmental effects of this activity. However, the LCA allows consideration of these emissions as part of the environmental effects.

The main conclusions and recommendations for improving the environmental and economic performance of the brewery are:

- Establish an environmental management system. To achieve this goal existing and new environmental information must be consolidated in one place so that it is readily accessible and allows identification of data gaps.
- Record data on energy use, natural resources (water) consumption, and emissions, both at the plant level and for each individual production line. By simultaneously monitoring and analyzing the resource consumption and waste generation of multiple processes, environmental and economical improvements can be identified.
- Redesign of the multipack cluster package should be considered for the product itself (beer in multipacks shipped on pallets) since the largest environmental impact is caused by production of cluster board.
- Optimize the product truck routes. Substantial environmental impact results from beer distribution. The multipack pallets should be designed to accommodate empty bottles returned to the brewery so the transport of empty crates is minimized.
- Search for local raw material alternatives. One of the main raw materials for the beer production is purchased and transported from Central Europe. Hence, the environmental impact of this long distance transport is significant.
- Separate bottles into individual crates at the shops. Approximately 9% of returned bottles are soft drink bottles, which must be separated at the brewery on a special process line. An entire process line could potentially be eliminated if the bottles were separated at the shops.

4. Case study iii: a novel process for the recovery of valuables from solid wastes [3]

A.N. Gokarn and S. Mayadevi

Chemical Engineering and Process Development Division, National Chemical Laboratory, Pune 411 008, Maharashtra, India

4.1. Introduction

The strategy used for waste elimination/minimization depends on the industry and the type of waste generated. A general approach to waste generation and its minimization for an operating industry is given in Fig. 4. Generation of waste can be reduced to a great extent by process optimization leading to improvement in the efficiency of operation. However, this process may not necessarily lead to "zero waste" processes. Effluents are unavoidable in most commercial

GENERATION AND REUSE OF WASTE

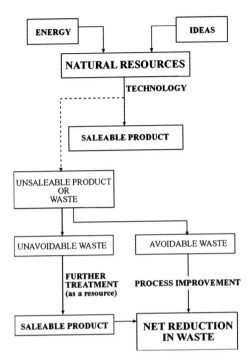

Fig. 4. Generation and reuse of waste: A general approach.

chemical processes; however, its treatment is typically an expensive, end-of-pipe process. Motivating industries to clean their effluents through the establishment and practice of stringent environmental norms by the government and regulatory agencies is not always successful. A more attractive way of motivating industries to clean up their waste, especially in developing countries, is to make waste reduction a revenue generating activity.

This strategy of generating revenue is termed value addition to the waste and has been investigated on a bench scale for two large waste streams: gypsum and ferrouso-ferric oxide. Gypsum is a waste generated in large quantity in the manufacture of phosphoric acid, in the neutralization of spent sulfuric acid, and in the manufacture of H-acid (1-amino-8-hydroxynaphthalene-3,6-disulfonic acid). These wastes are often disposed in landfills.

This case study describes high temperature processing of these two waste streams in the production of calcium ferrate and recovery of sulfur dioxide gas. It has been possible to achieve more than 90% recovery of sulfur dioxide in the laboratory. The other product, calcium ferrite, is used as a raw material in cement industry, pig iron refining etc. making this novel recovery process economically attractive and large-scale production would be a significant breakthrough in green engineering.

4.2. Generation of waste gypsum

Gypsum is a waste generated in large quantity in the manufacture of phosphoric acid, in the neutralization of spent sulfuric acid, and in the manufacture of many organic intermediates including the manufacture of H-acid. The main source of waste gypsum is the phosphoric acid plant. In the phosphoric acid plant, sulfur is converted to concentrated sulfuric acid and is used for digesting phosphate rock producing phosphoric acid and phosphogypsum ($CaSO_4$). In India, 4500–5000 kg of phosphogypsum are generated per ton of phosphoric acid. Roughly 30% of the imported sulfur that is used in the process ends up as waste phosphogypsum that is disposed in landfills. Waste gypsum produced by organic industries, as in the manufacture of H-acid, comes under the hazardous waste requirements due to the association with low levels of organic chemicals. Phosphogypsum is an unavoidable waste in phosphoric acid manufacture. In such a case, generation of negligible or zero waste is possible only by altering the process chemistry.

4.3. Generation of iron oxide sludge

Ferrouso-ferric oxide, Fe_3O_4, is generated as a solid waste in the reduction of organic nitro-group to produce amine-containing organic intermediates by the Bechamp reduction process. It is also a waste formed in the partial reduction of hematite in the iron and steel industry. In the manufacture of H-acid (1-amino-8-hydroxynaphthalene-3, 6-disulfonic acid), an organic intermediate useful for dyes, both gypsum and ferrouso-ferric oxide are generated. The raw materials requirement for the production of 1 ton of H-acid is given in Table 2. This process generates approximately 9000 kg of gypsum sludge (55–60% solids) and 2500–3000 kg of iron sludge (60–65% solids) per 1000 kg of H-acid produced. Both wastes are typically disposed in landfills.

4.4. Value addition to solid wastes – gypsum and iron oxide

The solid wastes gypsum and iron oxide can be reacted at elevated temperatures for the production of calcium ferrite and recovery of sulfur dioxide gas according to the following equation:

$$6CaSO_4 + 2Fe_3O_4 \rightarrow 3CaOFe_2O_3 + 6SO_2 + 3CaO + (5/2)O_2$$

The solid product calcium ferrite and the SO_2 gas are useful commodities.

In the process, stoichiometric proportions of these solid wastes in the form of dry powder were thoroughly mixed and shaped into cylindrical pellets using

Table 2
Raw material requirements per 1000 kg of H-acid produced (Basis: 1000 kg of H-acid) [4]

Raw materials	Quantity (kg)
Naphthalene	1000
Oleum	3000
Sulfuric acid	2300
Nitric acid	900
Soda ash	1100
Calcium carbonate	4800
Iron	1000
HCl	2000
NaCl	2500
Caustic lye solution	2300
Sulfuric acid (40%)	10,600
Total	31,500

die-punch without the use of a binder. The pellets were then weighed and transferred into a silica boat. The boat was pushed into a horizontally mounted silica reactor maintained at the desired temperature. An inert atmosphere was maintained in the reactor by passing nitrogen gas starting at the beginning of the experiment. Sulfur dioxide gas evolved during the reaction was absorbed in a bubbler containing excess hydrogen peroxide solution. The sulfuric acid formed ($SO_2 + H_2O_2 \rightarrow H_2SO_4$) was analyzed quantitatively as a function of time by titrating against standard 0.5 N sodium hydroxide solution. The yield of sulfur dioxide could be obtained from the amount of sulfuric acid generated. A 90% yield of sulfur dioxide could be achieved with a reaction period of 30–120 min when the reaction temperature was between 950 and 1100 °C. The cooled solid product was black in color with characteristics of calcium ferrite. The process details are schematically represented in Fig. 5. This novel and useful process developed in laboratory scale has been patented [5] and can be commercially developed after pilot-plant trials and refinements.

Calcium ferrite has many industrial and technological applications. It can be used in anticorrosive pigments [6], enhance the reducibility of iron ore pellets [7], and save energy in cement manufacture [8]. Millions of tons of waste gypsum can become a source of sulfur and lime for commercial applications. This sustainable process could result in large saving of dwindling natural resources like lime stone and sulfur.

4.5. Conclusions

This section describes a potential process using two hazardous solid wastes generated by chemical process industries to generate valuable, commercially

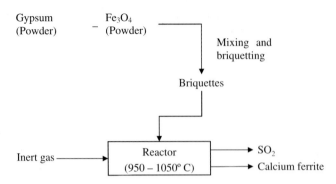

Fig. 5. Process details: Production of calcium ferrite and sulfur dioxide from waste gypsum and iron oxide.

viable chemicals. The products generated would have both technical and economic significance. Although the proposed process has been demonstrated only at a bench scale, the chemistry can be scaled to pilot—and full-size production.

It is possible to identify many similar systems where value addition to effluents can lead to products of commercial significance. The challenge is to have the vision and persistence to demonstrate the value of converting the wastes into resources to waste generators, environmental regulators, and commercial interests. The concept of sustainable engineering will be accepted in developing countries if positive economic impacts can be used as an incentive.

5. Case study iv: sustainable development of structural walls incorporating solid waste materials

Ahmed H. ElSawy and Joseph J. Biernacki

College of Engineering, Tennessee Technological University, Cookeville, TN 38505-0001,USA.

Sustainable development is a strategy that seeks economic industrial solutions that also benefit the environment and the quality of life. In response to the national goals for resources and energy conservation as well as efficiency improvement and reduction of greenhouse gas emission, sustainable cement-based composite construction blocks are being developed. The approach taken here is to develop high performance, cost-effective cement-based structural walls incorporating solid waste and by-product materials. A three-stage plan is proposed, which incorporates elements of full-scale demonstration, development of physics-based models and property characterization and economic analysis. A limited demonstration study as well as some characterization has been done at this time and is reported here along with a longer-term plan to develop a workable commercial technology.

5.1. Introduction

The objective of this research is to prove the feasibility of using different forms of solid waste materials and by-products of wood, glass (particulate and fiberglass), plastics, metals processing (spent foundry sand and blast furnace slag) and power generation (fly ash) to construct high-performance and cost-effective composite walls. The walls must be thermally efficient, sound proof, and have moisture and insect resistant structural members and be useful for both residential and industrial construction. High performance is defined here as having performance characteristics that are not typical of ordinary commercial building products. Since there are little data regarding the formulation of such non-traditional cement-based materials, this project identified and demonstrated one suitable model that can be used to provide guidance for next-generation engineered sustainable construction materials.

5.2. Sustainable wall demonstration project

Figure 6 illustrates the *sustainability cycle* for this project. Pine wood pallets are scrapped and stockpiled by the millions each year (a). These pallets can be chipped and mineralized to form a renewable coarse aggregate source. Different granulated solid waste materials (plastics, fiberglass, and glass) (b) were also

Fig. 6. Sustainability cycle for the sustainable wall demonstration project.

evaluated as partial aggregate substitutes for the fine aggregate (sand) typically used in Portland cement concrete mixtures, to produce new concrete composites [9,10]. Finally, these waste streams were utilized in the formulation of both concrete block (c) and structural concrete for filling block cavities (d) to construct a sustainable composite wall system.

Structural fill concrete was formulated by replacing a portion of the fine aggregate (sand) with one of three waste materials: plastic, fiberglass, or glass. Four different volume percentages of fine aggregate substitution (5, 10, 15, and 20%) were utilized for each of the three waste materials. A control, *ordinary* concrete, was also prepared. Strength tests were conducted after 28-days of curing. These tests included compression strength, splitting tensile, and flexural strength in accordance with ASTM standards.

The plastic aggregate was found to have reduced compressive, splitting tensile, and flexural strengths. On the other hand, the stiffness was almost the same as that of the control specimen. Glass-containing concrete composites had compressive and splitting tensile strengths comparable to that of the control specimen. In addition, the values of the modulus of rupture and elasticity of all the tested glass-containing concrete composites were almost the same as that of the control specimen. In the case of fiberglass-containing concrete composites, adding greater volume percentages of this aggregate reduced the compressive, splitting tensile, and flexural strengths of the concrete composite, yet, improved the stiffness.

The production of cement-based composite block [11] made from waste fiberglass in combination with mineralized wood chips, fly, ash and spent foundry sand as partial substitutes was also considered. The amount of fiberglass was increased in steps from 5% to 20% while decreasing the mineralized chips from 85%–65%. Proportions of fly ash and spent foundry sand were increased by 1% while decreasing the amount of Portland cement respectively to a max. of 5%. Combining fiberglass substitution in the range from 5%–20% with fly ash or spent foundry sand substitution in the range from 1%–5% was found to improve the mechanical properties over the control sample. Unfortunately, the use of solid waste substitution for wood chips increased the weight of the composite as much as 30% over the control specimen. The designed mixes that are considered here take advantage of the Pozzolanic nature of fly ash that consumes the undesired calcium hydroxide by-product of cement hydration and converts it into a dense cementitious phase. This should reduce water mobility and increase the strength of this product while at the same time improving the toughness (resistance to cracking) of these mixtures.

5.3. Commercial goals

Commercial goals for the proposed sustainable structural walls technology are driven by performance, production, and construction cost. This technology will result in significant energy savings, since R-values as high as 21 should be achievable and whereas conventional construction technology typically achieves an R-12 value. (A high R-value means greater resistance to heat flow.) Production cost of the sustainable wall-forms is expected to be less than for conventional concrete block, primarily due to the use of low-cost recycled materials. Finally, construction costs using the proposed wall-forms is expected to be competitive or lower than standard construction practices due to reduction in weight (handling costs) and ease of interfacing with other non-structural elements such as interior finishes. Proposed future projects will focus on quantifying each of these costs, providing data for projections and optimization, and are expected to lead to significant cost advantages for this sustainable technology [12,13].

5.4. Future work

There are two major challenges confronting the development of sustainable cement-based composites for construction: (1) What are the complex interactions between typical combinations of waste and waste by-products and Portland cement and how can we reliably formulate structurally sound and durable end products? (2) Which waste streams are consistent enough in composition and availability to be technically and economically viable?

The following are three tasks suggested to meet these challenges: Task I— conduct limited statistical optimization studies (similar to the case study presented here) to demonstrate feasibility; Task II—identify the major technical hurdles and define solutions; and Task III—conduct an economic assessment for production of cement-based structural walls using waste and waste by-products. As shown in Fig. 7, these three tasks must be supported by a three-fold approach

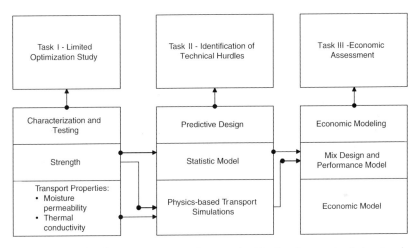

Fig. 7. Research and development strategy for sustainable development of structural walls incorporating solid waste materials.

of predictive design, characterization and testing, and economic modeling. The predictive design is driven by the use of both statistical and physics-based/first principle models. By combining computational tools, a predictive model for sustainable development of wall-forms could be developed in combination with focused experimentation to characterize mechanical performance (strength and toughness), microstructural interactions (physiochemical composite interactions) and transport processes (thermal conductivity, sound absorption and moisture permeability). Ultimately, economics are the driving force that enables this technology to be commercially successful. A comprehensive economic modeling strategy must be employed to integrate production and performance costs. The successful product will have an economic benefit realized as a reduction in manufacturing (production) cost and/or reduction in building costs by using the waste-based walls in construction (performance).

5.5. Acknowledgments

The authors would like to acknowledge the ORNL building thermal envelope systems and materials lab, the Tennessee Valley Authority (TVA), and Insul Holz-Beton International, Inc., the producer of commercial concrete construction components.

6. Case study v: engineering a solar house for a solar decathlon competition

Donald D. Liou

College of Engineering, University of North Carolina at Charlotte, 9201 University City Blvd., Charlotte, NC 28223-0001, USA

6.1. Introduction

The University of North Carolina at Charlotte (UNCC) students designed and constructed a solar house for the first Solar Decathlon Competition. This green building has 500 sq. ft. of conditioned floor area and is powered solely by the sun's energy. The house was one of the 14 entries in the competition, which was sponsored by the US Department of Energy, and was held in Washington, D.C.during the last week of September and the first week of October 2002 [14].

6.2. Engineering

6.2.1. Engineering goals
The competition is called Solar Decathlon because, for scoring purposes, the Department of Energy divided the project into ten different categories. The

categories included design and livability, design presentation and simulation, graphics and communication, the comfort zone, refrigeration, hot water, energy balance, lighting, home business, and getting around. In addition to addressing the Department of Energy requirements, the UNCC house was engineered with the following goals:

- to build a house that can be easily taken apart, easily transported to another location, and easily assembled for use in a new location;
- to build a house that uses only the sun's energy, and yet has enough energy to support normal living activities, such as washing, cooking, office work, and transportation;
- to build a house that not only requires no excavation, but also has enough strength to balance loads from nature and provide stability to the house itself;
- to design an energy-efficient house that not only has adequate insulation and moisture barrier, but also has enough lighting as well as other desired features of livability.

6.2.2. Course re-organizations

The project was developed to fit into the normal university course schedule for the fall semester of 2001, with progressive classes through fall semester 2002. Development of the project as individual classes for semesters during 2001 and fall semester 2002 precluded the normal design-bid-build process used by most commercial construction projects. The process had to be unique, one that was flexible enough to incorporate cash contributions and in-kind donations, and the fluctuation of faculty and student participations. With these objectives, the project advisors developed a combined engineering and architectural curriculum that was unprecedented. The contents of four upper division technical engineering courses were re-structured to incorporate various aspects of the project that were necessary to design and build the solar house.

6.2.3. Building envelope

While the architectural students developed concepts for the solar house, students in an *Electrical and Mechanical Building Systems* class were engineering the building envelope of the future solar house. Because the cooling load calculation for a building contains many components and is, therefore, much more complicated than the associated heating load calculation, it was decided that the basic design of the building envelope would be independent from the energy simulation. (The energy simulation was later performed by a mechanical engineering student team as a fulfillment of one of the Department of Energy requirements.) Therefore, the initial design was based on the heating load requirement alone.

A routine individual take-home assignment on heating load calculation was thus altered to facilitate the design process. Each student was required to estimate the heating loads, gather information on doors, windows and other building materials, and design the building envelope. Using a very sketchy floor plan, each student was required to solicit the cost associated with the building materials he or she used in the building envelope. Students were also encouraged to use the building materials, including building insulation panels, skylights, and Kal wall (a trade-marked transparent window panel), of those companies that exhibited interest in donating materials or providing engineering assistance to the project. After the information was gathered, two types of building envelopes were selected from the students' reports and summary sheets. The information and designs were subsequently passed on to engineering students in an individual-study class and a civil engineering technology capstone class, Senior Design Project. The participants of the solar house project were able to take all the classes sequentially or concurrently, which provided valuable continuity to the project.

6.2.4. Building systems

One or more student teams were organized within each of the class sessions. Initially each of these teams focused on the design or simulation of only one important building system without considering any system integration needs. During the actual design of the building systems performed later in the summer, the final design team adjusted the size of the originally conceptualized building systems by establishing a link between some of the units. For instance, a link was made between water heating and space heating and cooling. In this way, some of the excess heat exhausted from the house by the HVAC unit was captured and used to heat water in parallel with a small solar hot water panel on the roof of the solar house.

Certain key power-supply decisions needed to be made in the early stages of the project. These included the amount of solar power that would be provided by the solar panels, the type of current (i.e. direct current only or a combination of direct current and alternative current), and the associated voltages. After studying the issue, participating electrical-engineering technology students decided to use a roof-mounted solar photovoltaic (PV) power system that could provide 4.5 KW They estimated it could provide enough power for lighting, appliances, and charging an electrical vehicle during the competition. Ordinary appliances use 220 V and/or 110 V alternating current, while the PV panels produce direct current at different voltages. A conversion between direct current and alternating current is, therefore, usually required. The students tried to minimize the conversion need, contacting EQUATOR, a marine supplier, to provide a compact dishwasher, a dryer, and a clothes washer. These appliances require only 120 V direct current and would simplify the power requirements. A small 120 V stove was later installed. SEAWARD, another marine manufacturer provided a 120 V convection oven.

6.2.5. Design challenges

The engineering, architectural, and construction teams all worked in concert as they faced and overcame many challenges. They worked on a fast-track basis, meaning the design, fund raising, and construction of the solar house project were carried out at the same time. The most challenging aspect of the project was the need for the engineering team to make structural changes at the last moment.

The solar house was initially designed to use insulated walls and roofs panels produced by ThermaSteel Corporation and a beam-and-post timber system manufactured by a local timber company. The chassis and base plate of the house were delivered by R-Anell in early summer. The structural woods, including timber, joists, and rafters were due next. However, in early summer, the local timber company backed out. The company scheduled to deliver the insulated wall and roof panels also withdrew their delivery later in the summer. Modifications to the structural and building-envelope designs were quickly made to adapt to the new situations. Less expensive insulation panels made by Insulspan were identified and used. Wood beams and other structural elements were re-engineered and re-sized to adjust to the new loads. The revised design was based on the design manual by the Southern Pine Council. The substitution of the wall and roof panels' necessitated verification that the chassis design would accommodate the changes. The unit weights of the two panels were compared and it was safely concluded that the chassis would easily support the lighter Insulspan panels.

6.2.6. Transportation and foundation

The solar house was designed to accommodate multiple relocations, since it needed to be constructed in Charlotte, North Carolina, moved to Washington DC, for the competition, and then moved back to Charlotte. From a transportation viewpoint, one of the easiest ways to address this situation was to design the house as a mobile home. After consulting with manufacturing-home builders on several weight- and transportation-related issues, a decision was reached to design the house as an "irregular" mobile home. The design was irregular in that the chassis of the house would be longer and have more axles than most mobile homes, thus allowing more weight to be carried. The chassis would also be made of three connected sections, instead of one single section, allowing the tail two sections of the chassis to be uncoupled from the main section. This adjustment added flexibility to re-configure the house at a new location. Therefore, in the final design and actual construction, the chassis was made of three sections. The front section carried the main portion of the solar house, and the two tail sections the front porch and back sunroom. The temporary anchoring system used at the competition site was designed by the student team and was manufactured by a local steel manufacturing company. (At a permanent location, the solar house would be anchored similar to that of stationary mobile homes.)

6.2.7. Construction

The Solar Decathlon project provided a unique opportunity for students to work in a multi-disciplinary environment similar to that of the real world. This was particularly true during the construction phase of the project, which spanned from the summer session 2002 to the end of October 2002. Architectural and engineering students formed the core of the construction crews. The core crews planned, scheduled, and controlled some procurements, most material deliveries, and construction. Many building elements were donated to the project by outside companies. Other large items for the project were procured through regular university channels, as mandated by university rules. The teams were also aided by volunteer students.

Construction by the student team started with leveling of the steel chassis and floor plate of the solar house provided by R-Anell Homes. The insulated panels, provided by Panel Wrights Group and Jefferson Homes, were the first elements to be installed on the chassis. After the wall and roof panels were erected, the generative house began to take shape, which attracted positive attention and invigorated the students' enthusiasm. The VELUX skylights were installed next. This was followed by the placement of roof materials. After the roof was secured, the interior was framed, and some building systems were installed at the construction site. The balance, including the PV panels and water supply and circulation system, was later installed at the competition site.

6.3. Conclusion

This project clearly demonstrated the viability of sustainable residential buildings using commercially available products. Although many modifications to "normal" residential design concepts were necessary, the changes were relatively easy to incorporate.

The UNCC student teams successfully accomplished both the design and construction phases of the Solar Decathlon project. The completion of the UNCC solar house represents a significant accomplishment in the univeristy's sustainable engineering education. This unprecedented project was accomplished with extensive cooperation between various UNCC organizations and departments, the interest and dedication of the students and their professors, and the desire to tackle a complicated sustainable engineering challenge. Because the solar house was actually built as part of the course work rather than ending at the design stage, all who participated in the event have gained valuable "real world" experiences. This is especially true considering the extensive changes necessary to complete the project on schedule. The students were able to gain unique project management experience equivalent to that found in an innovative working environment. The students who participated in the Solar Decathlon represent the next generation of engineers. These engineers will be well prepared to pursue unlimited possibilities associated with sustainable engineering and design.

References

[1] S. Talve, Life Cycle Assessment of a Basic Lager Beer, Int. J. of Life Cycle Assess. 6 (5) (2001) 293–298.

[2] J. Seppälä, Decision Analysis as a Tool for the Life Cycle Impact Assessment, The Finnish Environment 123, Oy Edita Ab, Finland, Helsinki, 1997.

[3] A.N. Gokarn,S. Mayadevi, Treatment of industrial effluents for recovery of useful products, in: Proceedings of National Seminar on Recent Trends in Industrial Waste Treatments, Nagpur University, Nagpur, 13–14 January 2001, pp. 107–118.

[4] J.V. Hassler, W.E. McMinn, Ind. Eng. Chem. 37 (1945) 645.

[5] Indian Patent Application No. 1581/DEL/99, 28 December 1999.

[6] German Patent No. 2 827 638, 1979.

[7] Japanese Patent No. 89 136 937, 1989.

[8] M.S. Suvarna, S.N. Joshi, World cement, 18, 8 (1987) 342.

[9] I. Shehata, S. Varzavand, A. ElSawy, M. Fahmy, The use of solid waste materials as fine aggregate substitutes, in: Cementitious Concrete Composites, Semisequicentennial Transportation Conference Proceedings, Iowa State University, Ames, Iowa, May 1996. (http://www.ctre.Iastate.edu/pubs/semIsesq/session2/shehata/)

[10] I.H. Shehata, A.H. ElSawy, S. Varzavand, M.F. Fahmy, Properties of Cementitious Composites Containing non-Recyclable Glass as Fine Aggregate, Current advances, in: Mechanical Design and Production VII, Proceedings of the 7th International MDP Conference, Cairo-Egypt, (Eds) M.F. Hassan, S.M. Megahed, Pergamon Press, New York, February 2000, pp. 313–320.

[11] H. M. Vahradian, A. ElSawy, S. Varzavand, M. F. Fahmy, The Use of Solid Waste in the Development of Wood/Cement Blocks, in:Proceedings of the 7th International Conference on Production Engineering, Volume III, M. H. Elwany, M. Helaly, (Eds), Alexandria-Egypt, February 13–15, 2001 pp. 835–1846.

[12] R. Avula, Effect of Using Sawdust as Fine Aggregate Substitute on the Properties of Concrete Composite, M.S. Thesis in Chemical Engineering, Tennessee Technological University, December 2003.

[13] P. Muthukumaraswany, Thermal, acoustics and mechanical properties of concrete composites containing solid by-products, M.S. Thesis in Chemical Engineering, Tennessee Technological University, December 2004.

[14] D. Liou, Engineering a sustainable house for solar Decathlon 2002," Proceedings of International conference on sustainable engineering and science, Proceedings, Paper #25, pp. 1–10, Auckland, New Zealand, July 6–9, 2004.

Sustainability Science and Engineering: Defining principles
Martin A. Abraham (Editor)
DOI 10.1016/S1871-2711(05)01013-5

Chapter 13

Actively Engage Communities and Stakeholders in the Development of Engineering Solutions

L.G. Heine[a], M.L. Willard[b]

[a]*Green Blue Institute (GreenBlue), 600 E. Water St. Suite C Charlottesville, VA 22902, USA*
[b]*AXIS Performance Advisors, Inc., 2515 NE 17th Ave Portland, OR 97212, USA*

1. Introduction

1.1. What is green engineering?

According to the US Environmental Protection Agency (EPA), "Green engineering is the design, commercialization, and use of processes and products, which are feasible and economical while minimizing (1) generation of pollution at the source and (2) risk to human health and the environment. The discipline embraces the concept that decisions to protect human health and the environment can have the greatest impact and cost effectiveness when applied early to the design and development phase of a process or product" [1]. Green engineering has also been defined as "the design of systems and unit processes that obviate or reduce the need for the use of hazardous substances while minimizing energy usage and the generation of unwanted by-products" [2]. These definitions emphasize environmental protection via chemical engineering.

In May 2003, some 65 engineers convened in Sandestin, Florida and pushed the boundaries of green engineering beyond environmental considerations and chemical engineering to include all engineering disciplines and the broader scope of sustainability. They agreed that:

"Green Engineering transforms existing engineering disciplines and practices to those that promote sustainability. Green Engineering incorporates development and

implementation of technologically and economically viable products, processes, and
systems that promote human welfare while protecting human health and elevating
the protection of the biosphere as a criterion in engineering solutions" [3].

Having established this more expansive definition, it is logical to ask, how
does one practice green engineering? The nine principles that emerged from the
meeting in Sandestin, and continue to be forged, suggest how to think about
engineering problems and to design engineering solutions that promote
sustainability. While the terms "green" and "sustainable" are often used interchangeably, we consider the distinction between the EPA focus on pollution
prevention and risk reduction and the Sandestin focus on engineering that
promotes sustainability to be critical. Based on the 1987 report of the Brundtland Commission entitled, "*Our Common Future*", sustainability has come to be
defined as "meeting the needs of the present generation without compromising
the ability of future generations to meet their own needs." Operationally it
means synergistically optimizing environmental, economic, and social well-
being. The ideal of sustainability recognizes that the practice of focusing only on
environmental benefits to the exclusion of economic and social concerns, or on
economic and social concerns to the exclusion of environmental values, or any
combination not inclusive of all three concerns will result long-term failure, even
if there are limited short-term gains.

While engineering to protect the environment in the most economical manner
feasible has been within the purview of traditional engineering, especially in
environmental and chemical engineering, social well-being has historically not
been included as a design criterion. The US Declaration of Independence
declares we are endowed with certain "unalienable Rights, that among these are
Life, Liberty and the pursuit of Happiness". But the definition of happiness
varies between individuals, cultures, religions, ages, generations, genders, and
much more. The complexity of defining social well-being is well beyond the
scope of engineering, and so we often settle for environmental gains that are
economical and trust that by reducing health risks and environmental harm, we
are fulfilling our social responsibility. Protecting the health and safety of people
is a critical component, but only one of many criteria to be considered when
designing products for public use.

Sandestin Principle Nine states that in order to fully implement green engineering solutions, engineers should, "actively engage communities and stakeholders in
the development of engineering solutions". Stakeholders, according to R. Edward
Freeman, are defined as, "those groups who can affect or are affected by the
achievement of an organization's purpose"[4].

In this chapter, we discuss the drivers and benefits of engaging stakeholders,
including communities, in developing engineering solutions. We then present examples of how engaging stakeholders in different engineering and design-related
practices can promote movement toward sustainability. Examples include the use

of internal teams, eco-charrettes, supplier workshops, the development of environmentally preferable product standards, industry-working groups on materials flow management, and citizen advisory groups to improve corporate environmental performance. We also discuss how to manage these different group processes successfully. As the value of engaging stakeholders for developing sustainable solutions becomes more apparent, many more manifestations are likely to emerge.

1.2. Why engage stakeholders in the development of engineering solutions?

The reason to engage stakeholders is not evident if one limits the definition of green engineering to design for pollution prevention or risk reduction. But we believe that green engineering is engineering to promote sustainability and that there are a number of reasons to engage stakeholders in engineering and design practices:

1. Stakeholder engagement is ethically justified and therefore worthy of being elevated to the level of a principle rather than simply a strategy or practice. Engaging stakeholders is an effective way to clarify and prioritize the needs of society thus promoting sustainability by explicitly including the goal of human welfare in design.
2. Stakeholder engagement is a good strategic management practice at the corporate level. And during the design process, engaging stakeholders provides critical market information when it can have maximum impact, while at the same time creating a level of support important to the roll out or implementation of the project.
3. Pursuing sustainability raises system-level challenges that take place in the context of the supply chain, the market, and the natural environment — over the product lifecycle and various spheres of influence. Achieving sustainability requires cooperation among stakeholders to ensure both sustainable production and consumption [5]. Stakeholder engagement supports robust and sustainable engineering solutions and system design.

1.2.1. Ethically based and supportive of the social element of sustainability
Engaging communities and stakeholders in engineering decisions is a form of participatory democracy — "strong democracy" — a term borrowed from Benjamin Barber by R.E. Sclove [6]. It is based on the ethical principle of justice and supports the idea that "as a matter of justice, people should be able to influence the basic social circumstances of their lives"[6]. Technologies have significant political and cultural impacts and act as structures that shape behavior — physically or subconsciously. Technologies establish opportunities

or barriers for action and self-realization, affect non-users, shape communication, psychological development and many other aspects of culture [6]. Therefore, technologies cannot be considered politically or ethically neutral. For example, the Volvo Corporation experimented in manufacturing workplace reorganization at their automobile factory in Kalmar, Sweden by replacing a linear automobile assembly line with a system of independently movable electronic dollies. Working with the dollies required that workers organize into small teams of 15–20 members responsible for assembling an entire automobile subsystem. This organizational structure allowed flexibility to vary the nature, order and allocation of tasks, and within a narrow window, the pace of work. As a result, Volvo workers enjoyed increased fellowship and the chance to choose and vary their daily tasks. Volvo management benefited from reduced employee turnover and absenteeism [6].

Technical experts are not necessarily experts on important social questions. Good design should serve companies and communities in a way that respects relationships and desired ways of working and living. Stakeholder engagement allows us to consider social and environmental criteria in addition to economic criteria for assessing technologies. Stakeholder engagement supports democracy because it expands the influence of those who are affected by design decisions, allowing them to voice their own needs and definition of well-being rather than leaving those determinations solely to the engineer or designer.

1.2.2. Good business management

Organizations recognize that their success depends on their constituencies. The question of whether or how to include those constituencies, i.e. stakeholders, in business strategy development is more challenging [7]. Stakeholder theory is a system of organizational management that posits that organizations must attend to the interests and well-being of those who can assist or hinder the achievement of the organization's objectives. This means moving beyond a focus simply on maximizing shareholder wealth. While attention to stakeholders may lead to increased shareholder wealth, stakeholder theory posits that "attention to the interests and well-being of some non-shareholders is obligatory for more than the prudential and instrumental purposes of wealth maximization of equity shareholders" [7].

Engaging stakeholders at the level of the firm is practical because it can provide a number of valuable benefits, both directly and indirectly. These benefits, many of which are illustrated in the examples discussed throughout this chapter, include [8]:

- Building market share and advancing business objectives
- Increasing returns to shareholders and stakeholders
- Mitigating risk

- Reducing regulation
- Resolving conflicts, litigation, and ending boycotts
- Building relationships and reputation
- Enhancing media and public perceptions
- Driving innovation

Freeman notes that a stakeholder approach to strategic management is important not only to include stakeholder perspectives but to manage those with whom full agreement is impossible. If an organization chooses not to go "with" society, they should still manage relationships in order to deal with the resulting dissonance. For example, the tobacco industry must manage stakeholder opposition to the products they manufacture [9]. Stakeholder engagement at its least controversial is not very different from good marketing strategy. "In the free market, in nature, in every living system, feedback is the powerful catalyst that drives innovation and improvement via adaptation" [8]. Stakeholder engagement is a means of obtaining feedback and determining if the product, process, or system is something that the intended audience will accept and want.

Meeting stakeholder expectations can be profitable, just as misjudging the market can be costly. According to a British Telecom (BT) report, "Shareholders are the most influential group in any company. BT is convinced that the best way to create shareholder value and to add value to being a shareholder is by enhancing stakeholder value e.g. making it more worthwhile and satisfying to be a BT employee. Maximizing employee and customer satisfaction, partnering with suppliers for mutual gain, and being accountable for actions to society in general is as important as making a profit. Moreover, by addressing stakeholder expectations, BT creates shareholder value" [10].

Stakeholder engagement is also a form of risk management. Most expert engineering models focus on the physical aspects of the engineering system, and do not take into consideration the social and institutional system within which the technological system is embedded. Models developed by experts are affected by the cognitive limitations of the model-builders. Given that most large-scale engineering systems involve high stakes and large amounts of scientific and technical uncertainty, and will be executed within a larger social context, it is risky to use expert models that do not take into account stakeholder and decision-maker concerns and knowledge. To do so is to risk resistance and/or litigation by stakeholders who do not understand or agree with the recommendations [11]. As engineer Ian Alexander notes, "Stakeholders are sweeter if you talk to them first" [12].

Voluntary, consensus-based processes that engage stakeholders can reduce the need for regulation and foster broader support and buy-in for design decisions. The Performance Track and Sustainable Futures Programs at the US EPA are examples of programs by which stakeholder engagement can reduce the need for regulation. In these collaborative partnerships between businesses and the US EPA, companies demonstrate exceptional environmental management or new product development screening procedures that result in regulatory relief or expedited Agency review. Consensus-based stakeholder processes can circumvent traditional political processes by giving full voice to the under-represented, allowing for effective learning and information sharing, and building trust and commitment from a base of shared power, through processes based on principles of transparency and fairness [13].

The engagement of stakeholders can also help a company or organization create a vision, gain legitimacy and improve its performance in a community, thus making it less likely to run into expensive opposition to its operations. Engaging stakeholders upfront in decisions that affect the community or public perception can be critical to success.

In addition to the benefits stated above, stakeholder engagement and subsequent collaboration can lead to ongoing organizational learning that facilitates innovation. Good leadership means constantly monitoring environmental change, customer buying habits and motives, and providing the force necessary to organize resources in the right direction [14].

1.2.3. Robust solutions to system-level challenges

Engineering for sustainability requires engineers to address multi-disciplinary, system-based challenges. Engaging stakeholders in engineering decisions supports design within the complex systems of product lifecycles and material flows through society and the environment. As noted by Fiksel, a product cannot be sustainable in an absolute sense. It must be considered in the context of the supply chain, the market, and the natural environment. The challenge then is how to establish realistic boundaries for design within such a complex system [5]. Due to the interconnectedness of players within the enterprise, the supply chain, across material and product lifecycles, in the market, the broader society and the natural environment, there is a need for direct input from key stakeholders in order to better understand conditions, drivers, opportunities, and constraints. System-level change can be addressed by involving those who represent key aspects of the larger system and who are empowered to influence change within their organizations.

Sustainability involves the consideration of activities from the level of product and process development and manufacture within the firm, across material flow lifecycles, into the market and affecting society and ecosystems. Extending the boundaries of system design leads to increasing system scope and complexity and increasing relevance to global sustainability as illustrated in Fig. 1 [5].

In the next section, we discuss and illustrate a variety of multi-stakeholder processes that support the design and development of sustainable engineering solutions. In this treatment, stakeholders are considered from a product- and/or

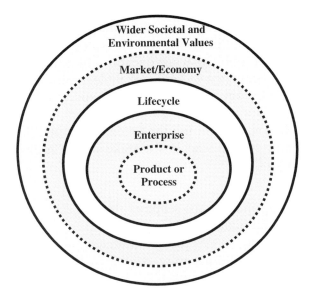

Fig. 1. Scopes of impact of engineering or design activities.

process-centric perspective, rather than from the perspective of the firm as a whole. Therefore, stakeholders may be viewed in different constellations. For example, shareholders are very prominent and influential from the corporate perspective while they may be less immediately relevant from the product development perspective. In contrast, relationships with suppliers and some key regulators are likely to be more pressing for product developers than for corporate managers.

2. Multi-stakeholder processes

There are many ways to engage stakeholders in product or process design, development, and deployment. We refer to these strategies together as "multi-stakeholder processes". A multi-stakeholder process (MSP) is used to bring together stakeholders in a new form of design, decision making and dialogue for the purpose of forming new partnerships, creating new solutions, and enabling decisions that are better informed and more widely acceptable.

Before creating a MSP, it is important to determine who might be considered a stakeholder. From the perspective of the firm, key stakeholders include *customers*, *employees*, *shareholders*, *vendors/suppliers*, and *community* members. The definition of community members may be expanded to include secondary stakeholders such as non-governmental organizations (i.e. not-for-profits, NGOs), government, and virtual communities who can affect or are affected by the achievement of the organization's purposes. Freeman carefully notes the importance of limiting the number of stakeholder groups defined. To expand indefinitely the boundaries that define stakeholders is to dilute the effectiveness and legitimacy of the process. So, while one may argue that CO_2 emissions have a global impact, it is important to limit stakeholder participation to those with direct involvement with key related issues, perhaps an NGO concerned with climate change [9].

Engineers and designers might use the following set of questions early in the design process to determine who has a stake in the project and should be engaged in its design:

- Who should be consulted on the scope of this project?
- Who has input to the budget of this project?
- Who can support or derail this project internally?
- Who has needed technical expertise related to the usability, functionality, qualities, and features of the product?
- Who can facilitate market penetration?
- Who will be impacted by the product over its expected life span (including production, use, and disposal)?
- Who can support or derail this project externally? [12]

Each of these questions suggests a different stakeholder or set of stakeholders that falls within one or more of the "scopes of impact" illustrated in Fig. 1. At the level of the enterprise, stakeholders may be individuals or departments within the firm working across functional boundaries to add multiple perspectives to the design process. At the level of the value chain or product lifecycle, stakeholders expand to include vendors, suppliers, and managers of product and material flows from origin to end-of-life. Many lifecycle stakeholders are not typically considered part of the supply chain (i.e. landfills, recyclers, etc.) unless one is considering sustainability. At the level of the market/economy, stakeholders such as standard setting organizations or government purchasing agencies may represent market pressures based on performance incentives or purchasing requirements; at the level of wider societal and environmental impacts, stakeholders may include NGOs and others who represent the well-being of current and future generations of human and non-human species.

The decision of whom to include in a stakeholder dialogue depends largely on two variables: the scope or reach of a project and the complexity of the endeavor. The examples that follow illustrate stakeholder engagement processes that map roughly to each scope of impact illustrated in Fig. 1, and describe the different kinds of outcomes that each type of MSP is intended to achieve. Note that due to the interrelatedness of roles and the complexity of human stakeholders, the boundaries are really quite porous and are used more for clarity of conception than to describe strict taxonomies.

2.1. Enterprise

2.1.1. Internal teams

At the core of the diagram in Fig. 1 are products and processes. Individual processes can be made more sustainable by internal collaboration between employees that represent different perspectives and areas of expertise and interest. Chemical company Rohm and Haas Co. frequently teams research, manufacturing, and marketing to ensure thoughtful and marketable design. When researchers partner with marketing professionals, they have a better idea of what customers expect. In turn both marketing professionals and scientists must be attuned to the regulations and other policies that affect their products and customers. This approach is at the foundation of the company's newly created Green Chemistry Laboratory at their Spring House, PA, facility, where they develop next generation sustainable technologies with the features and performance characteristics that customers want. Examples of greener technologies developed by Rohm and Haas via internal collaboration include water-based acrylics for paints and coatings, energy saving roof coatings that can be applied to asphalt roofs to reduce roof temperatures by nearly 30°C, and low monomer adhesives for flexible packaging to address European concerns about residuals in food packaging [15].

2.1.2. Creating common mental models

Employee training is another cross-organizational activity that can support sustainability and help guide business activities at the level of enterprise by cultivating shared mental models about what sustainability means. Organizations find it useful to base this training on a framework for sustainability such as The Natural Step, which presents a vision of a sustainable society through four System Conditions [16]. Cradle to Cradle Design is another framework that provides an effective mental model and strategy. Cradle to Cradle Design posits using renewable energy and viewing materials as *nutrients* that cycle through either *biological* or *technical metabolisms* [17]. The biological metabolism is made up of natural processes that circulate the pool of materials to support life on Earth. The technical metabolism mirrors natural nutrient cycles as a closed loop system in which valuable, high-tech synthetics, and mineral resources circulate in a continual cycle of production, recovery, and reuse. Following cradle-to-cradle principles, companies are creating products and materials designed as biological or technical nutrients, which either safely biodegrade or provide high-quality resources for subsequent generations of products. Related efforts such as the 12 Principles of Green Engineering also attempt to guide engineering practice to support sustainable material and energy flows [18]. Shared mental models can provide teams with clarity of direction and a clear commitment for design and development.

2.1.3. SC Johnson & Son Inc. and Greenlist^{TM}

SC Johnson used inter-divisional teams to pioneer the identification of preferred ingredients used in their products via their "Greenlist^{TM}". The development of the Greenlist^{TM} involved identifying key attributes of ingredients used in material categories (e.g. insect control active ingredients, surfactants, solvents, chelators, fragrances, dyes and colorants), and evaluating them with respect to related environmental impacts (biodegradability and environmental toxicity, regulations, human health, safety, and corporate sustainability goals). To ensure effectiveness, SC Johnson set up reward incentives for the chemists and designers who increase the use of those ingredients that are considered "best in class" and decrease use of those designated for "restricted use". As a result, the Greenlist^{TM} has been very successful in helping to focus and drive product development changes [19]. It is also having impacts on the supply chain by driving greater demand for preferred ingredients.

2.2. Lifecycle

The next ring represents the scope of supply chain and lifecycle impacts. There is a growing trend among manufacturers to include as design partners key vendors and suppliers. Companies interested in pursuing the development of sustainable products frequently must work closely with suppliers to find more benign ingredients and components, or more efficient and less resource-intensive packaging options.

2.2.1. Eco-charrettes

The building industry has long practiced a form of multi-stakeholder involvement through collaborative design intensives called "charrettes". These meetings, typically convened by the designing architects, bring together the contractors, engineers, clients, and other consultants to frame key planning and programming issues. This early "concept design" process sets the terms for the overall design and assures a whole systems approach to the creation of the project. Having all the critical expertise in the room together at one time enables the team to understand and efficiently manage the complex interdependencies of a building.

An "eco-charrette" is a special kind of design collaboration that focuses on sustainable building practices. Green building practices include reducing the need for energy and natural resources, eliminating toxic materials that cause "sick-building syndrome" and allergic/asthmatic reactions, treating storm water on site, and capturing rainwater for use in toilets, irrigation, or even drinking. The benefits of green building practices are compelling. They include improved occupant health and productivity, as well as energy efficiency and other potential cost savings. Accounting for all of these factors, the financial benefits of green buildings may on an average be ten times greater than additional up-front costs [20].

Maximizing these benefits, however, requires the collaboration of all the participants involved in the design and construction of a building. If window glazing is added and insulation increased, for example, what impact will this have on heating, ventilation and air conditioning systems? If piping is engineered to run as straight as possible, what will this do to the size requirements for pumps and relays and what will be the resulting impact on building layout and program requirements? Without all parties in the room, opportunities for maximizing the efficiency of the building would likely be missed.

2.2.2. Supply chain collaboration

In the same way that product design benefits from a collaboration of multi-stakeholders, so too does process design. According to a study done by Business for Social Responsibility, waste and inefficiencies across organizational boundaries can be staggering. Inefficiencies across the supply chain can waste up to 25% of a company's operating costs and a 5% reduction in waste throughout the supply chain can double a typical company's profit margin [21]. Organizations are responding by convening their entire supply chains to uncover waste and improve the process inputs. Sometimes called "supplier workshops", these customer-convened meetings bring first tier suppliers into a dialogue about how to minimize hand-off costs, identify new materials, and find improvements in quality.

Effective supplier workshops often have the following characteristics:

- The convening company represents a significant amount of business for the supplier. (If this is not true, it may be helpful to form a coalition with other customers of the supplier to have more leverage.)
- The supplier's product relates in some way to a company's goals for improvement. For example, there may be concern about toxic chemicals in subassemblies or the product itself, contributing to quality problems. Or perhaps packaging is excessive.
- All or most of a particular product is purchased from one vendor.
- The convening company has a trusting, collaborative relationship with the vendor.

General Motors, for example, discovered that by requiring their supplier of ignition sets to manufacture different versions for different cars, they inadvertently added significant costs for themselves and their supplier. Sometimes the solution is in changing the structural relationship. A number of automobile manufacturers have changed their relationships with chemical companies, for example, by paying Dow and DuPont to paint their cars instead of buying the paint directly. Now both parties are encouraged to minimize the use of this hazardous product [22]. Supplier workshops can be very positive forms of collaboration whereby barriers and opportunities are brought to light in a constructive way.

2.2.3. Sustainable Packaging Coalition

The Sustainable Packaging Coalition (SPC) is an example of a working group spearheaded by GreenBlue that is comprised of packaging professionals, ranging along the supply chain from manufacturers of materials such as paper and polymer resin, to converters, to consumer product companies, and retailers. The mission of the SPC is (1) to advocate and communicate a positive, robust environmental vision for packaging; and (2) to leverage innovative, functional packaging materials and systems that support economic and environmental health [23]. One of the first projects of the SPC is to define sustainable packaging by creating a common framework for the packaging industry that facilitates the design and development of sustainable packaging and systems. This definition is creating shared values among group members that will help guide product development, select future group projects, and educate the packaging industry about the Coalition's efforts.

2.3. Market/Economy

The next scope of impact includes market and other economic forces. These may be created via collaboration between industry members — perhaps even among competitors. Forming alliances with others within and across industry sectors can help to create markets for emerging materials that are more sustainable, and can enable the adoption of new industry standards. For example, Nike has joined forces with Patagonia and Norm Thompson to increase the demand for organically grown cotton, stimulating growth in availability [24]. This mutually beneficial alliance nets results that no single organization could achieve.

2.3.1. Intelligent materials pooling

Intelligent materials pooling (IMP) is a collaborative, market-based approach to managing materials in closed loops. Partners in an intelligent materials pool agree to share access to a mutually important material to generate a healthy system of closed loop material flows [25]. As partners share knowledge and resources, they develop a shared commitment to using the healthiest, highest quality materials in all of their products. Together they form a value-based business community built on cradle-to-cradle principles. The evolution of an intelligent materials pool follows the same steps as almost any kind of community or nation building: the community decides what it *does not* want; it chooses what it *does* want; and its members continually look for synergies where different organizations can work together for mutual advantage in pooling a number of resource investments:

- Intelligence and information — one company shares solutions not yet discovered by others
- Product development — companies share in the development and validation of possible solutions (with one or more suppliers)
- Purchasing — companies share in the bidding of "greener" replacements to get a more competitive price
- End of life — companies share in the development of systems for material recovery including the separation of products into feedstocks for technical and biological nutrient cycles.

The Society for Organizational Learning and its member companies are developing IMP communities for a number of different types of materials including cardboard, polypropylene and polyethylene, alternatives to hexavalent chromium treated metals, and more [26].

2.3.2. Design for the environment (DfE) green formulation initiative

GreenBlue has initiated a new collaborative project involving formulators that incorporates aspects of IMP as it aims to identify chemical ingredients that meet criteria for positive human and environmental health and safety performance, similar in aspects to SC Johnson's proprietary "GreenList^{TM}". The information will be used by formulators (i.e. "material choosers") to help promote the selection and use of such ingredients in product formulation. An MSP approach is being used to manage the challenge of protecting the confidentiality of raw material formulations and to establish the technical criteria and data requirements to identify ingredients with potential environmental benefits. Other advantages to collaboration include increased access to information, increased awareness of and subsequent demand for preferred raw materials, and the potential for economic benefit for raw material manufacturers and formulators by creating market opportunities based on verifiable claims.

2.3.3. US Green Building Council Leadership in Energy and Environmental Design (LEED)

The US Green Building Council (USGBC) is the nation's foremost coalition of leaders from across the building industry working to promote buildings that are environmentally responsible, profitable, and healthy places to live and work. Developed by the USGBC membership, the Leadership in Energy and Environmental Design (LEED) Green Building Rating System is a national consensus-based, market-driven building rating system designed to accelerate the development and implementation of green building practices. In short, it is a leading-edge system for designing, constructing, and certifying the world's greenest and best buildings. The full program offers training workshops, professional accreditation, resource support, and third-party certification of building performance [27].

LEED-registered projects currently represent close to 5% of the total square footage in US new construction. Rough approximations suggest that projects already registered for LEED certification can be expected to reduce key air contaminants (ozone and smog precursors) by about 170,000 tons and carbon dioxide emissions by more than 15 million tons during the next 50 years compared to conventional buildings. In terms of human and economic impacts, tangible boosts of labor productivity and well-being through improved working environments are expected [28]. The coalition approach to the development of the standards has been instrumental to their acceptance within the building industry.

2.3.4. Environmentally preferable purchasing

Environmentally preferable purchasing (EPP) is a federal government-wide program that encourages and assists Executive agencies in the purchasing of environmentally preferable products and services. All federal procurement officials are required by Executive Order 13101 and Federal Acquisition Regulation (FAR) to assess and give preference to those products and services that are environmentally preferable. Environmentally preferable products and services are defined as "... products or services that have a lesser or reduced effect on human health and the environment when compared with competing products or services that serve the same purpose ..." — Executive Order 13101.

EPA and its contractors and grantees have worked to support the development of criteria and products that meet EPP specifications. Related programs such as the EPA's Design for the Environment Formulator Initiative form direct partnerships with cleaning product formulators to review all ingredients, leading to formulation improvements with ingredients that are considered best in their class with respect to human and environmental health profiles [29].

2.3.5. Technology roadmapping

Technology roadmapping is a MSP involving industry, academia, and government collaborating to identify, select, and develop technology options to satisfy service, product, or operational needs 2–10 years into the future. Generally, this process is led by industry, and facilitated by government [30]. The main benefit of technology roadmapping is that it provides information to make better technology investment decisions by identifying critical technologies and technology gaps and identifying ways to leverage R&D investments. The roadmap identifies precise objectives and helps focus resources on the critical technologies that are needed to meet those objectives. This focusing is important because it allows increasingly limited R&D investments to be used more effectively [31]. Roadmapping processes with the greatest implications for sustainable development are issue-oriented roadmaps such as the U.S. Department of Energy roadmap to promote development of renewable bio-based products. Through such roadmapping exercises companies, research institutions, the academic community, and government agencies are motivated to align their research efforts with the agreed upon high-priority research needs to promote sustainable development objectives [32].

2.4. Wider societal and environmental values

The final ring represents the wider scope of impact on society and the environment and may be represented by citizens, advocacy groups, and advisory groups — both governmental and non-governmental. Emerging product responsibility legislation makes corporations increasingly accountable for a product's lifecycle impacts, thereby increasing the need to establish positive and reciprocal relationships with citizens and even countries in which a product is sourced, produced, used, and disposed. Emerging corporate commitments to social and environmental responsibility are changing the meaning of the term "license to operate" from compliance with local, national, and international regulation and legislation, to earning the trust and respect of diverse stakeholders with respect to company values and behavior [33].

2.4.1. Michigan Source Reduction Initiative (MSRI)

Stakeholder engagement can be a strategy for a company to resolve conflict and/or gain the support of members of the local or broader environmental community. When there is conflict, it is important to be inclusive of all potential stakeholders who could derail consensus. If key stakeholders are not included, then the conflict may not be resolved and any agreements arrived at during the process might not be accepted. A fine example of a MSP that resulted in significant improvements in environmental performance, in addition to strong relationship building and organizational learning, was a partnership called the Michigan Source Reduction Initiative (MSRI) in which Dow Chemical collaborated with stakeholders from the local and advocacy communities to improve environmental performance at their Michigan plant [33].

While Dow had already accomplished much in the area of pollution prevention, they were experiencing pressure from stakeholders to accomplish much more. Dow agreed to work with a broad range of stakeholders to review a list of chemicals of concern. The goals of the initiative were to gain approval for the use of capital that would achieve at least 35% reduction in the waste and emissions of 26 MSRI-priority chemicals at Dow; to foster institutional changes throughout Dow which would further shift the corporation's thinking from compliance to pollution prevention, and to further integrate health and environmental concerns into core business planning and decision making. Dow engaged a large contingency of stakeholders in the process that included internal leadership and technical resources, the Natural Resource Defense Council, Ecology Action Center, Lone Tree Council, Citizens for Alternatives to Chemical Contamination, Greenpeace, other local activists, and several external consultants and facilitators.

Throughout the project, there were detailed discussions of processes and emissions, and third-party validation of MSRI baseline waste and emissions. This enhanced both transparency and trust. As a result:

- Projects were implemented that reduced waste and emissions by 12 million lbs/yr
- MSRI wastes were reduced from 17.5 to 11 million lbs/yr — *a 37% reduction*
- MSRI emissions were reduced from 1 to 0.59 million lbs/yr — *a 43% reduction*
- Non-MSRI waste and emissions were reduced by 5 million lbs/yr
- 17 projects were identified with a combined return on investment of 180% — *a savings of $5.4 million/year*
- Almost every project met or exceeded the business criteria for return on investment.

Not only did Dow realize impressive, cost-saving results, but they also learned valuable lessons about using external stakeholders to recognize new opportunities for improvement and for creating a corporate culture of openness and innovation.

3. Managing the multi-stakeholder process

The examples in the previous section illustrate a few of the forms that multi-stakeholder processes can take from the engineering and design perspective. As the scope and complexity of projects increase, so too do the intricacies and challenges of managing multiple stakeholder engagements. Working with groups of diverse stakeholders is never a simple process, and is often complicated when participants hold very different views of an issue, have varying levels of knowledge of it, or hold strong positions or biases. The effort required to bring these collections of stakeholders to agreement usually pays off, though, because of the time saved later in redesign, and the ease of implementation within a supportive environment.

The authors of this chapter led a multi-stakeholder process whose mission was to identify criteria for sustainable industrial and institutional cleaning products. This MSP, called the Unified Green Cleaning Alliance (UGCA), used a structured facilitation process to allow broad participation from formulators, purchasers, users, advocacy-oriented NGOs, policy makers, and expert consultants. Despite their different interests, alliance members were able over several meetings to come up with a set of mutually agreeable attributes that green cleaning products should have. The UGCA process shed light on the strengths and limitations of current ecolabeling efforts and helped to bring about discourse on the value of complementary approaches to promoting sustainable products and practices in the industrial and institutional cleaning product industry. It also helped spawn new initiatives such as the DfE Green Formulation Initiative mentioned above. While every situation has its unique needs and characteristics, the UGCA experience illustrates several key steps that all multi-stakeholder processes share. These steps are explained below and organized into three phases: a project planning phase which describes the foundation building steps for a multi-stakeholder process, the convening phase which describes how to manage stakeholder meetings and processes, and a follow-up phase which makes recommendations about post project management.

3.1. Project planning phase

Before undertaking a multi-stakeholder process, it is important to lay a firm foundation. Taking the time to lay this foundation more than pays for itself by increasing the likelihood that subsequent meetings and decision forums run smoothly and achieve their goals.

3.1.1. Establish the scope and direction

As with any project, multi-stakeholder processes require clear direction and sense of purpose. The case could be made that this clarity is more important to a MSP because of the potential confusion multiple perspectives and positions can add to a project. To prevent "scope creep" and the unproductive pursuit of tangential issues that are common to MSPs, it is important to set up the process on a clear and focusing framing question. The question should clarify the outcome the process is intended to achieve or the problem it is designed to resolve in plain and measurable terms. The framing question acts as the compass for the process, to keep the effort on track and moving consistently toward the outcome. It also defines the measures against which the outcome will be assessed. A good framing question not only clarifies the purpose and outcome, but also determines the scope of the conversation and begins to suggest the key stakeholders who should be involved. While it sounds simple, framing a defining question can be challenging, as it should leave no room for confusion, ambiguity, misinterpretation, or diversion while at the same time avoiding any foregone conclusions.

The framing question at the foundation of the UGCA was, "What criteria should be considered in the evaluation of sustainable cleaning products"? As suggested by this question, the UGCA grappled with criteria that not only had environmental impacts, but because of our focus on sustainability, also considered impacts to society (including human health, jobs, and community impact) and to the economy (material costs and profitability potential for formulators).

3.1.2. Identify the convener

Ideally, every process should have a sponsor who launches it, convenes the group and holds them accountable for results. Depending on the sensitivity of the issues addressed in the framing question, it may be more or less important to choose a neutral party as convener, such as an NGO or a funding government agency. In any event, the sponsor or convener accepts responsibility for the outcome of the MSP and provides oversight to the project. Multi-stakeholder groups without a sponsor frequently fail to muster the necessary sense of shared responsibility needed to see a project through and run the risk of falling apart before a satisfactory conclusion has been achieved.

The UGCA was funded by a grant from the EPA that was managed by the Zero Waste Alliance (ZWA), a Northwest based non-profit. Because ZWA is held in high regard by both government and industry, it was a natural and logical sponsor for this project. ZWA was careful throughout the 6-month project to manage the process with neutrality and prevent individual organizations from undermining the process through "off-line" lobbying.

3.1.3. Determine stakeholder involvement

The scope and complexity of a project determine the appropriate reach of involvement. The framing question, however, suggests more specifically which stakeholder groups will have an interest in the particulars of the project. The challenge in establishing participation in a multi-stakeholder process is to be inclusive enough to get adequate representation, without including so many people that the process is burdened. Criteria for the selection of stakeholders should take into account the relationships they have with those they represent; the level of expertise or familiarity they have with the related issues; their availability to participate; and the level of decision-making authority they have.

The UGCA's framing question made the identification of stakeholders simple: cleaning product formulators, major purchasers, users (such as janitorial services), government agencies (in this case the EPA), and NGO's (ZWA, Center for a New American Dream, and Green Seal). The challenge for UGCA was creating a group large enough to be representative, but still small enough to fit within the constraints of the budget. Our solution to this dilemma was to include a broad reach of stakeholders but limit participation by geography inviting only organizations based in the Pacific Northwest. We were surprised by the attention and interest the project generated from around the country and were challenged to include people from outside the region in what we had defined as an open process. Limiting membership in the alliance created the concern that our conclusions would create a unique set of criteria and confound attempts by manufacturers to meet varying needs as well as concern from other groups that were trying to establish nationally recognized standards in the market place. In reality, no matter where the line is drawn in defining participation in a stakeholder process, someone is bound to feel left out. Our solution to the situation was to invite non-Northwest stakeholders to attend the meetings as presenters and discussion participants so that they had influence over the process, but limited voting and official input into the final report to our original Northwest members.

When stakeholders are invited to participate it is important to make clear the level of commitment that is expected and the role that each will play in the process. Will all participants, for example, have voting or decision-making authority? Is attendance at meetings required or can participants send representatives? How much time will the process require of each participant? What is the nature of the tasks or assignments expected of those who participate? Unless expectations are clearly conveyed at the beginning of the process, it will be difficult to fully engage participants and hold them accountable for outcomes. The UGCA made these expectations clear in its charter which detailed the Alliance's purpose, described the schedule and nature of the meetings, and defined voting procedures.

3.1.4. Design the approach and infrastructure

Dialogue is the heart of most multi-stakeholder processes. Dialogue among diverse groups can be robust or frustrating depending on the processes used to conduct them and the structures in place to support them. It is wise in most cases to choose a neutral facilitator experienced in designing processes to efficiently and effectively engage people in productive conversations, assure that those processes make good use of participants' time and guarantee that outcomes will be achieved within the agreed upon time frame. It is also important to allow sufficient time and resources to adequately plan, support, and conduct stakeholder meetings, as well as to put in place an infrastructure to manage communication, distribute documents and information, and record decisions and progress toward the goals.

The authors worked together to design meetings to achieve the UGCA's goals within the time frame stipulated in the contract. Critical to this process was identifying clear and measurable outcomes for each meeting, identifying, preparing and distributing support materials, sending notifications to participants and managing a web site for archiving meeting records and posting decision updates.

3.2. Stakeholder management phase

3.2.1. Identify stakeholder needs

Many stakeholders come to processes like these ready to advocate for a predetermined solution or course of action. Some stakeholders are, in fact, charged by the groups they represent to forward particular positions. However, at their best, MSPs are not about debates or convincing one party to agree with another. Rather they are about creating new solutions that meet the needs of everyone in the group. In order to achieve this, it is important to begin with a clear understanding of the needs represented by the participating stakeholders. Needs are distinct from positions. Positions are created in response to needs and frequently represent but one way to meet that need. Approaching MSP dialogues from the basis of participant needs enables the creation of multiple options and increases the likelihood that all stakeholders will be satisfied with the results. Beginning with the identification of stakeholder needs also facilitates the creation of meaningful project metrics. If it is clear what criteria define a satisfying outcome to those participating, then those criteria become natural yardsticks for measuring the success of the effort.

At the first meeting of the UGCA we grouped the participants by stakeholder representation (formulators/manufacturers, purchasers, users, NGOs, etc.) assuming that their interests in the project would be similar. We asked each group to generate answers to the question, "What would you describe as a successful outcome of this process"? This exercise was important for several reasons. It allowed participants to voice their needs and concerns right at the beginning. So often stakeholder processes get offtrack because participants feel unheard. This exercise not only gave each participant equal voice, but also helped them frame their needs in ways that were easier for the rest of the group to hear and address.

In addition, the answers to this question produced a vision that guided all subsequent conversations. The list of ideas became our "north star" by which we navigated. If the discussion took a tangent, we could use this list to determine whether or not it was moving us toward our vision of success. The process also provided us with the metrics for assessing the group's performance and the success of the project. We knew that the most successful outcome was one that matched the picture the group had generated.

3.2.2. Educate stakeholders

A diverse group of stakeholders usually means varying levels of familiarity with the issues under discussion. In providing information to MSP participants, the facilitator or convener must take into account three critical considerations. Firstly, there needs to be an accurate assessment of the group's learning needs. Presenting information that is irrelevant or redundant is a waste of valuable time. Secondly, determine the most efficient and effective means of delivering the information. Bear in mind that the most efficient means (distributing reading material before a meeting, for example) are frequently the least effective (there is no guarantee it will be read or interpreted reliably). Thirdly, the information presented should be free of bias so as not to prejudice perceptions or be seen as favoring any particular positions.

The issues under consideration for the UGCA were highly technical in nature. While a good portion of the participating members brought extensive knowledge, many were unfamiliar with the science and research behind the issues. We used several strategies for determining the learning needs of the group. The focusing question and stated objectives of the project suggested the need for training on certain topics, while a multi-voting process (see next section) allowed participants to reveal gaps in their knowledge. Using primarily these two methods we were able to craft an information plan suitable to the group.

While some information was posted on the web site for review by participants, we were reluctant to ask members to devote much time out of the meeting to reading. Our members were already volunteering significant time to the project and we were being careful not to exceed the time we had estimated. For these reasons and because we were more interested in the dialogue the learning would generate, we arranged for speakers to present during our meeting time. We mitigated the redundancy for some members by inviting them to present, leveraging the expertise we had in the group.

A good multi-stakeholder dialogue reflects the ideas and creativity of the participants. For this reason we were especially careful not to introduce bias into our education and information programs. We mitigated the potential for bias in several ways. First we based our assessment of the learning needs on the framing question, the tasks to be accomplished and participant requests to assure relevance and balance. Secondly, we chose a balance of presenters. For controversial topics such as certification options, we included three presenters representing three different perspectives. We also accepted proposals for presentations from members as well as from any interested party throughout the entire project. Thirdly, we used the presentation and ensuing dialogues to expose the limits in the knowledge base where they existed. As this is an emerging field, we were careful to distinguish between "known" facts and data and "suspected" or conjectured information.

3.2.3. Build strong working relationships
Honest, open and trusting relationships among participating stakeholders generate the best results with the fewest setbacks. Developing this type of relationship

with a large group of diverse stakeholders takes time and concerted effort. The UGCA used several strategies to achieve strong membership relationships. We built time into the process to allow people to get to know each other. We demonstrated respect for participants by designing processes that used their time well, honored their needs, and demonstrated commitment to the goal. We facilitated the meetings in such a way that assumptions were challenged and tested, participants felt safe speaking honestly, and information was accurate and up-to-date. We used a neutral facilitator to help create a climate of trust and assure that participation was balanced. Our ground rules further enabled healthy dialogue by forcing a conversation about what was considered acceptable behavior within the group and what would not be tolerated. Lastly, we assured that people fulfilled the commitments that they made to the group by following up on each assignment. Collectively these strategies enabled the UGCA to function effectively and produce the targeted results.

3.2.4. Get closure

Attention to all the preceding steps facilitates a successful result. A clear framing question obviates the outcome, measures provide the means for assessing the outcome, ground rules and agreements define the mechanisms for reaching agreement, and articulated stakeholder needs prescribe the level of individual satisfaction with the outcome. As a group prepares to document its conclusions, it should be sure to record the basis for the agreement, the assumptions upon which it was reached and the expectations for follow through and implementation. If agreement was not perfect, then the final report should include the conditions under which the agreement should be revisited. This involves documenting the concerns of members of the group and a commitment to monitor the ensuing actions to see if those concerns are realized. UGCA was fortunate in having broad agreement on the final report. In fact, the project generated so much enthusiasm, that a subsequent proposal was submitted and accepted by the EPA to continue the work into its next phase.

3.3. Post-project management phase

3.3.1. Follow-up

Too frequently multi-stakeholder processes end before their results are implemented. At the very least, the MSP should prescribe strategies for implementing the ideas, identify the methods and metrics for monitoring its progress, and establish schedules for reviewing and assessing success. Responsibility for monitoring follow-up logically lies with the sponsor (another reason having a dedicated sponsor is so important), but follow-up planning should include agreements for involvement throughout the full lifecycle of the project.

4. Conclusion

Engineering is a highly technical field that prepares professionals to tackle complex, highly integrated, system-based problems. Designing solutions that not only address technical problems, but also honor social and environmental concerns, requires the additional skill set to manage diverse stakeholder groups. Engaging stakeholders in the development of engineering solutions is good business management strategy and facilitates solutions to system-level challenges. Managing multi-stakeholder processes is a difficult task at best and one that is rarely addressed in engineering programs, but it is a skill that is becoming increasingly important to creating robust engineering solutions. This chapter makes the case for including this approach in a green engineer's repertoire and takes a first step at describing the processes and potential application to a variety of situations.

Acknowledgment

The authors gratefully acknowledge the assistance of Phil Storey.

References

[1] http://www.epa.gov/opptintr/greenengineering/ (accessed July 2004)
[2] P.T. Anastas, L.G. Heine, T.C. Williamson, Green Engineering, American Chemical Society, Washington DC, 2001.
[3] S.K. Ritter, Chem. Eng. News 81 (29) (2003) 30.
[4] R.E. Freeman, Strategic Management: A Stakeholder Approach, Pitman Publishing Inc., Massachusetts, 1984.
[5] J. Fiksel, Environ. Sci. Technol. 37 (23) (2003) 5330–5339.
[6] R.E. Sclove, Democracy and Technology, The Guilford Press, New York, 1995.
[7] R. Phillips, Stakeholder Theory and Organizational Ethics, Berrett–Koehler Publishers Inc., San Francisco, 2003.
[8] Adapted from Future 500. 2004. Stakeholder Engagement Presentation. Courtesy of Alison Wise.
[9] R.E. Freeman, University of Virginia, Darden Graduate School of Business Administration, Charlottesville, VA, personal communication, June 2004
[10] From: http://www.thetimes100.co.uk/case_study.php?cID = 55&csID = 106&pID = 4.
[11] Mostashari, Ali and J. Sussman. 2004. Engaging Stakeholders in Engineering Systems Representation and Modeling. http://esd.mit.edu/symposium/pdfs/papers/mostashari.pdf.
[12] Alexander, I, Stakeholders – Who is Your System For? http://easyweb.easynet.co.uk/~iany/consultancy/stakeholders/stakeholders.htm.
[13] M. Hemmati, Multi-Stakeholder Processes for Governance and Sustainability, Earthscan References, London, 2002.
[14] S.R. Covey, The 7 Habits of Highly Effective People, Simon & Schuster, New York, 1989.
[15] J.M. Fitzpatrick, K.A. Gedaka. Environ. Sci. Technol. 37 (2003) 445A.
[16] http://www.naturalstep.org/learn/principles.php.

[17] W.A. McDonough, M. Braungart, Cradle to Cradle: Remaking the Way We Make Things, North Point Press, New York, 2002.

[18] P.T. Anastas, J.B. Zimmerman, The 12 Principles of Green Engineering. Environ. Sci. Technol. 37 (2003) 94A.

[19] http://www.csrwire.com/pdf/Greenlistcasestudy.pdf. Adapted from presentation at Environ-Design8. 2004. John Weeks of SC Johnson.

[20] G. Kats, "Are Green Buildings Cost-Effective?" .Green@Work, May/June 04.

[21] Business for Social Responsibility. Suppliers' Perspectives on Greening the Supply Chain: A Report on Suppliers' Views on Effective Supply Chain Environmental Management Strategies, June 2001.

[22] B. Willard, The Sustainability Advantage: Seven Business Case Benefits of a Triple Bottom Line, New Society Publishers, Bagriola Island, BC, 2002.

[23] http://www.sustainablepackaging.org.

[24] M. Willard, C. James, Choosing Greener Products, AXIS Performance Advisors Inc., Portland, OR, 2001.

[25] M. Braungart, from "Intelligent Materials Pooling: Evolving a Profitable Technical Metabolism", from Monthly features at www.mbdc.com, September/October, 2002.

[26] http://www.solonline.org/public_pages/comm_SustainabilityConsortiumCore/ (accessed 7/2004)

[27] http://www.usgbc.org/AboutUs/mission_facts.asp (accessed 7/2004).

[28] http://www.edcmag.com/CDA/ArticleInformation/features/BNP_Features_Item/ 0,4120,80479,00.htm (accessed 7/2004).

[29] http://www.epa.gov/dfe/projects/formulat/index.htm (accessed 7/2004).

[30] http://strategic.ic.gc.ca/epic/internet/intrm-crt.nsf/en/Home (accessed 7/2004).

[31] http://www.sandia.gov/Roadmap/home.htm#abstract (accessed 8/2004).

[32] http://www.oit.doe.gov/agriculture/tech_roadmap.shtml (accessed 8/2004).

[33] Presentation by J. Feerer, Dow Chemical, Collaboration and Sustainability: A Case Study. Presented at EMS Essentials IV Workshop, Portland, OR, 2001.

PART III:
APPLYING THE PRINCIPLES

Sustainability Science and Engineering: Defining principles
Martin A. Abraham (Editor)
© 2006 Elsevier B.V. All rights reserved
DOI 10.1016/S1871-2711(05)01014-7

Chapter 14

Utilizing Green Engineering Concepts in Industrial Conceptual Process Synthesis

Robert M. Counce[a], Samuel A Morton III[b]

[a]*Chemical Engineering Department, University of Tennessee, 419 Dougherty Hall, Knoxville, TN 37996-2200, USA*
[b]*Chemical Engineering Department, Lafayette College, 266 Acopian Engineering Center, Easton, PA 18042, USA*

1. Introduction

Capstone design activities are described utilizing authentic industrial projects as a teaching mechanism for process synthesis and analysis; this approach is often referred to as experiential learning. The Capstone Engineering Program at the University of Tennessee's (UT), Chemical Engineering Department (ChE) has completed 51 projects involving 226 students in the period 1989–2004; it is one of several such departmental efforts at UT; bringing together students, faculty, and practicing engineers to work on authentic industrial projects [1–3]. Support of UT ChE Capstone Education efforts has been provided by INVISTA (formerly DuPont and DuPont Textures and Interiors), Eastman Chemical Company, Dow Chemical Company, and the Oak Ridge National Laboratory. Capstone projects in UT ChE have completed studies on far-ranging topics. It is similar to programs at other universities that regard the real-world solution of engineering projects as important learning activities [4]. It provides engineering students a realistic capstone conclusion to their undergraduate education under the direction of faculty and industrial "coaches". While there is some variation at UT between departments, these programs generally involve at least a two-semester sequence: some instructor-led elements, with transition to an industry-provided project for each team of four to six students.

In UT ChE Capstone Design projects, teams of students work on carefully selected industrially sponsored projects. Each project has an academic as well as an industrial coach. The team consists of typically four undergraduate students; however, graduate students have participated from time to time as team members or academic advisors and teams have varied from three to seven persons. These projects require the application of sound concepts of scientific and engineering principles, teamwork and project management to accomplish team goals. Protection of the sponsoring industries' confidential information is assured. In this paper, the underlying philosophy and special nature of this educational approach are presented and discussed.

2. Background

Experiential learning is not a new concept; historically, the apprentice working under a master was receiving an experiential education. Experiential learning is widely practiced in United States (US) engineering education. A study by Todd et al. [4] indicated a significant number of US engineering programs utilizing experiential learning through industrially supported capstone projects. The Engineering Clinic Program at Harvey Mudd College [5] and the Capstone Program at Brigham Young University [6] are well-known examples of the use of experiential learning. The capstone engineering projects performed by UT Chemical Engineering students are typically at a conceptual design level and result in the development of a conceptual design report; in other engineering disciplines the project deliverable may be a working prototype.

A comparison of attributes of experiential versus traditional learning models is presented in Table 1. It is likely that the contrasts presented here are never completely one or the other and may change according to the situation. Certainly students, who have not accepted the responsibility of self-(and continuous)education, are very disadvantaged in the rapidly changing global economy. Bright [5] indicates that the skills of abstraction and synthesis are key to self-education and are best developed by practice on open-ended problems under the guidance of an experienced problem solver. The role of a teacher in the area of capstone education is evolving from instructor to one of mentor and coach (guide and

Table 1
Model for experiential learning compared with the traditional model [5]

Student as self-instructor	versus	Student as novice
Faculty as coach	versus	Faculty as expert
Industry as partner	versus	Industry as consumer
Learning as collaboration	versus	Learning as acquisition

motivator). At times, however, the teacher may reassume the more traditional role as the "transmitter of knowledge." Certainly, the most important objective in these projects is to complete the student's undergraduate education. However, by careful attention to the design scope and objectives, the deliverable final project report should contain enough value on its own to justify the industrial participation. A final observation from Table 1 is that a successful project involves the individual acquiring knowledge as a collaborative effort in partnership with the other team members as they work to satisfy the ultimate goal and produce a unified engineering study.

The Capstone Engineering Program in UT Chemical Engineering proceeds through some typical steps leading to a conceptual design study or to the development of a prototype [2]. For a conceptual design study leading to preliminary process/product synthesis and evaluation, a typical pathway is presented in Table 2. The deliverable of this activity in Chemical Engineering is a final report with recommendations.

The deliverable may be different in other engineering departments where the capstone design project leads to the development of a prototype. The function of the faculty and industrial coaches is to provide the necessary conditions and support for a student process design team to function effectively. For most projects, students sign a "limited term" secrecy agreement with the industrial sponsor. Faculty members are covered under more inclusive secrecy agreements.

This capstone project is one of the few academic experiences involving a team rather than an individual effort. The students have typically had limited exposure to environmental regulations or waste management operations. Two to three hours per week of scheduled group meetings with their faculty and industrial advisors present are typical; in these meetings, goals are formulated, accomplishments presented and reviewed, and an occasional faculty lecture is presented. A useful project management tool is to have project milestones that

Table 2
Typical pathway for preliminary process/product synthesis and evaluation

---

1. Feasibility study
2. Narrowing the field of alternatives
3. Project initiation
4. Preliminary design report and presentation
5. Flowsheet development
6. Estimation of capital and operating costs
7. Selection of most promising alternative(s), and
8. Final report and presentation

are also draft sections of the final report; for example, the draft "Introduction" and "Design Objectives" are early milestones. A typical outline of final report milestones is presented in Table 3. These milestones correspond to the final report chapters and are correlated with the typical project steps in a logical manner. In general, the students contribute a great deal of time to successful conclusion of these projects, similar to that required of a typical engineering capstone design experience.

3. Project structure

3.1. Project selection

Internal discussions with the industrial sponsor begin several weeks in advance of the initiation of the project, with the faculty advisors typically involved in the final selection of the project; the selection criteria are based on their educational benefits, value to corporate sponsor, and on the possibility of being completed in one semester. Usually the individual or group providing the industrial problem will function as the industrial advisor(s) for the team.

3.2. Group membership selection

The students are selected for the "honors" or standard capstone design course based on their academic achievements and completion of an informal interview. Providing equal opportunity for all chemical engineering students having appropriate prerequisite course work is an important consideration. Various options for group determination exist, self-selection, lottery, and instructor pre-selection. All these forms have merit and the individual course instructor is tasked with determining the best method for his/her class.

3.3. Initial classroom meetings

At this point, groups are selected and projects presented to and accepted by the various teams. During this classroom time the confidentiality forms for each of the student teams are signed and witnessed by other members of the class.

After some introductory remarks by the instructor, a discussion of characteristics of "good" teams is initiated. One group recalled an experience where the team had a good time working and laughing together. Another student recalled a Navy team where organization was clear, project needs were known, and responsibility was clearly understood and accepted. Another student recalled a poor team experience where responsibility was unclear and subject to change throughout the experience. The discussion shifted to what individual

Table 3
Project milestones (report sections and outline)

1.0 Introduction
- Purpose of the manuscript
- Brief background information (*save detail for Background section*)
- Rational for the project
- Definition of the design problem
- Contributions of others
- Scope of the manuscript/indication of report contents
- Design objectives (*this should be a specific subsection*)
  - Defines the problem
  - Presents rudimentary material/energy balance considerations
  - Briefly indications the approach to the problem
  - Indicate economic ground rules
  - Spells out the expect deliverables

2.0 Background
- Overall process or design situation (includes schematic drawing)
- Details of specific project (includes more detailed schematic drawing)
- Chemical Reactions
  - Stoichiometry, temperature, and pressure of each reaction
  - Correlation of product distribution to conversion and temperature
  - Catalyst(s) for each reaction
  - Desired phase condition for each reaction step

- Products
  - Desired production rate and product purity
  - Product price and/or price versus purity comparison
  - Value of all by-products as marketable chemicals or fuel

- Raw materials
  - Composition, temperature, and pressure of all raw material streams
  - Prices of all raw material streams and/or price versus purity

- Constraints
  - Explosive limits and safety considerations
  - Coking limits, polymerizations or decomposition limits, etc.
  - Environmental regulations
  - Corporate policy

- Plant and site data
  - Utilities — fuel, steam levels, cooling water, refrigeration, etc.
  - Waste disposal facilities and costs

- Other physical and/or chemical properties for reactants, products and important intermediaries
- Other details/information to support the project

Table 3 (*continued*)

---

    o Technology potentially useful to goals of project
    o Related physical/chemical phenomena for these technologies

- Narrow the field of alternatives for further study
- Brief path forward

3.0 Material and energy balance details of selected alternatives
- Figures to present flow sheets (*number streams*)
- Tables to present materials flow rate(s) and other information
- Explain flow sheets in text in detail

4.0 Analysis of selected alternatives
- Profile all relevant assumptions
- Tables to present capital investment estimates
- Tables to present annualized operating/cost information
- Tables to present environmental metrics (Tiers 1, 2, etc)

5.0 Results
- Present proposed design
- Present materials vial to Conclusions and Recommendations sections

6.0 Conclusions (*briefly interpret results*)

7.0 Recommendations
- Be concise and clear
- Approach depends on purpose of report

8.0 Summary
- Introductory statement
- What was done and what the report covers
- How final results were obtained
- Important results, including
    o Quantitative information
    o Major conclusions and recommendations

- Often the only part of the report that is read
- Give reader entire contents of report in 1 or 2 pages

---

students want to get out of this capstone activity. These comments are recorded in Table 4 and with slight variation have been consistent throughout the course of the UT ChE Capstone Design Program. Following this, the discussion moved to team expectations, shown in Table 5, and team guidelines in Table 6. The class concluded with setting dates for milestones and determining weekly meeting times for each team.

Table 4
Individual student expectations for UT ChE Capstone Design Course from Spring 2003

- "Bring all stuff together from previous semesters"
- "Do something meaningful for a company rather than laboratory tests"
- "Gain industrial experience"
- "Graduate at end of semester"

Table 5
Team expectations for UT ChE Capstone Design Course from Spring 2003

- Participation from each member
- Everyone has assigned roles and is accountable for performance
- Complete milestones on time
- Group defines appropriate leadership for team
- Everyone has input
- Leaders get opinions from everyone (facilitates)

Table 6
Student developed team guidelines for UT ChE Capstone Design Course from Spring 2003

- Appreciate deadlines
- Being dependable
- No whining
- Follow through with individual commitments
- Bath daily
- Honesty and ethics in work
- Strive for excellence
- Try to work out problems within team
- Flexibility in scheduling and responsibilities
- Rules are subject to change

3.4. Project initiation

Various formats for project initiation have been utilized during the development of the UT ChE Capstone Design Program. In earlier projects, the initiation occurred at a production site where the facility was toured. Later projects have been initiated on campus, sometimes with a visit from the corporate sponsor. In the initiation discussions, production information may be provided and the ground rules for the project established. Sometimes alternative designs are suggested by the industrial sponsor in the initial meeting and supporting information provided when available. The supporting information may include desired product purity, relevant reaction rates and yields, reaction and phase equilibria information, by-product formation data, operating and pilot plant data, and safety and toxicity information. It is common that more supporting information be available for some alternatives than for others.

Due to limited travel budgets and minimal travel time available to the students, the current project initiation generally occurs on campus. When the students visit the related production facility later in this activity, they are more knowledgeable about the facility and perhaps gain more from their visit than otherwise. It is typical for the students to visit and present a briefing at the conclusion of their feasibility study or later in this activity when flow sheeting is complete and capital and operating costs are known. Attendance generally includes industrial and faculty advisors as well as interested industrial personnel.

3.5. Design objective

Early in the project activity a draft of the "Introduction" section, including design objectives, is required. This is a critical step in that it assures that there is a common understanding of the project goals and deliverables. It is not uncommon for the initial series of communications between the industrial sponsor and the student team to focus on agreeing on the official design objectives. This ensures that the industrial sponsor is engaged from the outset and that the students are working toward a mutually acceptable goal.

3.6. Feasibility study

The feasibility study provides information on the appropriate design alternatives. It is usually done in conjunction with production of the "Background" section of the final report. It insures that the students have the necessary information to make their own decisions as to the appropriate technology. A computer search of the literature greatly expedites this project phase and is a required student team activity. There is often extensive communication between the students and industrial project advisors via e-mail, FAX messages, and telephone conferences. Videoconferences are more expensive and technical issues complicate their use, however, they can be used when available, as they are still lower in cost than personal visits. The combination of faxed documents, speakerphones, and conference calling provide low-cost and effective means of communication involving a number of people at different sites. At the conclusion of the feasibility study, the students present and discuss their findings with the corporate sponsor. These are important discussions in insuring that the objectives of the design study are being met and that all feasible options are identified.

3.7. Narrowing the field of alternatives

At the conclusion of the feasibility study the alternatives are screened to insure that the most appropriate options are considered. Screening criteria based on the Tier 1 analysis from Allen and Shonnard [7] can be used to eliminate undesirable alternatives. At this phase in the project, material, and energy balance flowsheets have not been developed and no capital and operating costs have been determined. The results of this screening step are typically presented to the corporate sponsor simultaneously with feasibility discussions.

3.8. Preliminary design report and presentation

The result of the feasibility study and the alternative-screening step are provided in the form of an oral and a written report. The written report usually forms the basis for the "Introduction" and "Background" sections of the final report. The written report required at this point in the project spreads the report writing tasks over a greater portion of the semester than would be the case if only a final report were required. Essentially all of the material in the preliminary report will become a part of the final report. The preliminary report and all sensitive communications may be treated as business confidential.

3.9. Flowsheet development

Identifying waste streams in the early stages of process design is expedited by considering waste streams to be intrinsic or extrinsic. Intrinsic waste streams are those that are inherent to the process configuration while extrinsic waste streams are those that are associated with the operation of the process [8]. Some waste streams may be identified from the macroscopic material balances. Identification of waste streams not apparent in the macroscopic material balance may be difficult at an early stage of process development. Identification of intrinsic wastes may, at times, require experimental data. Identification of extrinsic wastes usually requires experience. A great deal of the nature and the input–output structure of alternative flowsheets [9] may be found from an examination of the reaction step(s). Information on some waste streams requires discussions with knowledgeable industrial sponsor personnel in order to get a total view of the wastes generated in an operating process. This designation of waste streams is discussed further by Berglund and Lawson [8].

The window for creativity in these activities comes after the students understand the process and its constraints and are formulating their flowsheets. The semi-structured brainstorming activities of the flowsheet formulation phase may take a considerable amount of time but are critical for the opportunity they offer for creativity.

3.10. Estimation of capital and operating costs

For preliminary estimates of fixed capital investment by persons other than an expert, the factored approach has generally proven reliable and is generally the method selected. In this method of cost estimating, the purchased cost of the major equipment items is estimated and the total fixed capital investment is estimated by applying a multiplier (Lange factor) to the purchased cost of the major equipment items [10]. When time for this activity has been compressed, an approach has been utilized based on the method of Zevnik and Buchanan [11]. This method must be carefully applied and calibration of the procedure using actual cost data is recommended. Specific operating cost information, product, and raw material costs should be consistent with those used within the industrial sponsor's organization.

3.11. Selection of most promising alternative(s)

Selection of the most promising alternatives occurs when capital and operating estimates have been completed. Again, the students are encouraged to use existing, or develop their own criteria that include green engineering metrics. This is only possible when the students have a good understanding of the principles and concepts of green engineering prior to this phase of the project. This understanding is fostered through the presentation of the principles of green engineering in the main lecture session and through discussion during the team project.

3.12. Final report and presentation

The final design report from the project is a business confidential document. As mentioned earlier, students typically sign a "limited term" secrecy agreement with the industrial sponsor. The agreement to hold findings of these projects and related information secret is important if the students need access to proprietary information in order to provide a useful and thorough study. The final report is reviewed first by the academic advisor(s). After these comments are addressed by the students, the report is reviewed a second time by both academic and industrial sponsor advisors. A final oral report of the design effort and pertinent findings is presented by the student design team at the conclusion of the project. The audience for this is typically selected by the industrial sponsor and normally occurs at the facility where the design problem originated.

4. Incorporating green engineering in capstone design

Recently the EPA has promoted the principles of green engineering. As a result a number of risk assessment tools, such as EPISUITE, are available and can be used in conjunction with the text by Allen and Shonnard [7] to augment the classical chemical engineering design curriculum. The principles of green engineering as developed by Anastas and Zimmerman [12] are shown in Table 7.

These principles were introduced to the students using the team meeting structure and were supplemental to the traditional chemical process design curriculum. The students were expected to analyze and discuss these principles and various published green engineering examples. The students then approached the design project in a manner consistent with the traditional design process. The students performed the process design with these principles in mind during the "Narrowing of the Field" and "Selection of Most Promising Alternative(s)" steps. To accomplish this most effectively, metrics must be used that can categorize the alternatives and rank them based on economic and environmental impacts. The most common metrics for green engineering are designated as Tier 1, 2, and 3 Environmental Metric [7]. Tier 1 is useful during the earliest stages of process synthesis and includes evaluation of toxicity of raw materials and products and evaluates environmental persistence and bioaccumulation. Tier 2 analysis is a reexamination of environmental performance after preliminary process flowsheets are developed where estimated emissions from process units, fugitive sources, and utilities are important information. Tier 2 culminates in estimation of Environmental Performance metrics such as material, energy, water, and pollutant indices, usually expressed per mass of product. The Tier 3 analysis utilizes environmental and health indices to assess the potential impacts of the nearly complete chemical process designs. The indices

Table 7
The 12 principles of green engineering

---

1. Risk: Inherent as opposed to circumstantial.
2. Prevention rather than treatment.
3. Design process for separation simplification.
4. Maximize resource efficiency.
5. Output-pulled instead of input driven.
6. Conserve complex productions through recycle and reuse.
7. Design products for durability instead of immortality.
8. Don't create excessive capacity during design.
9. Minimize material diversity in multicomponent products.
10. Maximize use of available material and energy resources.
11. Design products with post-use concerns in mind.
12. Maximize use of renewable resources.

---

typical to Tier 3 are for global warming, stratospheric ozone depletion, acid deposition, smog formation, inhalation toxicity and carcinogenicity, ingestion toxicity, carcinogenicity, and fish aquatic toxicity. The students may choose to apply these metrics at specific points along the design path; however, they make decisions regarding equipment, components, and processes based on their understanding of economic and environmental impacts throughout the project.

5. Green engineering design example

In the fall of 2003, a group of senior students performed a green engineering-centered design project. The project was concerned with the Scotch whisky process as employed at Bruchladdich Distillery, Islay, Scotland. The distillery provided information related to process performance, waste products, and technical information and support via telephone conversations and e-mail. While the distillery was not motivated by pressing environmental regulations, they were open to working with the student teams as part of a growing relationship between our department and the distillery. The students were expected to present the results of their project to the distillery manager when he visited the university later in the semester.

During the "Project Initiation" phase of the project, the students were assigned individual literature research by the main academic advisor. This would maximize the team's information while minimizing the time required for its acquisition. Additionally, it allowed for natural team development as the students presented and discussed the literature information during team meetings. The students were tasked with reviewing the structure and arrangement of the whisky distillation process. This process and a general review of raw materials, products, by-products, and wastes are shown in Fig. 1.

At this time the students were able to formulate an initial flowsheet, which detailed the processing stages and any pertinent inputs and output, as shown in Fig. 2. The project objectives were formulated and agreed upon by all involved parties. In this phase, the students were generally given activities by the team advisor rather than generating activities within the team. At this point, the students performed a combined "Feasibility Study" and n "Narrowing of the Field" to select the optimal waste stream(s) for this project. The students researched the selected waste stream; carbon dioxide generated during the fermentation process stage, and developed a preliminary listing of waste minimizational alternatives. During this phase of the project, the students began to direct the project activities and reported progress to the academic advisor.

As the design effort progressed, the students developed a design strategy for the project, shown in Fig. 3, and prepared the preliminary design report and

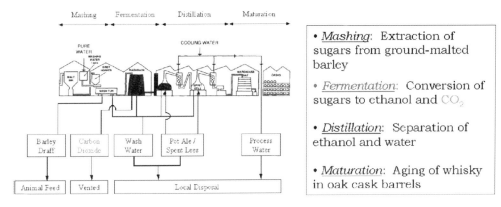

Fig. 1. Scotch whisky process overview.

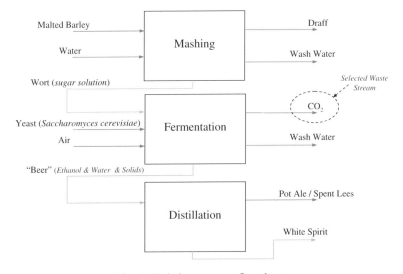

Fig. 2. Existing process flowsheet.

presented it to the academic advisor. When all parties were in agreement with the path forward, the team began to revise the process flowsheet with respect to the individual input, output, and storage requirements for the waste utilization alternatives. The students determined that a CO_2 recovery system followed by a CO_2 liquefaction system would be required. The main justification for this was that only one of the four selected alternatives, Algae Growth, required gaseous CO_2 while the others required liquefied CO_2. A two-stage process would allow for accurate project costs while limiting the creation of excess capacity. The proposed process design scheme along with preliminary equipment capital and operating costs can be seen in Fig. 4.

Proposed Process Flow Diagram:

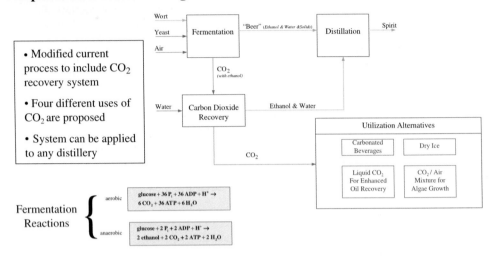

- Modified current process to include CO_2 recovery system

- Four different uses of CO_2 are proposed

- System can be applied to any distillery

Fermentation Reactions {
aerobic $glucose + 36\,P_i + 36\,ADP + H^+ \rightarrow 6\,CO_2 + 36\,ATP + 6\,H_2O$

anaerobic $glucose + 2\,P_i + 2\,ADP + H^+ \rightarrow 2\,ethanol + 2\,CO_2 + 2\,ATP + 2\,H_2O$
}

CO_2 Utilization Alternatives:

- **Algae Growth**
  - Enhances plant growth to be used for livestock food (similar to barley draff), fertilizer, or health food supplements
  - Additional costs: $26,800 (includes blower, algae tanks, and artificial lights)

- **Carbonated Beverages**
  - Supplied to bottling companies and restaurants for carbonation of soft drinks
  - Additional cost[‡]: $224,000 (includes 700 CO_2 transport cylinders at 100-lb each)

    [‡] Additional cost will be zero if Bruichladdich chooses to use CO_2 on site. The distillery currently has a bottling facility.

- **Enhanced Oil Recovery**
  - Pumped into oil wells to drive out remaining oil that is otherwise impossible to extract
  - Additional cost: $24,000 (includes 3-ton CO_2 capacity tanker truck)

- **Dry Ice Production**
  - Converted into solid refrigerant to be used in the areas of fresh meat processing, pharmaceuticals, sport fishing, etc.
  - Additional cost: $36,000 (includes dry icemachine)

Fig. 3. Projected design strategy and list of selected alternatives.

CO₂ Recovery System:

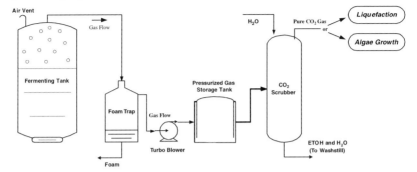

Installed Equipment Costs[†]:
- Foam trap → $40 (*Literature estimation*)
- Turbo Blower → $10,800 (*Literature estimation*)
- CO₂ Pressurized Tank → $93,500 (*Quote from Roy E. Hanson Jr. Mfg.*)
- Scrubber Column → $50,100 (*Literature estimation*)

Operating Costs:
- Blower Electricity → $5/year
- Scrubber Water → $284/year

[†] **Design is based on industrial size equipment. Probable accuracy of cost estimates is ± 25% (Douglas p. 7)**

CO₂ Liquefaction System:

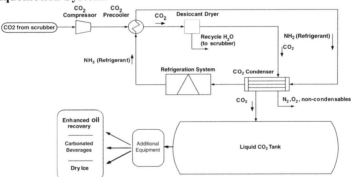

Installed Equipment Costs[†]:
- CO₂ Compressor → $10,800 (*Literature estimation*)
- Desiccant Dryer → $53,200 (*Quote from Fluid Energy*)
- Refrigeration System → $510,000 (*Literature estimation*)
- Liquid CO₂ Tank → $163,800 (*Quote from Cryotrader*)

Operating Costs:
- Compressor Electricity → $12/year
- Refrigeration System Utilities → $1331/year

Fig. 4. Proposed design scheme — equipment and operating costs.

Table 8
Comparison of economic potentials

CO_2 utilization	Installed cost ($)	Operating cost ($/year)	Product value	Payback period (years)
Carbonated beverages	1,115,000	1631/	$0.50/lb CO_2 or $498,000/yr	2.2
Dry ice	928,240	1866/	$0.25/lb CO_2 or $258,000/yr	3.6
Algae growth	181,240	498/	$0.003/lb CO_2 $2988/yr	60.7
Enhanced oil recovery	916,240	1666	$0.008/lb CO_2 or $7531/yr	122

At this point the design portion of the project approached completion. A finalized summary of the economic potentials for the CO_2 recovery system, the CO_2 liquefaction system, and any additional costs related to individual waste use alternatives can be seen in Table 8. A milestone for the project was the arrival of Jim McEwen, Bruichladdich Master Distiller, at UT for a series of lectures and student discussions. The student team presented the project results, received feedback, and engaged in an informative discussion with Mr. McEwen. They discussed the feasibilities of the various CO_2 alternatives specific to Islay and Bruichladdich. The students learned that all the Islay distilleries participate in a collective organization focused at improving the conditions and environment of Islay. As a result of these discussions the students developed a rough extension of the economic potentials relative to the whisky production levels of the various island distilleries (Fig. 5). The team then performed a Tier 1 environmental analysis as described by Allen and Shonnard. The results can be seen in Fig. 6. This analysis essentially indicated that the process alternatives exhibited similar environmental impacts.

As a final requirement for completion of the project, the students prepared a poster and a project report for dissemination to the involved parties. The students entered this poster in the Undergraduate Poster Competition at the American Institute of Chemical Engineers, 2003 Annual Meeting in San Francisco California [13], where they earned a first place award in Green Engineering and a second place award in the Environmental Division [13].

It was evident from the reception of the students work by Jim McEwen, the awards at the AIChE Meeting, and the responses of the students to the project, that they have developed an excellent concept of green engineering. During this project the students were able to satisfy a number of principles shown in Table 7. The capture and conversion of CO_2 into a valuable product limited the circumstantial risks of the process (1), was designed with the separation system

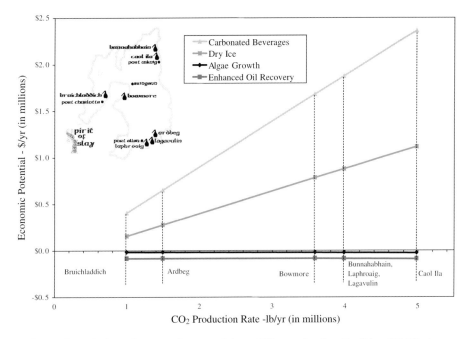

Fig. 5. Correlation of economic potentials to CO_2 production for Islay Distilleries.

Component			
Component	PEL (Permissible Exposure Limit index)	Aquatic Half-life (BDF - Biodegradation Factor)	Persistence (Atmospheric Half-life index)
CO_2	5000 ppm (1/5000)	Days - Weeks (3.10)	11.5 days (1)
EtOH	1000 ppm (1/1000)	Days - Weeks (3.25)	6 hours (0)
NH_3	50 ppm (1/50)	Days - Weeks (3.16)	2 months (3)
$CaSO_4$	0.015 ppm (1/0.015)	Weeks - Months (2.89)	months (3)
Totals	Environmental Index = 0.0002*	Average BDF = 3.13* (days –weeks)	Rating Index = 1* (weeks)

*Stoichiometric coefficients of NH_3 and $CaSO_4$ are assumed to be zero

CO_2 Utilization Alternative	
Use of CO_2	Environmental Index
Carbonated Beverages	0.0002
Dry Ice	0.0002
Algae Growth	0.0002
Enhanced Oil Recovery	0.0002

Fig. 6. Tier 1 environmental analysis.

fully considered (3), was output pulled (5), limited the excess capacity during thedesign effort (8), and was developed with the potential uses and future fate of the CO_2 product being considered (11).

6. Conclusions

Important benefits of the Capstone Design Program are: (1) valuable interaction with interdisciplinary student design teams working on projects that are important to the participating company; (2) access to faculty who have expertize in technical areas related to specific projects; (3) significantly improving the education of the next generation of engineers; and (4) visibility in identifying and recruiting new engineers.

Currently, in the Department of Chemical Engineering at UT there is some variation in our industrially sponsored capstone design activities. In general, these projects focus on process synthesis in response to an authentic industrial need. If the industry fully funds a project team, then they usually insist on confidentiality. Other situations are negotiable, with some projects having no confidentiality concerns. In general, the student teams do utilize a more extensive set of metrics than economics in developing a case for their recommendations.

References

[1] R.M. Counce, J.M. Holmes, E.R. Moss, R.A. Reimer, L.D. Pesche, Chem. Eng. Educ. 29 (1994) 116.

[2] R.M. Counce, J.M. Holmes, S. Edwards, C.J. Perilloux, R.A. Reimer, Chem. Eng. Educ. 31 (1997) 100.

[3] R.M. Counce, J.M. Holmes, R.A. Reimer, Int. J. Eng. Educ. 17 (2001) 396.

[4] R.H. Todd, S.P. Magleby, C.D. Sorensen, B.R. Swan, D.K. Anthony, J. Eng. Educ. 84 (1995) 165.

[5] A. Bright, Proceedings of Advances in Capstone Education, Provo, Utah, 1994, p. 113.

[6] S.P. Magleby, C.D. Sorensen, R.H. Todd, 21st Frontiers in Education Conference, 1992, p. 469.

[7] D.T. Allen, D.R. Shonnard, Green Engineering: Environmentally Conscious Design of Chemical Processes, Prentice-Hall, Upper Saddle River, NJ, 2002.

[8] R.L. Berglund, C.T. Lawson, Chem. Eng. 98 (1991) 120.

[9] J.M. Douglas, Conceptual Design of Chemical Processes, McGraw Hill, Boston, MA, 1988.

[10] M.S. Peters, K.D. Timmerhaus, R.E. West, Plant Design and Economics for Chemical Engineers, McGraw-Hill, Boston, MA, 2003.

[11] F.C. Zevnik, R.L. Buchanan, Chem. Eng. Prog. 59 (2) (1963) 70.

[12] P.T. Anastas, J.B. Zimmerman, Environ. Sci.Technol. 37 (2003) 94A.

[13] J. Campbell, D. McCollum, O. Melnichenko, C. Tyree, 2003 Annual Meeting of the American Institute of Chemical Engineers, San Francisco, CA, 2003.

Sustainability Science and Engineering: Defining principles
Martin A. Abraham (Editor)
DOI 10.1016/S1871-2711(05)01015-9

Chapter 15

Clean Chemical Processing: Cleaner Production and Waste

K.L. Mulholland

Kenneth Mulholland & Associates, Inc., 27 Harlech Drive, Wilmington, DE 19807, USA

1. Introduction

Processes that produce waste reduce their potential profitability. Waste reduction programs, such as pollution prevention, waste minimization, and cleaner production programs, can reduce waste generation by 40–50% with a 200% internal rate of return (IRR) [1,2]. Waste is defined as an unwanted byproduct or damaged, defective, or superfluous material of a manufacturing process. Unfortunately the education and training of chemical engineers has resulted into a thought framework, or "box of thought", that leads to process designs that produce excessive waste and thus have higher investment and operating costs. The engineer needs to be shown how to think outside the "box" and discover process improvements that traditional engineering problem-solving techniques do not find (Fig. 1).

Consider the person who walks along a circular path in the same direction every day. The first time everything is new. By the 30th time, the walker only notices the unusual. The slow deterioration of the walkway or cumulative minor changes in the landscape are not evident. However, if on the thirty-first walk the person walks in the opposite direction, all of a sudden everything is new. It is still the same scenery, but now it is seen from a different perspective.

The same holds true for a manufacturing process. Instead of starting from the front of the process, if you start with the waste streams and move backward through the process, and ask different questions of the same process information, then your view changes completely.

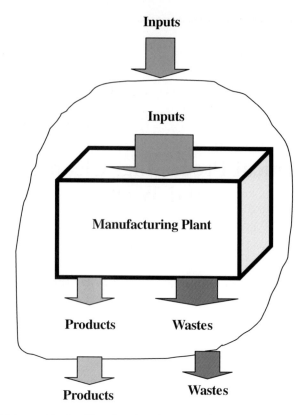

Fig. 1. Think outside the "box," reduce inputs and wastes and increase profits.

The basic skills and knowledge of the engineers, operators, mechanics, scientists, and business people are available to solve waste generation problems. The information is available. All that is needed is a different approach to looking at the information.

The book *Pollution Prevention Methodology, Technologies and Practices* [3] contains more detail on the cleaner production/pollution prevention engineering methods and practices described in this chapter. The book's information and technology were developed into the manual *Identification of Cleaner Production Improvement Opportunities* [4]. The manual was tested at The Dow Chemical Company and further refined with the Korean Institute of Industrial Technology. The manual is scheduled to be published through AIChE/Wiley in 2005.

In summary, this chapter contains a tested-engineering method that will lead business and design personnel to discover process improvements that will reduce waste generation, reduce resource requirements to manufacture product(s), and increase revenues to the business. Even though these engineering methods and practices were developed for chemical processes, the same techniques can be applied to any manufacturing process.

2. Waste

A waste is defined as an unwanted byproduct or damaged, defective, or superfluous material of a manufacturing process. Most often, in its current state, it has or is perceived to have no value. It may or may not be harmful or toxic if released to the environment. A secondary source of waste generation is the excess energy required to process and treat any waste that is generated.

Manufacturing processes produce three classes of waste:

- Process wastes are solid, liquid, vapor, and excess-energy-generated wastes resulting from transforming the lower-value feed materials to a higher-value product(s).
- Utility wastes are solid, liquid, vapor, and excess-energy wastes resulting from the utility systems that are needed to run the process, e.g. steam, electricity, water, compressed air, waste treatment, etc.
- Other wastes result from start-ups and shutdowns, housekeeping, maintenance, etc.

2.1. Process waste

Process chemistry often produces waste. For example:

$$A + B + C \rightarrow I1 + BP1$$

$$I1 + D \rightarrow P1 + BP2 + BP3$$

where:

- A, B, C, and D are feed materials.
- I1 and P1 are an intermediate and product, respectively.
- BP1, BP2, and BP3 are byproducts that have no value and require treatment or landfill.

To reduce the amount of byproduct waste, an economic value must be found for the byproducts (they would then be products), or the process chemistry has to be changed, for example:

$$A + E \rightarrow P1 + P2$$

Process waste is the *most costly waste*, because

- a process that produces no waste would not require costly waste treatment and monitoring systems,

- process waste requires more or larger process equipment to move, heat, cool, and separate from the product or feed materials.

If a business has limited resources, then the reduction of process waste should be the top priority. Lower process-waste generation results in:

- A lower cost-of-manufacture per pound of product, e.g., higher conversion of feeds to product(s), lower consumption of energy to move, store and process wastes, and so on.
- Reduced investment requirements per pound of product, i.e., lower process investment is required to process waste, or existing capacity that was required to process the waste is now made available. The investment required to move, store, separate, and treat waste ranges from 10% to 35% of the total plant investment. For example, if air is being used to remove a waste (contaminant) from a stream, the end-of-pipe treatment investment is a function of the airflow rate. If the contaminant does not need to be removed, then this investment is no longer required.
- Decreased cost of manufacture in other areas, e.g., lower solvent loses, reduced energy use, lower manpower requirements, lower testing requirements, and so on.

2.2. Utility waste

The major waste streams from utility systems are water, air, nitrogen, and energy. The utilities required are a strong function of the manufacturing process, that is, the amount of waste that the process generates and how inherently safe it is. Examples of excess waste or losses from utility systems are:

- Venting of steam, inefficient steam traps, and a low level of condensate return.
- Inefficient boiler operation and steam distribution system heat losses.
- Compressed air and nitrogen leaks.
- Inefficient chilled-water and brine systems.
- Excessive use of chemicals and energy for waste treatment systems.
- Waste treatment sludges, incinerator particulate, and acid gas control system discharges, excess energy use in incinerators and thermal oxidizers, contaminated carbon, and so on.
- Insufficient or compromised equipment insulation.

2.3. Other waste

As with utility waste, the quantity of other wastes, such as uptime losses, maintenance waste, and storage losses, is strongly dependent on the amount of process and utility wastes required to run the process. Examples of other waste include maintenance materials such as rags, filter cartridges, oils and greases, worn-out parts, gaskets, and so on. Uptime losses include lost product, solvents, and catalysts.

2.4. Pollution

Pollution is any release of waste to environment (i.e. any routine or accidental emission, effluent, spill, discharge, or disposal to the air, land, or water) that contaminates or degrades the environment.

Thermodynamic fact — A compound introduced to, such as a solvent, or created in a manufacturing process, such as a byproduct, will escape as a waste or will not be completely destroyed by an end-of-pipe treatment device. If the compound is flammable, toxic, bio-persistent, or bioaccumulative, the health and safety of humans and the ecology of the environment will be adversely affected.

3. Why waste?

The standard college education, normal process design procedures, and on-the-job training are three contributing factors as to why manufacturing processes produce excessive waste (Fig. 2). The education, design and training, focus has been on the product.

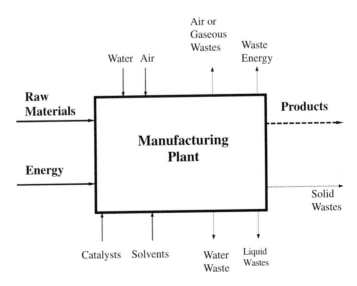

Fig. 2. Plant with pollution.

Until recently, that is within the last 5–10 years, when the vast majority of plants now operating within the US were designed, the teaching of process design focused on:

(1) Lowest possible investment,
(2) Lowest operating cost,
(3) Highest throughput possible especially first pass yield.

These three major objectives of process design were developed in an environment which now seems both innocent and naïve because of the beliefs that:

(1) Air was an infinite sink — that is the gases disappeared and were not even considered as harmful or at worst were easy to treat,
(2) The lines that left the flow-sheet for treatment were someone else's problem, and
(3) Energy costs were not significant.

Further exacerbating the problem was the fact that society was not aware of the problems being caused by waste. Jobs were more important than any harm to the environment or discomfort to the workers, for example, coal miners and coal dust, mill workers and ergonomics, smelter workers and fumes, farm workers and pesticides, and so on.

The normal design process has the following steps.

(1) Researchers develop a new product.
(2) Engineers synthesize a design using their previous experiences around the criteria of minimum investment, minimum operating cost, and maximum throughput.
(3) The design is further refined to make maximum profit.
(4) Just before final design any necessary treatment facilities to meet regulatory requirements are then considered, almost as an afterthought.
(5) Construction and finally startup.

Once the plant is running, the businessperson directs the plant manager to run the process to offset startup costs and generate revenue. The plant manager cannot take time to optimize the process — the business needs the cash.

Now comes the training of the engineers to solve process problems. An experienced engineer mentors an engineer new to the process. The mentor starts with the feed system and discusses the problems associated with the feed system, safety etc. The mentor explains how the reactor operates, temperatures, pressures, catalysts, hazards, etc. Then the mentor describes the separation systems, packaging, and shipping. Finally, the mentor states that the process does generate some waste and this or that is the waste treatment system. This system is described almost as an afterthought.

The "new" engineer is asked to solve some problem to see how well he/she will handle the challenge. Having no preconceived ideas, the engineer uses his/her full range of skills and does a good job. Now it is 10 years down the road. The now "experienced" engineer solves the problems tomorrow the same way as the problems were solved yesterday. Over time while solving problems the engineer acquires blinders on all other happenings, almost as if they were at the Colorado River of the Grand Canyon instead of being on the upper plateau and being able to have a greater view of everything.

There are two general categories of root causes for process waste — operational and fundamental. Operational root causes arise from how the process *is operating* versus how it *should operate*. Examples of operational root causes are:

- poor understanding of the process (the following callout box illustrates this concept),
- control problems,

Process understanding

> At a DuPont site, tars were plugging a distillation column and feed preheater [1]. The tar buildup resulted in plant shutdowns every 3 months to clean the preheater and every 9–12 months to replace the column packing. This cost the business hundreds of thousands of pounds of lost production each year. A second preheater was installed in parallel with the existing heat exchanger to allow cleaning without shutting the plant down. In addition, lab and plant tests were run to better understand the mechanisms behind the tar formation.
>
> As a result of these tests, the plant discovered that the tar formation reaction was pH-sensitive. By tightening Standard Operating Procedures and installing alarms on upstream crude product washers, the tar formation was virtually eliminated. This provided the business with approximately $280,000 per year in additional after-tax earnings.

- not following operating procedures,
- maintenance problems (the following callout box illustrates this concept).

Maintenance

> Methylene chloride is used as a coating and cleaning solvent in the manufacture of several graphic arts and electronic photopolymer films [3].

In the late 1980s, a site released more than 3 million pounds of chlorinated solvents into the air. The coating solution preparation areas accounted for the majority of the air emissions. Coating solutions are prepared batchwise in agitated vessels using a blend of polymers, monomers, photoinitiators, pigments, and solvents. These atmospheric mix tanks were not well sealed, resulting in large fugitive and point source emissions of methylene chloride.

In an effort to enclose the batch vessels as much as possible, the mix tanks were fitted with bolted, gasketed lids. The vessels were also designed with pressure/vacuum conservation vents to allow the vapor pressure to rise to 3 psig before the tanks breathed. By sealing up the process, the site was able to reduce air emissions by 40% and save $426,000 per year in methylene chloride costs.

Fundamental root causes arise from the chemistry, thermodynamic, and engineering limitations of the process. Examples of fundamental root causes are:

- chemistry route picked that requires toxic solvents (the following callout box illustrates this concept),
- catalyst selection and byproduct formation,
- reactor operation and byproduct formation,
- not understanding the functions and thermodynamic principles of the separation processes (the second callout box illustrates this concept),
- inadequate engineering of the process equipment.

Any unwanted material introduced to or created in a process will escape as a waste or will not be completely destroyed by an end-of-pipe treatment device.

Control of the reaction pathway

In hydrocarbon oxidation processes to produce alcohol, there is always a degree of over-oxidation [3]. The alcohol is often further oxidized to carboxylic acids and carbon oxides which are wastes. If boric acid is introduced to the reactor, the alcohol reacts to form a borate ester, which protects the alcohol from further oxidation. The introduction of boric acid terminates the byproduct formation pathway and greatly increases the product yield. The borate ester of alcohol is then hydrolyzed, releasing boric acid for recycle back to the process. This kind of reaction pathway control has been applied to a commercial process, resulting in about a 50% reduction in waste generation once the process was optimized.

Thermodynamic review

> In distillation, the conventional wisdom is to remove the low-boiling material first [3]. After the low-boiling material was removed in a DuPont batch process, the remaining mixture was very difficult to separate, because of azeotropes and pinch points formed by the remaining compounds. The separation difficulties resulted in about one-third of the production run having to be incinerated.
>
> The vapor–liquid equilibrium data for the compounds was reexamined — especially the binary interaction parameters. This reexamination revealed that the low-boiler could extract the product from the remaining compounds. A pilot plant test confirmed the concept, and a continuous extraction process was designed and constructed. The new process reduced the lost product from 200,000 lb/yr to less than 2,000 lb/yr, and the impurities in the final product were decreased from 500 ppm by weight to less than 1 ppm by weight.

The engineer's education, the design process, and training at their job result in a thought framework or "box of thought" that result in process that produce excessive waste, thus require higher operating costs (Fig. 1). The need is to show the engineer how to think outside the "box" and discover process improvements that standard problem-solving techniques do not find.

4. Waste reduction program experience

The following two callout boxes give brief summaries of two major studies that illustrate the value of waste reduction programs.

Cleaner production analysis defines improvement opportunities

U.S. EPA and DuPont Chambers Works Waste Minimization Project [1]

> In May 1993, the US EPA and DuPont completed a joint 2-year project to identify waste reduction options at the DuPont Chambers Works site in Deepwater, New Jersey [1]. The project had three primary goals as conceived:
>
> • Identify methods for the actual reduction or prevention of pollution for specific chemical processes at the Chambers Works site.
> • Generate useful technical information about methodologies and technologies for reducing pollution, which could help the US EPA assist

other companies implementing pollution prevention/waste minimization programs.
• Evaluate and identify potentially useful refinements to the US EPA and DuPont methodologies for analyzing and reducing pollution and/or waste generating activities.

The business leadership was initially reluctant to undertake the program, and was skeptical of the return to be gained when compared against the resources required. After completing a few of the projects, however, the business leadership realized that the methodology identified revenue-producing improvements with a minimum use of people resources and time, both of which were in short supply.

The pollution prevention program assessed 15 manufacturing processes and attained the following results:

• A 52% reduction in waste generation.
• Total capital investment of $6,335,000.
• Savings and earnings amounting to $14,900,000 per year.

o Reduced treatment costs 11%
o Recovered product 15%
o *Process improvements 74%*

11 of the 15 manufacturing processes identified waste reduction opportunities that would require less than $50,000 capital investment and could be completed within 6 months.

The key to the site's success was following a structured methodology throughout the project and allowing the process engineers' creative talents to shine through in a disciplined way.

Preventing industrial pollution at its source — Michigan Source Reduction Initiative [2]

The Natural Resources Defense Council, Dow Chemical, and a group of community activists and environmentalists initiated a project to reduce waste and emissions of 26 priority chemicals at Dow's Midland site by 35% using only pollution prevention techniques. [2]. The project exceeded its goals and reduced:

• targeted emissions by 43%, from 1 million to 593,000 pounds and
• targeted wastes by 37%, from 17.5 million to 11 million pounds.

The cost savings and process improvements were significant where the waste reductions will be paid for in less than 1 year, a 180% overall rate of return. For example, one project required $330,000 and will return $3,300,000 per year in raw material savings. The 17 projects revealed important insights into pollution prevention:

- Majority of projects required relatively small amounts of capital. Ten were $50,000 or less. Five required no capital at all.
- Some of the greatest reductions in waste cost the least amount of money. Good reduction opportunities were found in almost every production process even though Dow businesses doubted that they would be found.
- Opportunities were broadly available in the various businesses in the plant.
- Several projects focused on basic process changes, yet were implemented in a relatively short period of time.

5. Cleaner production and sustainable manufacturing

The traditional approach to process design is to first engineer the process and then to engineer the treatment and disposal of waste streams. However, with increasing regulatory and societal pressures to eliminate emissions to the environment, disposal and treatment costs have and will continue to increase. As a result, capital investment and operating costs for disposal and treatment have become a larger fraction of the total cost of any manufacturing process. For this reason, the *total system* must now be analyzed simultaneously (process plus treatment) to find the minimum economic option.

Experience in all industries teaches that processes that minimize waste generation at the source are the most economical. For existing plants, the problem is even more acute. Even so, experience has shown that waste generation in existing facilities can be significantly reduced (greater than 30% on average), while at the same time reducing operating costs and new capital investment [1–4]. Also, processes, which generate waste could require 10-35% more investment. The higher investment is required to store, heat, move, and separate the waste streams from the product(s) and recyclable feed materials, solvents, catalysts, and so on.

Cleaner production engineering provides tools, which address the problems of negative environmental impact, loss of materials to waste, and increased process investment to deal with wastes. The application of these tools will result in a manufacturing process evolving from a typical process shown in Fig. 2 to the desired process shown in Fig. 3.

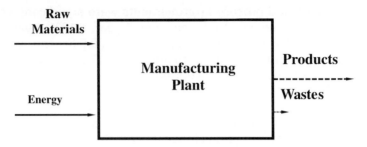

Fig. 3. "ZERO" waste generation facility.

The most common definition for sustainable development is development that "meets the needs of the present without compromising the ability of future generations to meet their own needs" [5]. For a manufacturing facility the definition translates to a process that requires the minimum amount of resources, both material and energy, and produces a minimum amount of waste for a given chemistry, that is, Fig. 3.

The concepts behind the methodologies presented in this chapter are described in the book "Pollution Prevention: Methodologies, Technologies and Practices". The last chapters of that book describe technologies and practices to reduce the waste from processes ranging from ventilation problems to reactors to pH control [3].

In contrast to Fig. 2, Fig. 3 depicts a manufacturing facility with an absolute minimum or "zero" amount of waste being generated. Since most processes require some level of solvents, catalysts, and other materials and normally produce byproducts in the reactor, to attain a "zero" waste process requires fundamental understandings of the process and how to implement changes inside the pipes and vessels (the following callout box illustrates this concept).

"ZERO" waste

Solvents used in the manufacture of intermediate monomers were incinerated as a hazardous waste. Alternative non-hazardous solvents were considered and rejected. However, the intermediate monomers were found to have the dissolution capacity of the original solvents and could replace them. By utilizing existing equipment, realizing savings in ingredients' recovery, and reducing operating and incineration costs, the project achieved a 33% IRR and a 100% reduction in the use of the original solvents.

6. Methodology to achieve cleaner production processes

Identification of opportunities to reduce waste from a manufacturing facility requires contributions from everyone, from process engineers and chemists, operators, mechanics, business people, and so on. To identify the opportunities, a structured methodology is required that

- uses minimum resources (time and money) to define process improvement opportunities and to conduct a process baseline analysis,
- uses existing process information to define the process improvement opportunities,
- defines the process characteristics, i.e. process changes required inside the pipes and vessels to optimize the process,
- has been proven to work.

The methodology described here has been used to identify process improvement opportunities for more than 50 processes, which include:

- pharmaceutical intermediates,
- elastomer monomers and polymers,
- polyester intermediates and polymers,
- batch processes such as agricultural products and paints,
- chlorofluoro-hydrocarbon and chlorocarbon processes,
- specialty chemicals and waste water treatment facilities,

and has in all cases identified new process improvement opportunities that minimize/reduce waste and in 75% of the cases identified improvements that cost less than $50,000.

The methodology is based partly on the observation that the volumetric flow of an air or gaseous waste stream and the volumetric flow and organic loading of a wastewater stream (Fig. 2) determine the required end-of-pipe treatment investment and operating cost. Manufacturing plant investment and cost of manufacture are also influenced by the same gaseous and water flows. A second observation is that end-of-pipe treatment is required only because the streams contain components that have to be abated or removed.

There are two general problem definition approaches that provide the data and information to identify process improvement opportunities. These approaches help the engineer to think outside the "box". One approach (Waste stream analysis) follows the waste from its point of emission back through the process to its point of origin. As one progresses backward, one asks a series of what, how, and why questions to either eliminate the waste or reduce its amount. The second approach (Process analysis) looks at the process holistically. This approach helps identify overall causes of waste and synergies between various parts of the process that the first approach might not reveal.

6.1. Waste stream analysis

The first step in implementing waste minimization is to identify the options that will eliminate or minimize the waste streams' volumetric flow (this has the greatest influence on investment and operating costs) or eliminate/minimize the components of concern (which are the reason to treat the stream). To uncover the best options, each waste stream should be analyzed as follows:

(1) List all components in the waste stream, along with any key parameters. For instance, for a wastewater stream these could be water, organic compounds, inorganic compounds (both dissolved and suspended), pH, etc.
(2) Identify the components triggering concern, e.g. hazardous air pollutants (HAPs), carcinogenic compounds, wastes regulated under the Resource Conservation and Recovery Act (RCRA), etc.
   • Determine the sources of these components within the process.
   • Then develop process improvement options to reduce or eliminate their generation.
(3) Identify the highest volume materials — often these are diluents, such as water, air, a carrier gas, or a solvent. These materials frequently control the investment and operating costs associated with end-of-pipe treatment of the waste streams and have a significant impact on the process cost of manufacture.
   • Determine the sources of these high-volume materials within the process.
   • Then develop process improvement options to reduce their volume or eliminate them.
(4) If the components identified in step 2 are successfully minimized or eliminated, identify the next set of components that have a large impact on investment and operating cost (or both) for end-of-pipe treatment. For example, if the aqueous waste stream was originally a hazardous waste and it was incinerated, eliminating the hazardous compound(s) may permit the stream to be sent to the wastewater treatment facility. However, this may overload the biochemical oxygen demand (BOD) capacity of the existing wastewater treatment facility, making it necessary to identify options to reduce organic load in the aqueous waste stream.

Waste stream analysis and process analysis

In a sold-out market situation, a DuPont intermediates process was operating at 56% of its peak capacity [3]. The major cause of the rate limitation was traced to poor decanter operation. The decanter recovered a catalyst, and fouling from catalyst solids caused its poor operation. Returning the process to high utility required a 20-day shutdown. During the shutdown, the vessel was pumped out and cleaned by water washing. The solids and hydrolyzed catalyst were then drummed and incinerated. A *waste stream analysis* identified three cost factors — the volume of wastewater that had to be treated, the cost of the lost catalyst, and the incineration cost.

An *analysis of the process* and its ingredients indicated that the decanter could be bypassed and the process runs at a reduced rate, while the decanter was cleaned. A process ingredient was used to clean the decanter, enabling recovery of the catalyst ($200,000 per year value). The use of the process ingredient cut the cleaning time in half, and that, along with continued running of the process, eliminated the need to buy the intermediate on the open market. The results were a 100% elimination of a hazardous waste (125,000 gallons/yr) and cash flow savings of $3,800,000 per year.

6.2. Process analysis

The only materials that are truly valuable to the business are the raw materials for reaction, any intermediates, and the final products. Aside from the feed, other input streams (e.g. catalysts, air, water, etc.) are required because of the designers' limited knowledge of how to manufacture the product without them. The function of these input streams is to transform feed materials into products. To reduce or eliminate them, the feed materials, intermediates, or products must serve the same function as those input streams, or the process needs to be modified to eliminate them.

For either a new or existing process, a process analysis consists of the following steps:

(1) List all feed materials reacting to salable products, any intermediates, and all salable products. Call this List 1.
(2) List all other materials in the process, such as non-salable byproducts, solvents, water, air, nitrogen, acids, bases, and so on. Call this List 2.
(3) For each compound in List 2, ask "How can a material from List 1 be used to do the same function as the compound in List 2?" or "How can the process be modified to eliminate the need for the material in List 2?"
(4) For those materials in List 2 that are the result of producing non-salable products (i.e. waste byproducts), ask "How can the chemistry or process be modified to minimize or eliminate wastes (for example, 100% reaction selectivity to a desired product)?"

6.3. Sample questions

When coupled with the application of fundamental engineering and chemistry practices, examining a manufacturing process by these waste stream and process analyses' techniques will often result in a technology plan for a minimum-waste-generation process.

  Typical questions are:

- If the solvent is "bad," can another more benign solvent be used?
- Can the process be modified to eliminate the solvent?
- If water is used for example to dissolve a salt of the intermediate, why use water? Why not transfer the intermediate as a solid?
- If multiple solvents are used, why?
- If a homogeneous catalyst is used and has to be separated, why not use a heterogeneous catalyst?
- If any reactions are energetic, is there a different reaction pathway?
- If air is being used as a source of oxygen, why?
- If some reactions are slow and some are fast, is the proper type of reactor being used? Fast reactions in a pipe reactor and slow in a batch reactor.
- If byproducts are produced, is there a reaction pathway or set of conditions that do not result in certain byproducts?
- If toxic or hazardous solvents or feed materials are required, can they be produced in situ to eliminate shipping and storage?
- Do separations unit operations use no solvents, require minimum energy, and produce minimum byproducts?

7. Application of the methodology

A single person could do the entire program [4]. However, the most efficient means to identify the largest number of improvement opportunities is to use structured brainstorming. The team of business personnel necessary to discover a larger number of improvement opportunities should consist of:

Person	Role
Business leader	Provides direction, resources, and support
Process team leader	Leads the entire program
Team members	3 to 4 members comprising of process engineer, process chemist, lead operator/mechanic, and possibly an environmental specialist or project engineer
Cleaner production	Advises the process team on the required expert information to assemble for the brainstorming and facilitates the brainstorming session

Scribe(s)   Records ideas during brainstorming session

Brainstorming team   8 to 15 people. The process team plus other members of the business and outside experts

7.1. Cleaner production expert

The cleaner production expert provides the guidance for an effective program. The major program objectives are to:

- Obtain the business leader's commitment to the cleaner production program. The business leader
  - provides resources to develop and implement the program,
  - defines the objectives of the program,
  - provides support for the business personnel involved in the program.
- Educate the core team leader on the cleaner production program.
  - acquaint the leader with the sequence of events in the program,
  - ensure the leader of support to make the program a success,
  - enlist the leader's active and enthusiastic support for the program.
- Educate the core team on the cleaner production program.
  - acquaint them with the sequence of events,
  - acquaint them with the data requirements,
  - enlist their support for the program.
- Facilitate the brainstorming to identify improvement opportunities.
  - acquaint everyone on their duties and responsibilities,
  - ensure that an effective brainstorming session occurs,
  - be a resource for the business leader and core team.

To fulfill the desired objectives the cleaner production expert should

(1) meet with the business leader,
(2) meet with the core team leader,
(3) provide for the core team a three-hour seminar/discussion session on the cleaner production program,
(4) become acquainted with the process by having the core team describe the process and be given a tour of the facilities,
(5) discuss data requirements and sources,
(6) develop a timeline for events.

8. Summary

The cleaner program developed in Korea (described in the callout box below) illustrates the value of the methodology. When first started the scientists and engineers at the Korea National Cleaner Production Center (KNCPC) and Hanwha Chemical knew that no new improvement opportunities would be identified. Hanwah Chemical is a very well run chemical company and is very proud of their knowledge and running of their processes. They had just finished an energy conservation program, thus they knew that no new improvements were possible at that time. However, using the methodology described in this chapter, the engineers and scientists were guided to view their process differently. More than 80 new improvement ideas were generated in the 3-day brainstorming session which resulted in a 35% decrease in waste generation and total of $9,000,000 increase in business revenues (cost reduction plus more product).

The discovery of process improvements that reduce waste generation is a **human** problem **not** a technical problem. Our education, process design methodology, and on-the-job training causes plants to be built and operated that generate billions of pounds of excess waste that is estimated to incur hundreds of billions of dollars of excess cost of manufacture not to mention threat of public relations debacles. One approach is for the engineer to think outside the "box," that is, to look at the manufacturing process from the back to the front end of the process.

The EPA/DuPont [1], Michigan Source Reduction Initiative [2], KNCPC program [4], and other cleaner production projects showed that: (The Waste Reduction Experience section gives summaries of the first two studies and callout box below a short summary of the KNCPC program.)

- product changes are not required,
- new technology is not required,
- any new investment has greater than 100% IRR,
- waste generation reduced by 40–50%,
- the knowledge resides in the business,
- everyone in the business can contribute, that is engineers, chemists, operators, mechanics, and so on, and
- the design and operation of low waste generation processes is a human not a technical problem.

Proven success in South Korea

To accelerate the introduction of cleaner production technologies to various industries throughout South Korea, in 1999 the Ministry of Commerce, Industry, and Energy formed the KNCPC as a division of the Korea Institute of Industrial Technology. Hanwha Chemical Corp. recently worked with KNCPC and Kenneth Mulholland and Associates to develop cleaner production technology for complex processes.

Hanwha Chemical manufactures polyvinyl chloride (PVC), low-density polyethylene (LDPE), linear low-density polyethylene (LLDPE), and chlor-alkali. It is headquartered in Seoul, has a research and development center in Daejeon, and operates two manufacturing plants — Yosu and Ulsan. Hanwha's policy is to follow stricter environmental management guidelines than required by regulation, and it has adopted the ISO 14001 Environment Management System. Even though Hanwha had just completed a massive energy conservation program, the company's leader-ship understood the value of cleaner production technologies and volun-teered to work with KNCPC to develop such technologies for complex processes.

The project focused on the Ulsan vinyl chloride monomer plant, since it had a higher manufacturing cost and environmental load. Plant personnel were trained on how to view their process by focusing on the waste streams instead of the product streams. A brainstorming team, consisting of experts in chemistry, engineering, environmental control, electricity, piping and equipment, and operations, met and generated more than a hundred process improvement ideas. The top ideas underwent further technical and economic analyses, and an implementation plan was for-mulated for the best ideas. The process improvements required no new technology, only better use and understanding of existing technologies, and ranged from revised operating procedures to major process modifi-cations that can be patented. The improvements, which are expected to be completed by the end of 2003, involve a capital investment of $2,600,000 and will achieve:

- a 35.7%, or 5232 m.t., waste reduction
- a cost reduction of $3,200,000
- additional revenue generation of $6,000,000.

Hanwha Chemical is expanding the program to all of its processes at the Ulsan site. KNCPC is actively implementing and promoting cleaner pro-duction technologies for all industries. It is working to: develop metrics by industry type; disseminate technology through collaborations with uni-versities, institutes and research teams, as well as via forums such as roundtables; and provide support for businesses developing and imple-menting cleaner production technologies.

Kenneth L. Mulholland, Kenneth Mulholland and Associates
Kwiho Lee, Korea National Cleaner Production Center
K. H. Hyun, Hanwha Chemical Corp.

References

[1] US Environmental Protection Agency, DuPont Chambers Works Waste Minimization Project, EPA/600/R-93/203, US EPA, Office of Research and Development, Washington, DC, November 1993.
[2] Natural Resources Defense Council and the Dow Chemical Company, Preventing Industrial Pollution at its Source, The Final Report of the Michigan Source Reduction Initiative, Meridian Institute: Dillon, CO, September 1999.
[3] K.L. Mulholland, J.A. Dyer, Pollution Prevention: Methodology, Technologise and Practices, AIChE:, New York, 1999.
[4] K.L. Mulholland, K. Lee, Identification of Cleaner Production Improvement Opportunities, KNCPC and Kenneth Mulholland & Associates, August 2001.
[5] A. AtKisson, Believing Cassandra, An Optimist Looks at a Pessimist's World, Chelsea Green Publishing Company:, White River Junction, 1999.

Sustainability Science and Engineering: Defining principles
Martin A. Abraham (Editor)
© 2006 Published by Elsevier B.V.
DOI 10.1016/S1871-2711(05)01016-0

Chapter 16

Role of Chemical Reaction Engineering in Sustainable Process Development

C. Tunca, P.A. Ramachandran, M.P. Dudukovic

Chemical Reaction Engineering Laboratory (CREL), Washington University, St. Louis, MO, USA

1. Introduction

Achieving sustainable processes that allow us at present to fully meet our needs without impairing the ability of future generations to do so, is an important goal for current and future engineers. In production of new materials, chemicals, and pharmaceuticals, sustainable processes certainly require the most efficient use of raw materials and energy, preferably from renewable sources, and prevention of generation and release of toxic materials. Advancing the state of the art of chemical reaction engineering (CRE) is the key element needed for development of such environment friendly and sustainable chemical processes.

Current chemical processes depend heavily on the non-renewable fossil-based raw materials. These processes are unsustainable in the long run. In order to make them sustainable, chemical technologies must focus on employing renewable raw materials as well as preventing and minimizing pollution at the source rather than dealing with end-of-pipe treatments. New technologies of higher material and energy efficiency offer the best hope for minimization and prevention of pollution. To implement new technologies, a multidisciplinary task-force is needed. This effort involves chemical engineers together with environmental engineers and chemists since they are predominantly in charge of designing novel chemical technologies.

Pollution prevention problem can be attacked via a hierarchical approach based on three levels as outlined in the book by Allen and Rosselot [1]. Each level uses a system boundary for the analysis. The top level (the macro-level) is the largest system boundary covering the whole manufacturing activity from raw material extraction to product use, and eventually disposal. These activities involve chemical and physical transformation of raw materials creating pollution or wastes as shown schematically in Fig. 1. The scope of the macro-level analysis is mainly in tracking these transformations, identifying the causes of pollution and suggesting the reduction strategies.

The next level is the plant level or the meso-scale, which is the main domain of chemical engineers. The meso-level focuses on an entire chemical plant, and deals with the associated chemical and physical transformations of non-renewable resources (e.g. petroleum and coal, etc.) and renewable resources (e.g. plants and animals) into a variety of specific products. These transformations can result in a number of undesirable products which, if not checked, can result in pollution of the environment. The challenge then for modern chemical engineers is to improve the efficiency of existing processes, to the extent possible, and design new cleaner and more efficient processes. While in the past the effort was focused on end of the pipe clean up and remediation, the focus now is on pollution prevention and ultimately on sustainability. Mass and energy transfer calculations as well as optimization of processes that result in less pollution can be performed at this level. This plant scale boundary usually consists of raw material pre-treatment section, reactor section and separation unit operations (see Fig. 2). Although each of these sections are important, the chemical reactor forms the heart of the process and offers considerable scope for pollution prevention. The present paper addresses the scope for pollution prevention in the

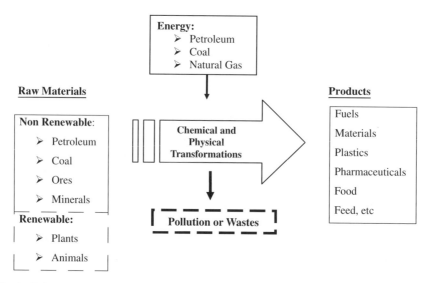

Fig. 1. Schematic of chemical and physical transformations causing pollution or wastes.

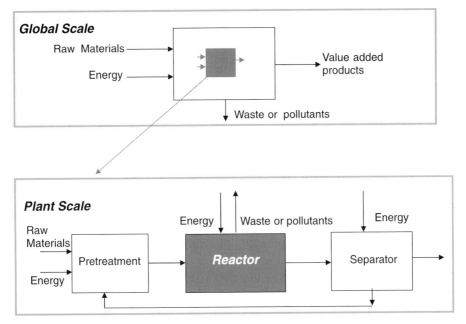

Fig. 2. Schematic of system boundaries. The first figure shows the global scale and the second one shows the plant scale. The reactor itself constitutes another boundary that chemical reaction engineers focus on.

reactor unit and sustainable development. It may however be noted that the reactor and separator sections are closely linked and in some cases, improvements in the reactor section may adversely affect the separation section leading to an overall increase in pollution. Hence any suggested improvement in the reactor section has to be reevaluated in the overall plant scale context.

The third level indicated by Allen and Rosselot [1] is the micro-level that deals with molecular phenomena and how it affects pollution. Analysis at this scale includes synthesis of benign chemicals, design of alternative pathways to design a chemical, etc. CRE plays a dominant role at this level as well and some applications of CRE at the micro-level pollution prevention are also indicated in the paper.

Waste reduction in chemical reactors can be achieved in the following hierarchical manner: (i) better maintenance; (ii) minor modifications of reactor operation; (iii) major modifications or new reactor concepts; and (iv) improved process chemistry, novel catalysts and use of these in suitable reactors. The items (i) and (ii) are usually practiced for existing plants while items (iii) and (iv) above are more suitable in the context of new technologies or processes. Also items (iii) and (iv) involve a multidisciplinary R&D activities which can be costly. But since long-term sustainability is the goal this is a worthwhile effort and increasing activities are expected in this direction in the future. The items (i) and (ii) are addressed in a paper by Dyer and Mulholland [2] and this paper focuses mainly on (iii) and (iv).

It is also appropriate at this point to stress the multiscale nature of reaction engineering. Figure 3 shows the multilevel CRE approach. We have at the tiniest scale the catalytic surface up to macro-scale of huge 10 m high chemical reactor and hence the task of pollution prevention in chemical reactor is a formidable task and has to address phenomena at all these scales. At the molecular level, the choice of the process chemistry dramatically impacts the atom efficiency and the degree of potential environmental damage due to the chemicals produced. We will first review how some micro-level concepts such as atom efficiency (Section 2) and optimum catalyst development (Section 3) affect reactor choice and pollutants generated. At the meso-level, optimizing the catalyst properties (Section 3) and choosing the right media (Section 4) significantly reduces the adverse environmental impact of chemical processes. Clearly, proper under-standing and application of the principles of multiphase reaction engineering are very important for proper execution of truly environmentally benign processes since almost all processes involve more than one phase. At the macro-level, it is crucial to estimate the hydrodynamic effects (Section 5) on reactor performance as this will lead to the selection of the right reactor. Furthermore, novel ap-proaches to reactor design using process intensification concepts (Section 6) can be implemented to improve efficiency and minimize pollution. In addition, en-vironmental impact analysis (Section 7) of the developed process has to be

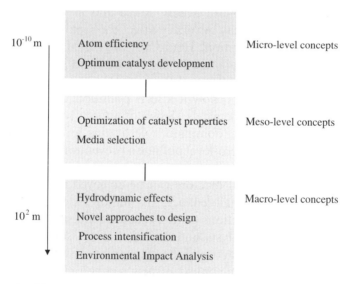

Fig. 3. Schematic of integrated multilevel CRE approach. All the above levels are important in design and operation of a successful reactor.

evaluated to see if green chemistry conditions are met and to assess the overall global impact of these changes. As can be seen, CRE is a marriage of multidisciplinary and multiscale efforts with the aim of operating at sustainable green chemistry conditions to reduce pollution and maximize efficiency. The summary of these efforts and conclusions are given in Section 8.

2. Raw materials selection

The aim of benign synthesis is to reduce the amount of side reactions generating undesired by-products, i.e. waste. Alternative direct synthetic routes are therefore advantageous from the point of waste reduction and have an economic advantage. The organic chemistry is rich with reactions where large quantities of inorganic salts are generated as wastes that lead to poor atom efficiency. These can be avoided by selecting raw materials appropriately. Therefore, using environmentally benign raw materials can make a big impact on pollution prevention and thus enhance sustainability. Atom and mass economy calculations, that measure how efficiently raw materials are used, are the key decision-making concepts in selection of raw materials for a given process.

To give an example, two raw materials, benzene and *n*-butane can be employed in the production of maleic acid. The following reactions are possible routes to maleic anhydride (the first step to produce maleic acid).

Benzene route:

$$2C_6H_6 + 9O_2 \xrightarrow{V_2O_5MoO_3} 2C_4H_2O_3 + 4H_2O + 4CO_2$$

n-Butane route:

$$C_4H_{10} + 3.5O_2 \xrightarrow{(VO)_5P_2O_5} C_4H_2O_3 + 4H_2O$$

In the benzene route, there are four carbon atoms in the product maleic anhydride per 6 carbon atoms in benzene. Therefore, the atom efficiency for the carbon atom is $4/6 \times 100\% = 66.7\%$. In the *n*-butane route, there are four carbon atoms in *n*-butane per four carbon atoms in maleic anhydride thus giving 100% atom efficiency.

If mass efficiency is considered, then the mass of the product is compared to the mass of the raw materials. The molecular weight of maleic anhydride is 98. For the *n*-butane route, we need 1 mole of *n*-butane (molecular weight, 58) and 3.5 moles of oxygen (total mass of $3.5 \times 32 = 112$). Thus, the total mass of raw materials needed is 170. The mass efficiency of the *n*-butane route is therefore 98/170 or 57.6%. By a similar calculation, we can show that the mass efficiency of the benzene route is only 44.4%.

As can be seen, *n*-butane is favorable to benzene in comparison of both atom and mass economy. In addition, benzene is expensive and toxic. Therefore n-butane has replaced benzene in maleic anhydride production since the 1980s. A detailed case study comparing the two routes is given in the Green Engineering book [3].

Choosing benign and efficient raw materials is the first step in designing environment friendly processes. Once the process route is selected, the process yield and efficiency can be further improved, and pollution can be minimized more effectively by the proper choice of catalyst and solvents used. Issues related to catalyst selection and development are discussed next.

3. Catalyst selection, development and reactor choice

Catalysts are being used extensively in chemical and fuel industries. Hence, selecting and developing the right catalyst has a huge impact on the success of a proposed process route. However, it may be noted that the development of a novel catalyst has to be often combined with novel reactor technology for the process to be economically viable and environmentally beneficial. Hence catalyst development and reactor choice often have to be considered in unison.

As a general heuristic rule, the processes that replace liquid routes by solid catalyzed routes reduce pollution significantly especially when some of the liquid reactants are toxic. For example, many liquid acids such as H_2SO_4 and HF used as catalysts in petroleum refining industry impose a significant environmental hazard as they are highly corrosive and toxic. Therefore, based on environmental concerns these catalysts are now being replaced by solid-acid catalysts.

Once selected, to further tailor the catalyst to get optimum yield and selectivity, physical properties of the catalyst have to be determined and improved. Experimental methods, namely NMR, spectroscopy, kinetic measurements are available to study important properties such as the surface topology of the catalyst, the adsorption sites and how the molecules are adsorbed on the catalyst surface. Computer simulations of pore structure are also becoming increasingly popular to study the transport behavior of the catalysts. Detailed micro-kinetic modeling using molecular dynamic simulations are also useful to guide the design of a new catalyst.

Examples of catalyst development and appropriate reactor selection are illustrated in the following discussions by consideration of oxidation and alkylation, which are two common reaction types in organic synthesis.

3.1. Oxidation

Oxidation reactions are important in producing many fine chemicals, monomers and intermediates. O_2, H_2O_2 and HNO_3 are common oxidants used in these reactions, with oxygen being the most benign oxidant. We will particularly talk about *n*-butane oxidation to maleic anhydride in this section and discuss the reactor choice for vanadium phosphorous oxide (VPO) catalyst.

Initially, fixed bed reactor configurations were used for this process. Fixed bed catalytic reactor is one of the most utilized reactors in the petrochemical and petroleum refining industry. These reactors use solid catalysts in pellet or granular form and they can be visualized as shell and tube heat exchangers. There are several limitations in employment of fixed bed catalytic reactors. The reactors are expensive and only up to 2% *n*-butane can be used in the feed [4]. Yield is around 50% with 70–85% conversion and 67–75% molar selectivity to maleic anhydride [4]. Moreover, since the reaction is exothermic, hotspot formation must be avoided. Reactor designs have advanced to a point where some of these issues can be addressed effectively. Catalyst development has also progressed and the yield and selectivity have improved over the years by controlling the chemical composition and morphology of the catalyst. However, it may be noted that, the mechanical properties of the catalyst is equally important. The catalysts used in packed beds are usually supported metals from 1 to 10 mm in size. These must have adequate crushing strength to carry the full weight of a packed bed.

In order to minimize hotspots, fluidized catalyst beds are preferred over fixed catalyst beds. The advantages of fluidized catalyst beds include the ease of temperature control, superior heat transfer and lower operating temperatures compared to fixed catalyst beds. In a fluidized catalyst bed, higher butane concentrations (up to 4%) are handled reducing operating costs [4]. The disadvantage of the fluidized catalyst bed is the rapid reduction of the catalyst surface, catalyst attrition and carry over of fines leading to air pollution. Again the mechanical properties of the catalyst play an important role. In fluidized beds, much smaller (compared to packed bed) catalyst particles on support (20–150 μm) are used and these must exhibit outstanding attrition properties. Hence, extensive catalyst development was required to move butane oxidation from fixed to fluidized beds.

In the mid-1990s, another improvement by DuPont de Nemours [5] in reactor design introduced circulating fluid bed (CFB) technology to maleic anhydride production. The incentive for this change was provided by the realization that much higher productivity and selectivity can be obtained by using the catalyst in transient rather than steady-state operation. CFB provides an ideal reactor set-up for such cyclic transient operation. In this reactor configuration, the chemistry is executed in a fluid bed-riser combination to accommodate successive oxidation and reduction of the catalyst. In the riser, *n*-butane gets converted into maleic anhydride while the catalyst gets reduced from V^{+5} to V^{+3} given by the scheme as:

$$V^{+5} \xrightarrow{HC} V^{+4} \xrightarrow{HC} V^{+3}$$

The reduced catalyst then circulates to the regenerator where it comes in contact with air and gets oxidized back to V^{+5} given by the scheme as:

$$V^{+3} \xrightarrow{O_2} V^{+4} \xrightarrow{O_2} V^{+5}$$

In this way, *n*-butane and oxygen are not in direct contact and this leads to minimizing side reactions and higher maleic anhydride selectivity (up to 90%) is therefore obtained [4,6]. Figure 4 shows the CFB reactor configuration. Again, to enable the use of CFB extensive catalyst development took place to introduce a highly porous but extremely attrition resistant shell on the VPO type catalyst.

VPO catalyst has also been subject to detailed investigation to further optimize its physical properties. For example, Mota et al. [7] investigated modifying VPO catalyst by doping with Co or Mo to operate under fuel-rich conditions (i.e. $O_2/C_4H_{10} = 0.6$). The authors state that Co-doped VPO catalyst performed better than the Mo-doped VPO catalyst and did not deactivate as the original VPO catalyst.

3.2. Alkylation

Hydrocarbon alkylation reactions are important in petroleum industries for producing high-octane gasoline stocks. Traditional routes use liquid phase acids such as HF or H_2SO_4 or Lewis acid metal halides such as $AlCl_3$ and BF_3. In these reactions, stoichiometric quantities of acids and/or halides are often needed and generate massive corrosive and toxic effluents. Alkylation reactions are, therefore, excellent targets for new cleaner chemistry.

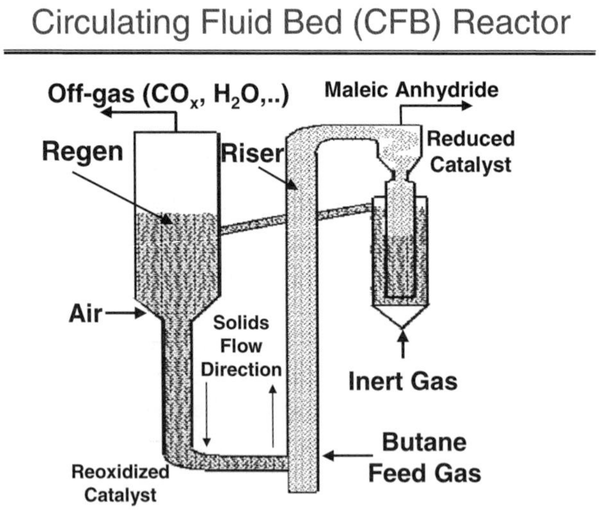

Circulating Fluid Bed (CFB) Reactor

Fig. 4. Circulating fluid bed reactor used for maleic anhydride production.

There has been a significant development in the reactor design in liquid phase processes to lessen the environmental risks. In the original process, stirred tanks (mixer settlers with heat exchanger), operated in parallel to keep olefin concentration low, were used with HF being the catalyst. Figure 5 shows the schematics where HF is recycled with an external pump. Due to HF use and leaky seals on the pump and reactor mixing shafts, this process is environmentally unfriendly. The newer reactor solved this problem by utilizing an HF internal recycle. This design accomplished mixing by utilizing the buoyancy force created in the mixture of a heavy (HF) and light phase (hydrocarbon paraffin–olefin mixture). Unfortunately, the process is still environmentally unfriendly due to presence of HF. The challenge is to develop a stable solid catalyst that will be effective and regenerable as the conventional HF/H_2SO_4 catalyst that is still employed worldwide.

Common types of solid acids are the Beta zeolites, ion exchange resins, such as silica, supported nafion and heteropoly acids such as tungsto-phosphoric acids. However, these catalysts are easily deactivated and must be reactivated each time. Hence, complex reactor types must be designed so that the reaction and regeneration activities can be combined. CFBs, packed beds with periodic operation, stirred tanks with or without catalyst baskets and chromatographic reactors are types of reactors that have been considered. Figure 6 shows a CFB, which is similar in concept to the reactor used in maleic anhydride production.

In order to select the best reactor among these reactor types, reactor models based on hydrodynamics, kinetics and pore diffusion must be accounted for since transport resistances may play a significant role in reactor performance. Therefore, proper understanding of these factors is a must to interpret the product selectivity and extent of formation of waste products. An additional discussion of transport resistance is in Section 5.

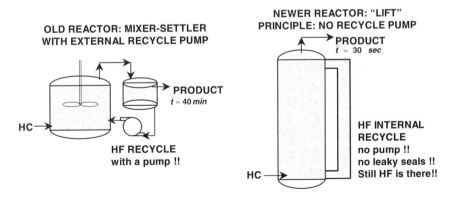

Fig. 5. Reactor types for liquid phase alkylation process.

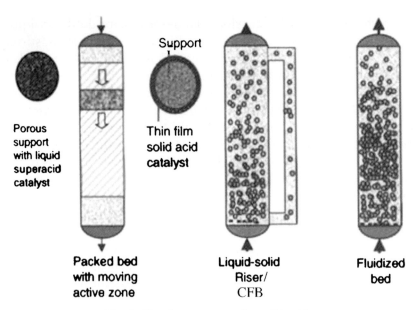

Fig. 6. Novel reactor type for solid acids.

4. Solvent selection

As a general guideline, for an improved process reactor design, the use of solvents should be reduced and benign solvents should be employed if necessary. With these guidelines in mind, much attention has been given to "green" solvents such as supercritical CO_2 ($scCO_2$) and ionic liquids with the hope that they will replace the current solvents that cause pollution.

$scCO_2$ is environmentally friendly as it is non-toxic, unregulated and non-flammable. It is also preferable as it is ubiquitous and inexpensive. $scCO_2$ has been extensively used in oxidation reactions. However, $scCO_2$ based oxidations are limited by low-reaction rates. Further the homogeneous catalysts [8] needed in the reaction have limited solubility in $scCO_2$. Hence, the use of expanded advents is being advocated for homogeneous catalytic oxidations, and currently there has been a shift in the research to use the CO_2-expanded solvents for many organic processes. Advantage of CO_2-expanded solvents is that the process pressure can be significantly lower compared to $scCO_2$. By changing the amount of CO_2 added, it is possible to generate a continuum of media ranging from the neat organic solvent to pure CO_2 [8,9]. Wei et al. [8] have studied the solubility of O_2 in CO_2-expanded CH_3CN and found that it was two times higher compared to neat CH_3CN, resulting in maximizing oxidation rates. The authors also reported that conventional organic solvent was replaced up to 80%. Hence CO_2-expanded solvents look promising in replacing the current solvents that are not environmentally friendly. More research is still needed in this area to enable scale-up and commercialization.

Ionic liquids are solvents that have no measurable vapor pressure. They exhibit Brϕnsted and Lewis acidity, as well as superacidity and they offer high solubility for a wide range of inorganic and organic materials [10]. The most common ones are imidazolium and pyridinium derivatives but also phosphonium or tetralkylammonium compounds can be used for this purpose. Classical transition metal catalyzed hydrogenation, hydroformylation, isomerization and dimerization can be all performed in ionic liquid solvents [11]. The advantages of ionic liquid solvents over conventional solvents are the ease of tuning selectivities and reaction rates as well as minimal waste to the environment [11]. As with scCO$_2$, more research is needed in scale-up and commercialization as well as in investigating different types of ionic liquid solvents for other catalytic reactions.

5. Reactor design

Choice of reactor type should be made in the early stages of process and catalyst development. It should consider the kinetic rates achievable and their dependence on temperature and pressure, the transport effects on rates and selectivity, the flow pattern effect on yield and selectivity as well as the magnitude of the needed heat transfer rates. Then plug flow or perfect mixing is identified as ideal flow pattern that best meets the process requirements in terms of productivity and selectivity. The final reactor type is chosen so as to best approach the desired ideal flow pattern and provide the needed heat transfer rates. The reactor should not be overdesigned to reach the desired product selectivity and to minimize waste generation. Therefore, sophisticated reactor models have been developed as essential tools for reactor design and scale-up. For these models to be accurate, information on volume fraction (holdup) distribution, velocity and mixing of the present phases has to be known. This hydrodynamic information is then coupled with kinetics of the reaction and deactivation to develop a sophisticated reactor model.

There are several experimental measurement techniques to get information on velocity and turbulence parameters in gas–solid, gas–liquid, liquid–solid, and gas–liquid–solid systems. Computed Tomography (CT) and Computer Automated Radioactive Particle Tracking (CARPT) are non-invasive measurement methods well suited for providing information needed for validation of CFD codes and reactor model development. Figure 7 gives the schematics of these experimental techniques. CT experiments are used to obtain density distribution and CARPT experiments to obtain the velocity field and mixing information [12,13].

Computer Tomography (CT)
Provides Solids Density Distribution

Radioactive Particle
Tracking (CARPT) Provides
Solids Velocity and Mixing
Information

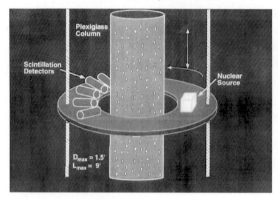

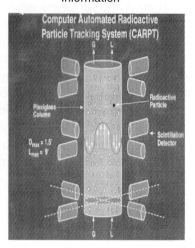

Fig. 7. Computer Automated Radioactive Particle Tracking (CARPT) and Computer Tomography (CT) experiments.

Hydrodynamic information obtained from CT and CARPT experiments can then be applied to reactor models. For example, for a liquid–solid riser, there are four reactor models at different sophistication levels: heterogeneous plug flow model (as the simplest), 1-D axial dispersion model, core annulus model and 2-D convection–dispersion model as the most complex. Each model requires an appropriate set of hydrodynamic parameters. The CT and CARPT provide these and combining this data with the kinetics of reactions and deactivation of the catalyst, one can develop detailed models to guide the selection and design of catalytic riser reactors.

6. Process intensification

Process intensification involves design of novel reactors of increased volumetric productivity and selectivity. The aim is to integrate different unit operations to reactor design, meanwhile operating at the same or better production rates with minimum pollution generation. To meet these goals, the practice involves utilization of lesser amount of hazardous raw materials, employing efficient mixing techniques, using microreactors, catalytic distillation, coupling of exothermic and endothermic reactions and periodic operations. As examples, consider catalytic distillation and coupling of exothermic and endothermic reactions.

6.1. Catalytic distillation

Catalytic distillation has been employed successfully for ethylacetate, H_2O_2, MTBE and cumene production. The catalytic distillation unit consists of rectifying and stripping sections as well as a reaction zone that contains the catalyst. The column integrates separation and catalytic reaction unit operations into a single unit therefore reducing capital costs. It is also beneficial in the fact that separation reagents that are toxic are no longer required. Since unreacted raw materials are recycled, less amount of feed is converted into products, thus reducing the use of harmful raw materials and waste generation. Significant energy savings due to utilizing the heat released by exothermic reaction for distillation is also another factor that contributes to this unit being environment friendly.

An application to ketimine production scheme using a catalytic distillation unit is illustrated in Fig. 8. Condensation of ketones with primary amines results in the synthesis of ketimines given by the reaction scheme:

$$R\text{-}NH_2 + \overset{O}{\underset{\displaystyle \bigwedge}{\|}} \longleftrightarrow R\text{-}N = \!\!\!< + H_2O$$

The reaction is reversible and therefore equilibrium limited. In the conventional process, the reversible reaction can be pushed forward to products by

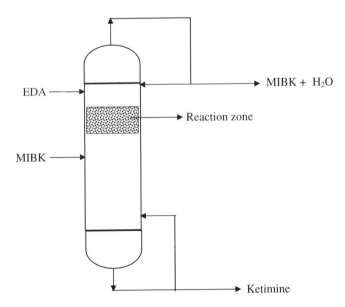

Fig. 8. Ketimine production by catalytic distillation. MIBK: ketones; EDA: amine.

employing drying agents such as $TiCl_4$ [14], $BuSnCl_2$ [15], Al_2O_3 [16] and molecular sieves [17], which remove the product water. By using catalytic distillation, ketones and amine react to produce ketimine. Excess ketones are recycled back and the product water is separated via *in situ* separation. No drying agents are needed thereby making the process cost efficient and environment friendly.

6.2. *Coupling of exothermic and endothermic reactions*

Using heat integration is another important guideline for improved reactor design and is another example of process intensification. Exothermic and endothermic reactions can be combined together in a reactor configuration to maximize energy conversion in the process. As an example, steam reforming of methane, an endothermic reaction (Reaction 1) can be combined with partial oxidation of methane, an exothermic reaction (Reaction 2).

$$CH_4 + H_2O \Leftrightarrow CO + 3H_2$$ (Reaction 1)

$$CH_4 + \frac{1}{2}O_2 \rightarrow CO + 2H_2$$ (Reaction 2)

The coupled reactions can be carried out in a shell and tube exchanger design that improves the energy efficiency of the process. This leads to a compact design. By contrast conventional steam reforming require huge furnaces leading to large capital cost and loss of energy.

Additional examples of process intensification such as microreactors, disk and plate reactors and reactive membranes can be found at Tsouris *et al.* [18] and Reaction Engineering for Pollution Prevention book [19].

7. Environmental impact analysis

Once new chemical processes are developed, catalyst, solvent and reactor type are selected using CRE methodology, the impact on the environment of the new chemical processes must be compared with the impact of conventional processes before implementation of the new process. Several tools such as Waste Reduction Algorithm (WAR) [20,21], Life Cycle Analysis (LCA) [22] and Environmental Fate and Risk Assessment Tool (EFRAT) [23] are available for environmental impact analysis. Much research is devoted to improve the accuracy of these tools and to develop a standard methodology.

The WAR algorithm is used for determining the potential environmental impact of a chemical process based on nine different impact categories listed in Table 1.

Potential environmental impact is a conceptual quantity that arises from energy and material that the process takes from or emits to the environment [24].

Table 1
Potential environmental impact categories

Physical potential effects	Acidification
	Greenhouse enhancement
	Ozone depletion
	Photochemical oxidant formation
Human toxicity effects	Air
	Water
	Soil
Ecotoxicity effects	Aquatic
	Terrestrial

The impact categories listed in Table 1 are weighted according to local needs and policies, and the scores are normalized to eliminate bias within the database. WAR is particularly useful for comparison of different existing processes but does not provide modifications that would minimize the waste. Application examples of WAR studies include methyl ethyl ketone production from secondary butyl alcohol [24,25], ammonia production from synthesis gas [25], and reactive distillation for butyl acetate production [26].

LCA has been developed to understand and characterize the range and scope of environmental impacts at all stages within a product or process [27]. LCA basically evaluates the process based on the boundaries of the assessment. Life cycle inventories for the inputs, products and wastes are evaluated for the system boundary. The results then can be compared for different chemical processes. LCA is very useful for global analysis shown in Fig. 1.

EFRAT is a simulation package developed by the EPA. It is used to estimate the environmental and health impacts of chemical process design options through a combination of screening-level fate and transport calculations and risk assessment indices [23]. EFRAT is a powerful simulation tool as it provides the process design engineer with the required environmental impact information, so that environmental and economic factors may be considered simultaneously [23].

CRE together with green engineering principles [28], provide the key concepts in designing and operating chemical processes at sustainable conditions. The improved process, however, must also be examined under close scrutiny by using the tools mentioned above. The results must be compared to conventional processes to see if the overall environmental impact has been reduced.

8. Summary and conclusions

In our road to achieving sustainability in production of materials and chemicals we must strive to eliminate pollution at the source, improve material and energy efficiency of our processes and use renewable resources. The best way to prevent pollution is at the source. Thus, if we want to have high-tech processes that are "sustainable" and "green", we must use CRE concepts to the fullest extent. The days when the chemist found a magic ingredient (catalyst) for a recipe and the chemical engineer tried in earnest to get its full potential expressed in an available "kettle", must be replaced by the coordinated effort of the chemist to select the best catalyst and the chemical engineer to provide the best flow pattern and reactor. This effort requires the multiscale CRE approach consisting of molecular, particle/eddy and reactor scale considerations.

Since the last decade, the chemical reaction engineers have been re-focusing on developing new technologies that prevent or minimize pollution rather than dealing with "end-of-pipe" treatments. In order to develop such technologies, a quantitative understanding of reaction systems and transport properties on the reaction rates is a must. Furthermore, the physical properties of the catalyst and media are also determining factors in choosing the "right" reactor for an environmentally benign process. Hence, it is a multidisciplinary task combining chemistry, reaction engineering, environmental impacts and economics. This chapter outlines the multiscale nature of the CRE approach starting from the molecular level at the atom efficiency to the process level at the scale-up of a reactor. Each scale is important in design and operation of a sustainable process. The combination of CRE approach with "green processing" principles should lead to the development of a sustainable chemical industry with minimal waste production.

Acknowledgements

The authors would like to acknowledge the National Science Foundation for their support through the NSF-ERC-Center (EEC-0310689) for Environmentally Beneficial Catalysis that allowed us to examine this topic.

References

[1] D.T. Allen, K.S. Rosselot, Pollution Prevention for Chemical Processes, Wiley, New York, 1997.
[2] J.A. Dyer, K.L. Mulholland, Chem. Eng. Prog. 94 (1998) 61.
[3] D.T. Allen, D.R. Shonnard, Green Engineering, Environmentally Conscious Design of Chemical Processes, Prentice-Hall, PTR, Upper Saddle River, NJ, 2002.
[4] S. Mota, M. Abon, J.C. Volta, J.A. Dalmon, J. Catal. 193 (2000) 308–318.
[5] R.M. Contractor, European Patent Application 01899261, 1986.
[6] M.J. Lorences, G.S. Patience, F.V. Diez, J. Coca, Ind. Eng. Chem. Res. 42 (2003) 6730–6742.
[7] S. Mota, J.C. Volta, G. Vorbeck, J.A. Dalmon, J. Catal. 193 (2000) 319–329.
[8] M. Wei, G.T. Musie, D.H. Busch, B. Subramaniam, J. Am. Chem. Soc. 124 (11) (2002) 2513–2517.

 [9] B. Subramaniam, C.J. Lyon, V. Arunajatesan, Appl. Catal. B-Environ. 37 (4) (2002) 279–292.
[10] K.R. Seddon, Green Chem. 1 (1999) G58–G59.
[11] J.D. Holbrey, K.R. Seddon, Clean Prod. Process. 1 (1999) 223–236.
[12] M.P. Dudukovic, N. Devanathan, R. Holub, Rev. Institut Francais Petrole 46 (4) (1991) 439–465.
[13] J. Chaouki, F. Larachi, M.P. Dudukovic, Ind. Eng. Chem. Res. 36 (11) (1997) 4476–4503.
[14] I. Moretti, G. Torre, Synthesis 141 (1970).
[15] C. Stetin, Synth. Commn. 12, 495 (1982).
[16] T.-B. Francoise, Synthesis 679 (1985).
[17] K. Taguchi, F.H. Westheimer, J. Org. Chem. 36 (1971) 1570.
[18] C. Tsouris, J.V. Porcelli, CEP Mag. 99 (2003) (October) 50–55.
[19] Reaction Engineering for Pollution Prevention, M.A. Abraham, R.P. Hesketh (Eds), Elsevier, Amsterdam, Netherlands, 2000.
[20] A.K. Hilaly, S.K. Sikdar, J. Air Waste Manage. Assoc. 44 (1994) 1303–1308.
[21] D.M. Young, H. Cabezas, Comput. Chem. Eng. 23 (1999) 1477–1491.
[22] R.L. Lankey, P.T. Anastas, Ind. Eng. Chem. Res. 41 (2002) 4498–4502.
[23] http://es.epa.gov/ncer_abstracts/centers/cencitt/year3/process/shonn2.html.
[24] H. Cabezas, J.C. Bare, S.K. Mallick, Comput. Chem. Eng. 21 (1997) S305–S310.
[25] H. Cabezas, J.C. Bare, S.K. Mallick, Comput. Chem. Eng. 23 (1999) 623–634.
[26] C.A. Cardona, V.F. Marulanda, D. Young, Chem. Eng. Sci. 59 (2004) 5839–5845.
[27] P.T. Anastas, R.L. Lankey, Green Chem. 2 (2000) 289–295.
[28] P.T. Anastas, J.B. Zimmerman, Environ. Sci. Technol. 5 (2003) 94A–101A.

Sustainability Science and Engineering: Defining principles
Martin A. Abraham (Editor)
DOI 10.1016/S1871-2711(05)01017-2

Chapter 17

Green Engineering and Nanotechnology

B.J. Yates[a], D.D. Dionysiou[b]

[a]*Department of Civil and Environmental Engineering, University of California, Berkeley, CA 94709*
[b]*Department of Civil and Environmental Engineering, 765 Baldwin Hall, University of Cincinnati, Cincinnati, OH 45221-0071, USA*

1. Introduction

In 1959, Richard Feynman gave a lecture at the California Institute of Technology during the annual meeting of the American Physical Society, which described humanity's first acknowledged attempt to manipulate matter at the nanoscale (10^{-9} m) [1]. In Feynman's speech, he foresaw future scientists developing methods to produce a more efficient catalyst or a more powerful computer and saying to themselves "why was it not until the year 1960 that we began moving in this direction?" Indeed, nanotechnology, over its short life, has already produced many new tools, and some specifically designed to improve the environment. Nano-sized catalysts have the potential to reduce by-products through more selective reactions; environmentally benign nanocomposites can be used for construction of optical and electronic devices; and novel biosensors can probe single cells to help understand deleterious processes *in vivo* as a step forward in the fight against disease. Nanotechnology has also made feasible the tools necessary for the next level of precision in environmental engineering practice. These tools include membranes with tailor-designed porosity, morphology, and surface chemistry; efficient catalysts for solar detoxification of water; and selective, extremely sensitive solid phase microextraction devices for monitoring of water and air quality. In many ways, then, nanotechnology has positive environmental implications.

At least equally important as the potential for nanotechnology to further the goal of a cleaner environment is the adverse impact that the production and implementation of the technology may have on the environment. It will be essential to go further than simple paradigms of pollution prevention and "reduce, recycle, reuse" to analyze nanotechnology completely according to The 12 Principles of Green Engineering described by Anastas and Zimmerman [2] and the sustainable engineering principles developed in Sandestin, FL [3]. For nanotechnology to beneficially impact the environment, it will be essential for the discipline to follow these principles during the production, use, and reuse of nanoparticles.

In this chapter, some of the current production methods of nanomaterials are presented from a green engineering standpoint, and shortcomings are noted. Examples of novel research are also presented that go beyond current and dominant technologies by explicitly incorporating into their designs elements to (1) reduce the use of harmful and depleting material inputs, (2) maximize space, time, and energy efficiency, and (3) holistically design nanoparticles through eco-effective solutions. Specific applications of green nanoparticle production for construction of ultrafiltration membranes, as well as some health concerns with prefabricated nanoparticles are also discussed.

2. Health concerns and risks

Before discussing ways to apply the principles of green engineering to the production of nanoparticles, it is necessary to consider what impacts these fabricated particles may have on human health and the environment once they are synthesized. Because nanotechnology offers the promise of more efficient materials and processes, and because the production methods are now within our reach and becoming less expensive, future human exposure to nanoparticles will increase dramatically. Environmental exposure will also increase as new nanomaterials are trusted to the hands of the consumer. In light of this, The National Science Foundation (NSF) and the US Environmental Protection Agency (EPA) have begun an initiative to develop technology to sense and characterize nanoparticles in the environment [4]. Already reports are being published on the adverse effects nanomaterials may have on human life and the sustainability of ecosystems.

In a recent Washington Post article, Rick Weiss compiled the current opinions of academic and private sector scientists intimately involved with the development of different nanomaterials [5]. Opinions are diverse, but mainly fall into three categories. In one category, activists against the science, or those who suggest an international moratorium on the production of nanoparticles, claim that nanomaterials will initiate human health and environmental problems of equal or greater magnitude to those that began during the industrial revolution. In the second category, those with the vision to believe in the potential for nanoparticles to benefit human health and the environment (and those in a position to benefit economically from the unregulated production of nanoparticles) remain optimistic that nanoparticles will prove to be relatively low-risk materials. Opinions in the third category state that a broad-reaching, scientific endeavor must be carried out to systematically uncover the aspects of nanomaterials that have the greatest potential to affect human health and environmental sustainability, in tandem with the regulated approval of new products and production methods. This endeavor is already underway as evidenced by the available literature concerning toxicity of nanoparticles, especially in relation to drug delivery systems.

Nanoparticle drug delivery systems are designed for the controlled release of pharmaceuticals at a specific anatomical site to improve the efficacy of a drug while reducing unwanted side effects, and are the subject of numerous studies, especially in the past 20 years [6–8]. Briefly, a drug is incorporated into a polymeric nanoparticle by dissolution, entrapment, or encapsulation or it is attached to the surface of the nanoparticle. Depending on the preparation method, nanoparticles, nanocapsules (vesicular systems in which the drug is confined to a cavity), or nanospheres (matrix systems in which the drug is physically and uniformly dispersed in the polymer) are formed [7]. These nanoparticles are delivered by inhalation or parenteral injection and are designed to target diseased organs or physiological compartments. Because the use of nanoparticles as drug delivery systems involves deliberate exposure of them to humans, it is possible to conduct controlled studies on the toxicological risk assessment associated with them. In the following discussion, it is worthwhile to emphasize that risk assessment hinges on the evaluation of two equally important and mutually exclusive factors: exposure and hazard [9]; anything that does not present a hazard to human health does not present a risk, no matter how great the exposure; conversely, anything to which there is no exposure does not present a risk, no matter how great the intrinsic hazard.

The three factors (apart from chemical composition) that most greatly affect the risk associated with *in vivo* nanoparticles are the size, method of preparation, and surface characteristics.

The diameters of the smallest human blood capillaries are 5–6 μm [8], thus giving nanoparticles the ability to pass through the entire venous and arterial networks. Particles that are larger than 800 nm, however, are recognized by the mononuclear phagocytic system (MPS) as a foreign object and are dealt with through macrophage phagocytosis [6,8]. Assuming that the particles themselves can be destroyed and removed from the body with no toxic effects (an issue pertaining to the intrinsic chemical composition of the particles), the hazard associated with these larger particles is less than in their smaller counterparts owing to their recognition by the body's immune system. Because the release of drugs from nanoparticle drug delivery systems hinges on the degradation of the nanoparticle itself [7], most drug delivery systems rely on this safe bio-degradation when the MPS is targeted. Some drug delivery systems, however, are designed to pass through the MPS to target specific anatomical sites (these particles are on the order of 100 nm [8]), or are designed to remain in the vascular system for long periods of time.

Nanoparticles may present a substantial pulmonary risk, as evidenced by the results of two studies not dealing with toxicity of drug delivery systems, but aimed at investigating the adverse effects on rodent lung tissue of single-walled carbon nanotubes [10,11]. In these studies, the formation of pulmonary was observed. Two very important conclusions were derived from these studies: Firstly, on an equal-weight basis, these nanotubes can be considered more toxic than some of the most destructive pulmonary health hazards (including quartz) [10]; Secondly, the pathogenic mechanisms appear to be different than those involved in other inhaled particles [11]. These conclusions imply that single-walled carbon nanotubes could represent a new serious health hazard in which extrapolating current toxicological data from *in vitro* studies is not an acceptable way to approach the problem. Care must be taken in interpreting pulmonary toxicology results in other organisms, especially rodents, as these results provide only an estimate of the possible effects in humans and some mechanisms may not even be relevant to human toxicity [12].

Another, possibly more familiar inhalation exposure of nanoparticles that highlights the effect particle size may have on toxicity is that of soot from combustion processes, especially the diesel engine. The particles that are emitted from the diesel engine are on the order of 30–80 nm and are especially toxic because they can enter smaller lung air spaces than larger particles. While the US-EPA cites PM-10 and PM-2.5 as criteria air pollutants, the particles that fit into these classifications are still 2–3 orders of magnitude larger than nanopar-ticles. Inside the lung tissue, particles of nano-size are not recognized by alveolar macrophages [13], and thus, are able to penetrate the epithelium and enter the pulmonary vasculature. Once inside the vessels, nanoparticles may place oxi-dative stress on the vascular tissues (because of the ability of ultrafine particles to produce many free radicals [14]) and lead to chronic tissue inflammation. The adverse health effects of these soot particles have been evident through past air pollution episodes, but now that scientists and engineers propose to create par-ticles of this size, methods for exposure control must be studied and developed. In addition to the direct health effects that inhaled nanoparticles may have, small atmospheric particles of Fe_2O_3, MnO_2, and FeS_2 greatly increase the formation of atmospheric ozone [15], a US-EPA Criteria Air Pollutant. In this way, atmospheric metal nanoparticles act as generators of secondary air pollution.

Depending on the chemicals used for processing, the method of preparation can also have an effect on the toxicity of nanoparticles. For example, the method for preparing polymeric nanoparticles for drug delivery systems affects their behavior and ultimate fate *in vivo*. Essentially, there are two methods for preparing polymeric drug delivery systems: (a) dispersion of preformed polymers and (b) polymerization of monomers [7]. The concern with methods of dispersion of preformed polymers is that many conventional techniques use organic solvents, which are evaporated to precipitate the nanoparticles. These solvents (such as dichloromethane/methanol mixtures and ethyl acetate [16]) may still exist as residues in the final product and are biologically as well as environmentally hazardous. Alternatives to the use of hazardous solvents are the supercritical anti-solvent method [17] and the gas anti-solvent method [18], both of which allow for the use of harmless supercritical solvents, which do not remain as residue in the final product. Problems of unwanted chemicals in the final product also occur in methods that directly polymerize monomers in the presence of stabilizers and surfactants. This problem can be overcome by selecting the correct starting polymer such as in the development of biodegradable polyesters, which self-assemble to form nanoparticles [19].

One of the most impressive aspects of nanoparticle drug delivery systems is their ability to target specific organs or tissues that are in greatest need of the administered drug. This targeting is generally best accomplished by chemical modification of the nanoparticle's surface to increase its affinity for specific cells. The most common modifiers are polyethylene glycol, polyethylene oxide, poloxamer, poloxamine, polysorbate, and lauryl ethers [7]. The surfaces are most commonly characterized by means of hydrophobic interaction chromatography [20], X-ray photoelectron spectroscopy [21], and evaluation of the surface zeta potential (ζ), an indication of surface charge characteristics. By changing the surface characteristics of these drug delivery systems in precisely quantifiable ways, researchers have shown that there is no physiological barrier to nanoparticle movement in the body. As examples, commercially available polystyrene nanoparticles have been coated with poloxamer 407 and poloxamine 908 for enhanced uptake by the spleen [22]; polystyrene-latex nanoparticles have been coated with lactosyl-polystyrene to enhance accumulation in the liver [23], and even the blood-brain barrier (and thus, the central nervous system) is permeable to some surface modified nanoparticles [24]. These examples illustrate the possibility that given the correct size and surface characteristics, specific nanoparticles have the possibility of entering specific physiological compartments in the human body.

One study was conducted to determine morphological, crystallographic, and chemical characteristics of atmospheric nanoparticles in El Paso, Texas, using transmission electron microscopy, selected-area electron diffraction, and analysis of the energy dispersive spectrum. The study discusses the fact that crystallinity and overall particle shape have an impact on nanoparticle toxicity [25]. The enhanced toxicity of crystalline nanoparticles is not only due to the fact that crystalline particles are responsible for mutagenic and carcinogenic effects [26], but also because of the induced oxidative stress (and hence epithelial damage) owing to the highly catalytically active surfaces of crystalline particles. This study also found that most atmospheric nanoparticles were agglomerates of smaller particles some of which were structurally unstable. For example, one particle viewed by TEM displayed layered features characteristic of graphitic materials [25]. These layers are easily cleaved forming even smaller, more hazardous particles. Thus, particle structure must be evaluated along with particle size in toxicity determination.

This small sampling of toxicological problems associated with nanoparticles serves to indicate that no production mechanism can be accepted as healthy — much less green — if the hazards of the proposed product are not considered. These hazards should be investigated through clinical studies seeking correlations between the physical/chemical properties of certain classes of nanoparticles and specific diseases. Once a firm scientific basis can be established regarding the hazards associated with nanoparticles it may be used to evoke legislation to reduce the exposure and hence, the risk associated with specific hazardous nanoparticles.

3. Methods of production

3.1. Metal and metal oxide nanoparticles

Metal and metal oxide nanoparticles are of interest for their use as solid electrolytes (cerium oxide), sensors (metallic silver and gold), and as very efficient catalysts (titanium dioxide and zinc oxide). In all these applications, the size of the material enhances properties that allow them to be used more efficiently. Producing metals and metal oxides in the nano-size range using traditional techniques usually requires the use of a solvent (as a medium for carrying out the synthesis reaction), a reducing agent (to produce elemental metals from metal salts), and a stabilizing agent (sometimes called a capping agent, to inhibit agglomeration of particles). Although in most cases these three ingredients cannot be completely eliminated, they should be environmentally benign and non-depleting, which may require that solvents and reducing and capping compounds interact weakly with nanoparticles so they may be decoupled from the product and reused.

3.1.1. Biomimetic processing: an alternative to sol-gel
Nature still produces the most intricate structure, the most practical use, and the most inspirational products of any on Earth. Engineers and scientists looking to make a more efficient process may want to look at metabolic pathways.

Metallurgists looking to synthesize a particular material for a particular environment would do well to find a natural environment that is similar and study the materials that remain. Often scientists and engineers do look to nature for inspiration, and this inspiration results in practical processes that in many cases are green and sustainable. This design process is known as *biomimetic processing*.

To produce metallic silver nanoparticles with particle sizes from 1 to 10 nm via a completely green approach, Poovathinthodiyil et al. [27] revamps the classical solvent/reducing agent/capping system by using three benign (and naturally produced) materials. Water is the solvent throughout the synthesis. With gentle heating to less than 40°C, a reducing β-D-glucose forms silver nanoparticles from a 0.1 M AgNO₃ solution. Because this sugar is found in natural metabolic pathways, it is non-toxic, renewable, and inexpensive. Starch (amylose) was chosen as a capping agent because its supramolecular structure allows for inclusion of newly reduced silver atoms into capsules formed by intramolecular hydrogen bonding. These capsules act analogously to micelles in traditional organic surfactant based sol–gel processing, but offer a green alternative to surfactant use. Apart from this, Poovathinthodiyil outlines several other advantages: (1) dispersions of starch in water are possible without the use of organic solvents and (2) the weak interaction between starch and the silver nanoparticles may allow reversible binding of silver to the starch capsules at moderately higher temperatures. This temperature-controlled reversibility would facilitate reuse of the capping agent.

Besides its use as a highly sensitive sensor, silver is also a strong metal catalyst employed in industrial chemical processing. The use of a nanoparticle catalyst greatly increases the rate of catalysis owing to the large specific surface areas of these materials. Two very important material properties must be controlled to make nano-catalysis effective: the particle size and particle size distribution of the catalyst. Photocatalysts illustrate the benefit of high specific surface area by control of particle size, and the biomimetic synthesis of one photocatalyst will be discussed next.

Titanium dioxide (TiO₂) is known for its photoactivity and ability to produce OH under excitation by near UV light, which can quickly and completely mineralize aqueous pollutants such as hydrocarbons [28,29], chlorinated phenols and pesticides [30], bacteria [31], biological toxins [32], proteins [33] and reactive dyes [34,35], as well as carbon monoxide [36], and organic aerosols (propionaldehyde) [37]. TiO₂ nanoparticles increase catalyst kinetics immensely, and the prospect of grafting these extremely active catalysts into ceramic membranes may offer a very efficient and resilient composite disinfection/separation system.

Production mechanisms for the synthesis of TiO₂ nanoparticles can be broadly classified as chemical or physical. In most cases, chemical production mechanisms begin with a toxic or otherwise hazardous titanium source.

Through a series of chemically harsh techniques carried out at elevated temperatures, the final TiO_2 nanomaterial product is separated from the chemical production matrix with the use of more chemical solvents. All of these lead to problems of waste production and subsequent expensive waste clean up to protect human health. Worse still, the original source of titanium for such operations is open mines, which not only scars but also pollutes the natural landscape and contributes to the depletion of natural deposits of this and other mineral ores.

One chemical production mechanism for TiO_2 nanoparticles employs mesoporous silica for the size control of grafted titanium nanoclusters [28]. The production and grafting mechanism requires the dissolution of a complex hexanuclear titanium cluster in tetrahydrofuran (THF), refluxing with the silica support for 20 h, washing with THF, and finally high temperature (550°C, 15 h) calcination. In attempting to graft TiO_2 nanoparticles in mesoporous alumina, Schneider et al. [38] used the "chemie douce" method — in which materials produced retain a "memory" of their precursors [39] — for the incorporation of nanoparticles into the alumina support. Although this production method avoids the energy intensive calcination step, it still employs large quantities of toluene, and two environmentally unfavorable precursors: bis(toluene)titanium^0 and an organometallic grafting reagent [$(\eta^5$-Cp)TiCl$_2$].

Energy intensive limitations plague physical techniques, such as flame synthesis and mechanical ball milling. In a large-scale operation of flame synthesis, gas streams of methane/oxygen/argon continuously deplete embedded entropy, so it offers little or no opportunity for recycle. Large ball milling operations feed on and degrade energy inputs, creating waste heat as well as air and water pollution.

The two most popular mechanisms to synthesize nanostructured TiO_2 are the sol–gel and vapor precipitation methods. The sol–gel technique requires that a soluble titanium precursor (usually titanium tetraisopropoxide) be blended with an organic solvent (in some cases, cyclohexane) before it is templated into a narrow size range nanoparticle by nano-scale emulsions. After the titanium has reacted with the templates, the emulsions must be destabilized and removed to reclaim the titanium product — a procedure that uses additional organic solvents and/or alcohols (isopropanol, ethanol). Use of vapor precipitation techniques have also been proposed for the production of a gold/titanium dioxide hybrid [40]. Although these vapor precipitation techniques avoid the use of large quantities of diverse, harsh chemicals used in sol–gel, they require ultrasonic treatment for spray preparation and high-temperature reactions (> 1000 K) for the size control of nanoparticles. However, the most unacceptable aspect of the vapor precipitation techniques is the common use of the precursor titanium tetrachloride. This extremely toxic compound readily causes burns [41, 42] and is carcinogenic under low level chronic exposure [43, 44].

Considering the above production mechanisms, a new green production mechanism for synthesis of TiO_2 nanoparticles is necessary and should obviate design flaws. Specifically, the new production mechanism should strive to use material inputs, which are inherently benign, reduce the complexity of nano-particle production (to increase efficiency), reduce or reuse materials and energy used in separation techniques, and integrate the entire production process with other industries to recycle technical nutrients and reduce the need to mine titanium ore.

A proposed alternative to these methods for producing TiO_2 nanoparticles was inspired by similar work with gold and silver published by Gardea-Torresdey [45–48]. This innovative approach exploits a natural biological process to produce metal nanoparticles, a design approach coined "Eco-Effectiveness" by McDonough and Braungart [49].

Gardea-Torresdey demonstrated that alfalfa and oat could precipitate nano-particles of elemental gold and silver inside their leaves, shoots, and roots. The exact method of uptake, translocation, and precipitation is unknown, but it is hypothesized that precipitation of the metal is accomplished by naturally occurring reducing sugars and supramolecular cavities that act as reducing agents and capping agents respectively. Precipitation is possible in live (as an analogue to phytoremediation) or inactivated biomass, in aqueous solution at room temperature, with no addition of chemicals. Nanoparticles produced in this way are on the scale of 14 nm, have narrow particle size distributions and are formed in a variety of shapes, including nanorods with multiple twinnings.

An application of this same principle has been applied by Unocic et al. for the production of anatase (TiO_2) nanoparticles from algae [50]. In this work, amorphous silica nanoparticles synthesized naturally by single-celled aquatic algae are used as templates for the halide gas/solid displacement reaction producing regular anatase nanoparticles by the following reaction:

$$TiF_{4(g)} + SiO_{2(s)} \rightarrow TiO_{2(s)} + SiF_{4(g)} \tag{1}$$

What is promising about this work is the biological aspect: the ability to cultivate algae to produce enormous numbers of similarly shaped structures for use as templates, thus, the templates are produced by a benign eco-effective process. The chemical aspects of this work cause several concerns: the precursor TiF_4 is a toxic, halogenated metal gas, and the production of another halo-genated gas (SiF_4) and fluorine gas (due to the conversion of an intermediate $TiOF_2$) represent environmentally dangerous byproducts of the process. The halide/gas solid displacement method for incorporating the titanium into the silica template structure must be modified if this process is to be consistent with green engineering principles.

Similar to the work done by Unocic et al. is the use of biomolecular structures such as microtubules or tube shaped viruses for the directed assembly of nanowires. This bottom-up fabrication of nanoparticles is less wasteful than its alternative (conventional top-down fabrication), where multiple steps involve the removal of the majority of deposited unstructured layers [51].

Microtubules are cylindrical structures found in many eukaryotes and are involved in life processes such as mitosis, cell motility, and transport of organelles. They are characterized by dimensions from nanometer to several micrometers, and possess molecular recognition capabilities due to the presence of many different functional groups. Behrens et al. have been able to exploit the highly oriented nature of these functional groups to produce silver nanowires about 1 μm in length and 50 nm in diameter [52]. The tubules were first fixed with glutaric dialdehyde in order to suppress the natural polymerization/depolymerization activity of such structures and subsequently exposed to silver in the form of $AgNO_3$ as well as the reducing agent, sodium borohydrate. Small spherical particles of silver were formed and densely bound to the microtubule surface forming a silver nanowire. In the absence of the microtubule, large silver agglomerates were formed, not silver nanoparticles. This indicates that the microtubules are essential for structure direction during the formation of the silver nanowires.

The ability of microtubules to direct the formation of silver nanowires is due to the highly oriented functional groups they possess. In this respect, helical RNA is similar and has been used for the formation of nickel and cobalt wires by tobacco mosaic virus (TMV) [51]. TMV is a stable tube shaped complex of helical RNA about 300 nm in length and consists of well-defined chemical groups, which act as ligands for metals. Using TMV for directed assembly resulted in continuous silver nanowires of 500 nm after reduction of metal salts by ascorbic acid, hypophosphite, and dimethylamine borane. Temperature and pH studies of the virus itself revealed that the TMV structure was largely unaffected when exposed to temperatures of 90°C and a pH range of 3.5–9.0 for several hours. The ability of RNA to recognize certain metallic nanopartices has also been shown in DNA strands where palladium, copper, and platinum nanowires have been formed [53–56].

3.1.2. Supercritical solvents

Recently, Beckman extolled the virtues of thermodynamic control of supercritical carbon dioxide solvents for oxidation reactions [57] as a green technology. Control of a system's thermodynamics to alter reaction rates and product characteristics is a green synthesis approach compared to chemical control. A thorough understanding of the thermodynamic behavior of a system and less reliance on its chemical behavior substantially reduces chemical consumption and byproduct formation. Carbon dioxide is a benign solvent in the liquid or supercritical state. It is naturally abundant and is relatively inexpensive. It seems desirable, then to use microemulsions present in supercritical CO_2/solvent systems to stabilize metal nanoparticles after their reduction from metal salts.

This has been the goal of research done by Ji et al. who report the synthesis of silver nanoparticles in water/supercritical CO_2 systems [58]. Although the natural reverse micelles and microemulsions formed from this solvent system stabilize the silver nanoparticles, they require a fluorinated cosurfactant (perfluropolyether-phosphate ether) as well as sodium borohydride as the reducing agent, illustrating a return to the problems associated with traditional sol–gel processing. Further, work should attempt to replace these harmful chemicals with more benign ones.

The obvious and outstanding benefit of using supercritical carbon dioxide is that it is a benign solvent. Solvents account for most of the (volumetric) waste from industrial processes, and their disposal would be much more acceptable if they had minimal effect on natural ecosystems. Water is the most unobtrusive solvent meeting this criterion.

Sue et al. reported continuous synthesis of zinc oxide nanopaticles in supercritical water with no calcination [59]. The supercritical water represents a clean synthesis environment, which is easily altered through control of certain thermodynamic properties. Using water at its critical or near critical state, the hydrothermal reaction rate and metal oxide solubility (two parameters which greatly affect the particle size and particle size distribution of the product) can be controlled by changing the temperature and pressure of the system. Sue et al. report that this reaction occurs in two steps: (a) zinc hydroxide sol formation (reaction 2) and (b) dehydration from the sol (reaction 3).

$$Zn(NO_3)_2 + 2KOH \rightarrow Zn(OH)_2 + 2KNO_3 \tag{2}$$

$$Zn(OH)_2 \rightarrow ZnO + H_2O \tag{3}$$

Because of the simplicity of the reactions, it may be possible to synthesize other metal oxide nanoparticles in a similar fashion.

Although the critical temperature of water is high (642 K), 92% conversion of $Zn(NO_3)_2$ into 31 nm ZnO is observed at temperatures as low as 573 K. This operation within a range of temperatures also sheds some optimism on the potential use of other supercritical solvents at lower temperatures.

By coupling the use of supercritical solvents with hydrothermal crystal growth, the need for cosurfactants can be eliminated. This is illustrated in a study that focused on the production of cerium oxide (CeO_2) nanoparticles [60].

Cerium oxide nanoparticles are used as polishing agents, solid electrolytes, diesel fuel additives, and as automotive exhaust catalysts. Of the reported methods for producing CeO_2 nanoparticles, hydrothermal crystallization can be viewed as green, because it does not include the use of organic solvents or require an energy intensive calcination step. Despite this, hydrothermal crystallization usually requires the use of at least one surfactant to inhibit agglomeration of the nanoparticles, which reduces its benefits as a green technology. Other practical problems such as precursor cost plague past attempts to produce cerium oxide nanoparticles via hydrothermal crystallization. However, a recent study by Masui et al. [60], which produces cerium oxide nanoparticles in ammonia water using citric acid as a capping agent has circumvented the practical problems of agglomeration and cost while simultaneously offering a green alternative to current hydrothermal synthesis.

Cerium chloride and citric acid are mixed in a 1:1 (mol/mol) ratio in an excess of ammonia water. These chemicals represent the entirety of the precursors for synthesis of cubic cerium(IV) oxide with particle sizes on the order of 3.9 nm. The small particle size can be attributed to the homogenous coating of citric acid on the surface of CeO_2 particles, which prevents the agglomeration of small particles during and after synthesis. Although the cerium source (a halogenated metal) is not benign, subsequent work indicates that cerium oxide with 5 nm particle sizes can be produced without agglomeration even if Ce(III) salts are used [60]. The entire reaction is carried out between 323 and 353 K.

3.2. Nano-membranes

Membrane science has given the engineer a means to separate solutes from gas and aqueous phase solutions with a high degree of selectivity for chemical processing, water treatment, or remediation. The approach to the problem of chemical separation on the molecular scale is a shift in focus to the nanoscale architecture of membrane surfaces. Design of filtration systems through careful control of pore size on the nanoscale allows levels of pollutant removal and chemical separation unmatched when the macrostructure of the membrane is the only consideration. Increased performance and efficiency of membrane separation may also be realized through the doping of fixed site carriers on the surface to produce "smart" polymeric materials. These fixed site carriers recognize particular pollutants, and preferentially pass them through the membrane, or reject them, hence removing the pollutant in the waste effluent.

Ceramic membranes are most suited for certain ultrafiltration applications because of their resistance to chemical and temperature degradation and their adaptability to use as supports for facilitated transport. But how green are the current ceramic membrane synthesis techniques?

DeFriend and Barron at Rice University have found an application of acetate-alumoxane nanoparticles (A-alumoxane) for the synthesis of "hierarchical" ceramic membranes [61]. This hierarchical structure — in which a thin effective layer is coated onto a mesoporous support — has been previously produced through the sol–gel technique [62–65]. In contrast, Barron and DeFriend use A-alumoxane nanoparticles as the effective layer, grafting them onto prefabricated supports [66]. This exploits the high specific surface area of these nanoparticles to increase flux through the membrane, while still allowing for rejection of nano-sized contaminants. The control of membrane pore size, pore size distribution, and membrane surface defects is completely dependent on the control of particle size and particle size distribution of the precursor A-alumoxane nanoparticles, and almost independent of sintering temperature. Thus, the challenge of creating a green, practical hierarchical ceramic membrane of this type is centered at the nanoparticle production.

Classically, the composition of alumina slurry similar to that synthesized for use in a hierarchical membrane involves alumina as the precursor powder, 1,1,1-trichloroethylene (TCE) as the solvent, menhaden oil as a deflocculating agent, poly(vinyl butyrol) as a binder, and poly(ethylene glycol) as a plasticizer [67]. These chemicals are harmful as aerosols — their ultimate fate in traditional ceramic processing — and TCE has been labeled a high priority toxic chemical targeted for source reduction by the US-EPA [68]. With all of this as motivation, Barron's group developed a method of synthesis for A-alumoxane nanoparticles with two design goals: (1) to minimize chemical emissions during processing and (2) to obviate the use of hazardous solvents and strong acids [69].

Drawing on newfound knowledge of the structure of alumoxanes as a three-dimensional cage, not as linear or cyclic chains, Barron's group formulated a synthesis method based on the interaction between boehmite (a naturally occurring aluminum containing mineral of similar structure) and carboxylic acid (reaction 4) [70].

$$[Al(O)OH]_n \xrightarrow{(HO_2CR)} [Al(O)_x(OH)_y(O_2CR)_z]_n \qquad (4)$$

This synthesis scheme was used as a guidepost for the synthesis of A-alumoxane particles in water using acetic acid as the source for the organic substituted coating; the A-alumoxane nanoparticles are water soluble, and have dimensions less than 30 nm. By grafting these particles onto an alumina me-soporous support, a hierarchical ultrafiltration membrane was synthesized. The report that A-alumoxane nanoparticles suitable for use in ultrafiltration membranes can be synthesized in water using acetic acid represents a desirable step away from dominant ceramic technologies that depend on plasticizers and solvents that must be vaporized. There are also reports of use of carboxylate-alumoxane nanoparticles [71] and ferroxane nanoparticles [72] for use in ultrafiltration membranes.

From a practical standpoint, this ceramic membrane production process using A-alumoxane nanoparticles is efficient, cheap, and productive. The ceramic yield of this process is high (about 75–80%), and size control of the nanoparticles is possible by altering the pH of the solution, carrying out the reaction in pure acetic acid instead of water, and carefully controlling the reaction time [66]. The low cost of acetic acid as the only solvent also makes the production economically attractive. It will be important, as the process becomes more refined, to find alternative energy sources with integration and interconnectivity of energy flows for the sintering step that currently requires temperatures of about 600°C.

By controlling the nanoparticle size and size distribution, a decrease in the molecular weight cut off (MWCO) of these hierarchical ceramic membranes from 100,000 g/mol to less than 1000 g/mol was observed [66]. This MWCO corresponds to an average pore size of 16 nm allowing for the possibility of effective removal of large organic molecules, pharmaceuticals, and synthetic dyes from industrial waste streams.

To separate even smaller solutes from waste streams simply by altering the pore size of the membrane may stimulate researchers to "chase the dragon" by increasing chemical and energy inputs to tailor the membrane pores smaller and smaller, with diminishing returns. This is a vicious design loop that is contradictory to the principles of green engineering. It may be more attractive to manipulate the membrane surface on the nanoscale by the incorporation of fixed site carriers — complexing agents that selectively pass or reject solutes through the membrane — to filter ultra-small contaminants.

Thunhorst et al. have described the incorporation of crown ether molecules into a polymeric support for the active transport of alkali metals [73]. This has application to metal recovery from waste and process streams as well as in-process recycling, two effective pollution prevention techniques.

Unfortunately, the fixed site carriers used in Thornhurst's study (benzo-18-crown-6 crown ether) are small (thus, bioavailable) and toxic. If leaching from the polymer support is possible during use of these membranes for water treatment, contamination is inevitable. Analysis of the structure/activity relationships [74] of the crown ethers and a similar class of compounds — pseudocrown ethers — with respect to incorporation into the polymeric support and their interaction with solutes, gave rise to a greener composite membrane. Elliott et al. were able to synthesize a pseudocrown ether network that is less toxic than its crown ether homologues for the synthesis of a polymeric membrane with incorporated fixed site carriers [75]. Also because the pseudocrown ethers are incorporated into the polymeric support during the polymerization of the support, the natural polymerization process (in this case photopolymerization) and the incorporation of the pseudocrown ether into the polymer is interconnected, maximizing synthesis efficiency. These membranes show good flux and selectivity (K/Na) in trial experiments [76] and illustrate the use of structure/activity analysis to make a greener composite membrane.

4. Conclusions

From nanostructured membranes to selective photocatalysts and quantum-sized sensors, nanotechnology has the potential to monitor and clean the environment in ways unmatched by macrotechnology. Nanotechnology is still a relatively new discipline, and often the incredible properties of nanoparticles can mislead us to think that benefits realized from the production of molecules outweigh the consequences of irresponsible fabrication. We must be careful that we do not fall into this trap as we did in the industrial revolution, when developers became blind to the adverse effects of depleting resources and unchecked expansion.

In 1959, Feynman seriously started considering the world of the nanoscale and wished that progress in this direction had been made sooner, so that in the year 2000 we could actually be producing and using nanotechnology to further science, humanity, and the quality of our planet. In the new millennium, examples of nanoparticles in practical use exist, indicating that we moved more quickly toward Feynman's goal than he had predicted. In our haste, however, we have developed production mechanisms that have the potential to retard, not further, the sustainability of our planet. Possibly, in the year 2040 we will look back on the dawn of the new millennium and wonder why it was not until now, that we began to seriously evaluate the environmental implications of producing, using, and reusing nanoparticles, and why it was not until now that we did not begin to consider a discipline of *Green Nanotechnology* [77].

5. Acknowledgments

The authors would like to thank the US Environmental Protection Agency (P3 Grant) and the National Science Foundation (OCE-0304171 Grant) for financial support on their work on Environmental Nanotechnology. D.D. Dionysiou also would like to thank the organizers of the Green Engineering Defining the Principles conference (May 18–22, 2003, Sandestin, FL) for providing financial support to attend the conference.

References

[1] R. Feynman, Eng. Sc. 23 (1960) 22.
[2] P. Anastas, J. Zimmerman, Environ. Sci. Tech. 37 (2003) 94A.
[3] M. Abraham, W.H. Sanders III (chairs), Conference on Green Engineering: Defining the Principles, Sandestin, FL, May 18–22, 2003.
[4] US Environmental Protection Agency (2003) EPA Nanotechnology and the Environment: Applications and Implications, STAR Progress Review Workshop, August 28–29, 2002, Arlington, VA. EPA Document Number: EPA/600/R-02/080.
[5] R. Weiss, Washington Post, February 1, (2004) A01.
[6] J. Kreuter (Ed.), Colloidal Drug Delivery Systems, Marcel Dekker, New York, 1994.
[7] K.S. Soppimath, T.M. Aminabhavi, A.R. Kulkarni, W.E. Rudizinski, J. Control. Release 70 (2001) 1.

[8] M.L. Hans, A.M. Lowman, Curr. Opin. Solid State Mater. Sci. 6 (2002) 319.

[9] D. Allan, D. Shonnard, Green Engineering: Environmentally Conscious Design of Chemical Processes, pp. 35–62, Prentice-Hall, Upper Saddle, NJ, 2002.

[10] C.-W. Lam, J.T. James, R. McCluskey, R.L. Hunter, Toxicol. Sci. 77 (2004) 126.

[11] D.B. Warheit, B.R. Laurence, K.L. Reed, D.H. Roach, G.A.M. Reynolds, T.R. Webb, Toxicol. Sci. 77 (2004) 117.

[12] D.B. Warheit, Mater. Today 7 (2004) 32.

[13] J. Heyder, International Workshop on Vehicle Exhaust Nanoparticles, Tsukuba, Japan, (January) (2003) 14–15.

[14] K. Donaldson, International Workshop on Vehicle Exhaust Nanoparticles, Tsukuba, Japan, (January) (2003) 14–15.

[15] R.R. Chianelli, M.J. Yacaman, J. Arenas, F. Aldape, J. Hazard. Subst. Res. 1 (1998) 1.

[16] D.T. Birnbaum, J. Control. Release 65 (2000) 375.

[17] S. Mawson, Macromolecules 28 (1994) 3182.

[18] M. McHugh, V. Krukonis (Eds), Supercritical Fluid Extraction: Principles and Practice, pp. 342–356, Butterworth-Heinemann, Boston, MA, 1994.

[19] T. Jung, A. Breitenbach, T. Kissel, J. Control. Release 67 (2000) 157.

[20] H. Carstensen, Int. J. Pharm. 67 (1991) 29.

[21] B.D. Ratner, J. Biomed. Mater. Res. (Appl. Biomater.) 21 (1987) 59.

[22] M. Demoy, Pharmaceut. Res. 16 (1999) 37.

[23] T. Shinoda, Drug Deliv 6 (1999) 147.

[24] U. Schroeder, B. Sabel, Brain Res. 710 (1996) 121.

[25] J.J. Bang, L.E. Murr, J. Mater. Sci. Lett. 21 (2002) 361.

[26] S. Momarca, Sci. Total Environ. 205 (1997) 137.

[27] P. Raveendran, J. Fu, S.L. Wallen, JACS Commun. 125 (2003) 1394013941.

[28] A. Tuel, L.G. Hubert-Pfalzgraf, J. Catal. 217 (2003) 343.

[29] C.B. Almquist, P. Biswas, Appl. Catal. A: Gen. 214 (2001) 259.

[30] D.D. Dionysiou, A.P. Khodadoust, A.M. Kern, M.T Suidan, I. Baudin, J.M. Laîné, Appl. Catal. B: Environ. 24 (2000) 139.

[31] H.D. Jang, S.-K. Kim, S.-J. Kim, J. Nanoparticle Res. 3 (2001) 141.

[32] G.S. Shephard, S. Stockenström, D. deVilliers, W.J. Engelbrecht, G.F.S. Wessels, Water Res. 36 (2002) 140.

[33] L. Muszkat, L. Feigelson, L. Bir, K.A. Muszkat, J. Photochem. Photobiol. B 60 (2001) 32–36.

[34] I. Arslan, I. Balcioglu, D.W. Bahnemann, Dyes Pigments 47 (2000) 207.

[35] Y. Xie, C. Yuan, Appl. Catal. B– Environ. 46 (2003) 251.

[36] K. Mallick, M.S. Scurrell, Appl. Catal. A: Gen. 253 (2003) 527.

[37] H. Yoneyama, T. Torimoto, Catal. Today 58 (2000) 133.

[38] J.J. Schneider, N. Czap, J. Engstler, J. Ensling, P. Gutlich, U. Reinoehl, H. Bertagnolli, F. Luis, L. Jos de Jongh, M. Wark, G. Grubert, L. Hornyak, R. Zanoni, Eur. J. Chem. 6 (2000) 4305.

[39] G. Chow, K. Gonsalves (Eds), Nanotechnology Molecularly Designed Materials, American Chemical Society, Washington, DC, 1996, pp. 237–249.

[40] L. Fan, N. Ichikuni, S. Shimazu, T. Uematsu, Appl. Catal. A: Gen. 246 (2003) 87.

[41] D.K. Chitkara, B.J. McNeela, Br. J. Ophthalmol. 76 (1992) 380.

[42] S.M. Paulsen, L.B. Nanney, J.B. Lynch, J. Burn Care Rehabil. 19 (1998) 377.

[43] W.E. Fayerweather, M. Elizabeth Karns, P.G. Gilby, J.L. Chen, J. Occup. Environ. Med. 34 (1992) 164.

[44] K.P. Lee, Toxicol. Appl. Pharm. 83 (1986) 30.

[45] J.L. Gardea-Torresdey, E. Gomez, J.R. Peralta-Videa, J.G. Parsons, H. Troiani, M. Jose-Yacaman, Langmuir 19 (2003) 1357.

[46] J.L. Gardea-Torresdey, K.J. Tiemann, G. Gamez, K. Dokken, S. Tehuacanero, M. Jose-Yacaman, J. Nanoparticle Res. 1 (1999) 397.

[47] G. Canizal, J.A. Ascencio, J. Gardea-Torresday, M. José-Yacamán, J. Nanoparticle Res. 3 (2001) 475.

[48] V. Armendariz, I. Herrera, J.R. Peralta-Videa, M. Jose-yacaman, H. Troiani, P. Santiago, J.L. Gardea-Torresdey, J. Nanoparticle Res. 6 (2004) 377.

[49] W. McDonough, M. Braungart, Cradle to Cradle, pp. 68–91, North Point Press, New York, 2002.

[50] R.R. Unocic, F.M. Zalar, P.M. Sarosi, Y. Cai, K.H. Sandhage, Chem. Commun. 2004 (2004) 796.

[51] M. Knez, A.M. Bittner, F. Boes, C. Wege, H. Jeske, E. Maiβ, K. Kern, Nano Letters 3 (2003) 1079.

[52] S. Behrens, Chem. Mater. 16 (2004) 3085.

[53] J. Richter, M. Mertig, W. Pompe, I. Mönch, H.K. Schackert, Appl. Phys. Lett. 78 (2001) 536.

[54] W.E. Ford, O. Harnack, A. Yasuda, J.M. Wessels, Adv. Mater. 13 (2001) 1793.

[55] M. Mertig, L.C. Ciacchi, R. Seidel, W. Pompe, A. de Vita, Nano Letters 2 (2002) 841.

[56] C.F. Monson, A.T. Wooley, Nano Letters 3 (2003) 359.

[57] E.J. Beckman, Env. Sci. Tec. 37 (2003) 5289.

[58] M. Ji, X. Chen, C.M. Wai, J.L. Fulton, J. Am. Chem. Soc. 121 (1999) 2631.

[59] K. Sue, K. Murata, K. Kimura, K. Arai, Green Chem. 5 (2003) 659.

[60] T. Masui, H. Hirai, N. Imanaka, G. Adachi, J. Mat. Sci. Lett. 21 (2002) 489.

[61] K.A. DeFriend, A.R. Barron, J. Membr. Sci. 212 (2003) 29.

[62] A.F.M. Leenaars, K. Keizer, A.J. Burggraaf, J. Mat. Sci. 19 (1984) 1077.

[63] J. Luyten, S. Vercauteren, A. De Wilde, C. Smolders, R. Leysen, Ceram. Trans. 51 (1995) 647.

[64] K. Lindqvist, E. Liden, J. Eur. Ceram. Soc. 17 (1997) 359.

[65] Y.S. Lin, K.J. de Vries, A.J. Burggraaf, J. Mat. Sci. 26 (1991) 715.

[66] K.A. DeFriend, M.R. Wiesner, A.R. Barron, J. Membr. Sci. 224 (2003) 11.

[67] J.A. Mangals, G.L. Messing, (Eds), Advances in Ceramics, Vol. 9, American Ceramic Society, Westville, OH, 1984.

[68] http://www.epa.gov/opptintr/3350/index.html.

[69] R.L. Callender, C.J. Harlan, N.M. Shapiro, C.D. Jones, D.L. Callahan, M.R. Wiesner, D. Brent MacQueen, R. Cook, A.R. Barron, Chem. Mater. 9 (1997) 2418.

[70] C.C. Landry, N. Pappe, M.R. Mason, A.W. Apblett, A.N. Tyler, A.N. MacInnes, A.R. Barron, J. Mater. Chem. 5 (1995) 331.

[71] C.D. Jones, M. Fidalgo, M.R. Wiesner, A.R. Barron, J. Membr. Sci. 193 (2001) 175.

[72] M.M. Cortalezzi, J. Rose, G.F. Wells, J.-Y. Bottero, A.R. Barron, M.R. Wiesner, J. Membr. Sci. 227 (2003) 207.

[73] K.L. Thunhorst, R.D. Noble, C.N. Bowman, J. Membr. Sci. 156 (1999) 293.

[74] D. Allan, D. Shonnard, Green Engineering: Environmentally Conscious Design of Chemical Processes, pp. 93–138, Prentice-Hall, Upper Saddle, NJ, 2002.

[75] B.J. Elliott, W.B. Willis, C.N. Bowman, Macromolecules 32 (1999) 3201.

[76] B.J. Elliott, W.B. Willis, C.N. Bowman, J. Membr. Sci. 168 (2000) 109.

[77] D.D. Dionysiou, J. Environ. Eng. (ASCE) 1307 (2004) 723.

Sustainability Science and Engineering: Defining principles
Martin A. Abraham (Editor)
© 2006 Published by Elsevier B.V.
DOI 10.1016/S1871-2711(05)01018-4

Chapter 18

Technology Assessment for a More Sustainable Enterprise: The GSK Experience

David J.C. Constable[a], Alan D. Curzons[b], Concepción Jiménez-González[c], Robert E. Hannah[a], Virginia L. Cunningham[a]

[a]*GlaxoSmithKline, Corporate Environment, Health and Safety 2200 Renaissance Blvd. Suite 105, King of Prussia, PA 19406, USA*
[b]*Southdownview Way, Worthing BN14 8NQ, UK*
[c]*Five Moore Dr., Research Triangle Park, NC 27709, USA*

1. Introduction

The achievement of truly sustainable products and services is likely to take some considerable time given the many technical, economic and social barriers that must be overcome. For truly sustainable chemical processes to be developed, a concurrent, integrated approach that systematically and holistically considers triple bottom line aspects of chemistry and technology must be taken. One of the main challenges of integrating sustainable or "green" thinking into process design is to get process designers to think differently. One approach to aid this change is to develop standardized, scientifically sound and comparative methodologies that would allow process designers to readily determine how 'clean,' 'green,' or sustainable a particular technology or a specific set of reactions are, so different process or unit operation alternatives can be identified and compared.

One aspect of GlaxoSmithKline's (GSK) Environment, Health and Safety (EHS) vision for environmental sustainability is to champion the research and implementation of increasingly sustainable technologies and processes. GSK is driving continuous improvement toward greater efficiencies in chemical

Nomenclature

ΔH_R^T	heat of reaction at temperature T
η	efficiency
v	change in stoichiometric coefficients
A	area
C_p	heat capacity at constant pressure
E	electricity requirements
m	mass
Q	heating or cooling requirements
Q'	refrigeration requirements
t	time
T	temperature of the system
T_1	initial temperature
T_2	final temperature
T_w	temperature of heating/cooling media
U	global heat transfer coefficient.

processing through innovative and cost-effective implementation of novel chemistries and technologies [1]. As part of an overarching Corporate EHS strategy to promote the implementation of more efficient and sustainable chemical processing practices, GSK Corporate EHS has developed unique, innovative and comprehensive programs to provide guidance while promoting and implementing Green Chemistry and Technology, life cycle thinking, environment, health and safety-conscious materials selection, Green Packaging and others [2–9].

While developing and implementing these programs, it became increasingly clear that the greatest short to medium term gains toward more sustainable practices would be realized at the interface of chemistry and technology. While GSK Corporate EHS had developed considerable understanding of fundamental pharmaceutical industry chemistry and chemical processing approaches to more sustainable practices, there was a lack of understanding about the materials and energy efficiency implications related to technology selection. As a result, a 'Green Technology Guide' (GTG) was developed for technologies and unit operations of interest to the pharmaceutical industry [6,7]. The Guide was developed as a module of the existing Web-based Green Chemistry Guide, and was designed to provide scientists and engineers with comparative assessments of unit operations from a sustainability perspective.

In this chapter, a summary of the methodology for comparing the technologies in a given case scenario basis is presented and illustrated with an application example.

2. Green Technology Guide (GTG) as part of GSK's Eco-Design Toolkit

The GSK Eco-Design Toolkit is available company-wide to scientists and engineers, and provides general and specific materials, chemistry, technology, life cycle and packaging guidance. The GTG component uses a transparent evaluation process that integrates technology with green chemistry principles to relatively rank key chemical processing technologies and provides scientists and engineers with comparative environmental and safety information and guidance on traditional and emerging technologies. The tool is user friendly, has clearly delineated alternatives, is modular, and can accommodate future growth. Figure 1 shows a screen shot of the GTG.

The GSK GTG is under continuous development and in its current configuration organizes technologies for each specific case scenario into a matrix as shown in Fig. 2. The framework for the matrix consists of case scenario comparisons of technology options that have been scored in four discrete categories: environment, safety, efficiency and energy. A user may follow the links provided in the matrix and explore the detailed information that was developed for each

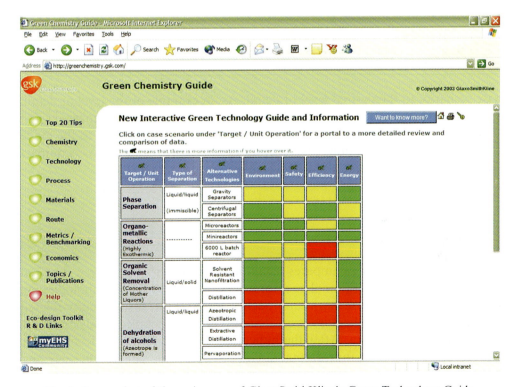

Fig. 1. Screen shot of the main page of GlaxoSmithKline's Green Technology Guide.

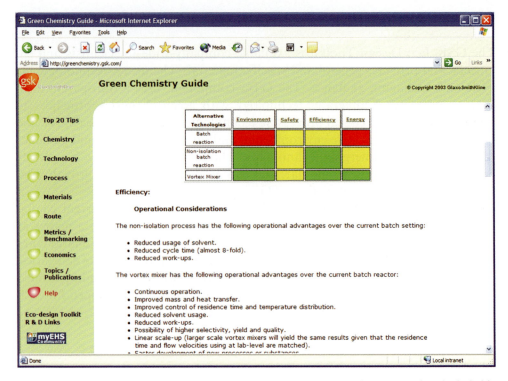

Fig. 2. Screen shot of one portion of the case-scenario comparison of a vortex mixer included in GSK's Green Technology Guide.

case scenario, to the desired level of detail. As an example of the level of detail, Fig. 2 shows a screen shot of the first part of a case scenario summary.

As time, data and resources permit, case scenarios will be added to the guide to expand the unit operations, the technology alternatives, and/or to encompass wider operational ranges for a given technology alternative.

3. Methodology

The GTG systematically compares technologies that perform a given function or achieve the same operational goal for a specific process. The technologies are compared using a case scenario approach where specific sustainability-related and life cycle-based indicators are calculated from a detailed analysis of the particular unit operation. After the indicators are developed, the technologies are relatively ranked in four discrete areas: environment, safety, efficiency and energy.

The process utilized for technology assessment follows four main steps:

- definition of scenario,
- data gathering,
- calculation of indicators and
- comparative ranking

as has been described in detail elsewhere [6,7]. The following paragraphs briefly describe the four steps used for technology comparison.

3.1. Definition of scenario

Because the **GSK** Green Technology methodology uses a life cycle-based approach, the crucial first step is to define the boundaries of the system, unit operation or process being evaluated.

The GTG is also intended to evaluate technologies that perform a particular function, objective or goal. The objective or goal may include an entire reaction system or it may be limited to the chemical processing function delivered by a single unit operation. In each case, the objective should be defined quantitatively and include any important constraints. For example, an objective might be to obtain 1 kg of product or to remove 99% of an impurity from a reaction mixture. One example of an added constraint might be that the process must be maintained at ambient temperature.

After the objective is clearly defined, suitable technologies are identified through literature reviews and/or discussions with internal and external scientists and engineers.

3.2. Data gathering

Once the objective, goal or function has been defined, and the technologies to be evaluated are identified, empirically derived data are collected from the literature (journal articles, books, etc.), through experimentation (research reports, pilot plant and laboratory records, etc.) and/or directly from scientists and engineers.

Full- or pilot-scale data are usually easier to obtain with in-house and traditional technologies and rather difficult to obtain for emerging technologies. In the case of emerging technologies, bench-scale data are generally the only information that is available. When only bench-scale data are found, there should be an assessment of the data quality for scaling-up to pilot or full scale. Data requirements include the information needed to perform a mass and energy balance for the system (i.e. flow diagrams, mass flows, operating temperatures, concentrations, physico-chemical properties, etc.), and hazard information on the process and materials such as calorimetry results, human and eco-toxicology test results, hazard assessments, etc.

3.3. Estimation of indicators

A set of core and complementary metrics that provided good differentiation between technologies were explored as indicators of the "greenness" of a given technology for a given case scenario. In addition to traditional indicators such as mass, energy, safety and efficiency, life cycle indicators were chosen and evaluated. The addition of life cycle indicators incorporates a broader view of technology evaluation and is intended to facilitate movement toward more sustainable business practices.

The mass indicators define environmental impacts and raw material utilization (e.g. emissions and mass intensity), while the energy indicators evaluate energy consumption for each of the alternatives. The safety indicators highlight potential hazards associated with the materials and processes used for the case scenario, and the need for occupational exposure controls. The efficiency indicators include mostly operational considerations such as conversion and complexity. The life cycle indicators account for life cycle environmental impacts from materials use, and emissions from extraction and production of raw materials, energy production and waste treatment.

3.3.1. Mass indicators

The calculation of process or unit operation inputs and outputs is performed using common chemical engineering principles. The denominator for the mass indicators should be based on the objective of the case scenario (e.g. production of 1 kg of product, treatment of 1 kg of waste, etc.) and should be the same for each technology alternative evaluated. All assumptions used in calculating the mass balance should be stated clearly. Once mass balances are completed, the mass indicators may be tabulated to compare each technology alternative.

Some of the mass indicators used are:

$$\text{Mass intensity} = \frac{\text{total mass input to the process excluding water}}{\text{basis of the mass balance calculations}} \tag{1}$$

$$\text{Solvent intensity} = \frac{\text{total solvent input excluding water}}{\text{basis of the mass balance calculations}} \tag{2}$$

$$\text{Waste intensity} = \frac{\text{total waste produced}}{\text{basis of the mass balance calculations}} \tag{3}$$

$$\text{Specific compounds released} = \frac{\text{amount of compound } i \text{ released as an emission}}{\text{basis of the mass balance calculations}} \tag{4}$$

3.3.2. *Energy indicators*

Energy for cooling, heating or power (electricity) is calculated for each technology option evaluated. The calculation of energy requirements should be on the same basis as is used for the mass balance, with all data and assumptions clearly noted. Equations (5)–(9) provide a sample of the calculation methods for energy requirements. An 85% heat transfer efficiency is assumed in all cases.

$$
\begin{aligned}
\text{Total heating requirements} = {}& \sum_n \left. \frac{mC_p(T_2 - T_1)}{\eta} \right|_{T_2 > T_1} + \sum_n \left. \frac{UA(T_w - T)t}{\eta} \right|_{T_w > T} \\
& + \sum_n \left. \Delta H_R^T \right|_{\text{endothermic}} - \text{heat recovered}
\end{aligned}
\tag{5}
$$

$$
\begin{aligned}
\text{Total cooling requirements} = {}& \sum_n \left. \frac{mC_p(T_2 - T_1)}{\eta} \right|_{T_1 > T_2} + \sum_n \left. \frac{UA(T_w - T)t}{\eta} \right|_{T > T_w} \\
& + \sum_n \left. \Delta H_R^T \right|_{\text{exothermic}} - \text{heat recovered}
\end{aligned}
\tag{6}
$$

$$
\text{Sensible heat} : Q = \frac{mC_p(T_2 - T_1)}{\eta}
\tag{7}
$$

$$
\text{Heating or cooling a vessel at a constant temperature} : Q = \frac{UA(T_w - T)t}{\eta}
\tag{8}
$$

$$
\text{Heat of reaction} : \Delta H_R^T = \frac{m}{\eta}\left(\Delta H_R^{298} + \int_{298}^{T} vC_p - dT \right)
\tag{9}
$$

It should be noted that the energy indicators above only include the energy requirements for each technology option used in the case scenario. The energy implications of utilities (e.g. steam, syltherm, etc.) supplied to the system or unit operations are covered in the life cycle portion of the methodology.

As shown in Eq. (5), the total heating requirement is a summation of all energy that needs to be added to the system minus any heat recovery. Added energy includes the heat of reaction for endothermic reactions, the sensible heat required to elevate the temperature, and the heat required to maintain a system above ambient temperature over a certain period of time.

Likewise, the total cooling requirement is a summation of all energy that needs to be removed from the system, with a final temperature not lower than

25°C. As shown in Eq. (6), the calculations discount any heat recovery when applicable. When a process stream is cooled below 25°C, it is generally assumed that a refrigeration system is used.

Electricity needed for mechanical devices such as pumps, compressors (e.g. refrigeration cycle), agitators, process controls and monitoring equipment, etc. is calculated using common chemical and mechanical engineering methods and has been described elsewhere [10–13].

3.3.3. Simplified life cycle approach

A simplified approach to life cycle inventory is used as an additional comparator of the technology alternatives in the case scenario. This approach first requires that life cycle information from available databases and literature is collected to estimate the life cycle impacts of the process that result from raw material and energy production [15–19]. The emissions from energy and raw materials production are then added to the emissions associated with a given unit process to estimate the total life cycle emissions associated with the unit operation. Because this is a simplified approach to life cycle inventory, life cycle impacts from transportation and waste treatment are not included in the evaluation.

In some cases, life cycle inventory information for raw or ancillary materials is not considered in the comparison because there are no significant differences in material or energy usage between technology options, or there are significantly smaller quantities of materials used for a given option, or there is a lack of data. Data gaps or exclusions should be clearly noted when they occur.

3.3.4. Operational considerations

Significant quantitative and qualitative operational differences between technology options affecting the efficiency and quality of the desired outcome must be identified and assessed. Efficiency considerations would include factors such as first pass yield, isolated yield and mass and separation efficiencies. Quality considerations would include such factors as product/impurities ratios or percentage purity. Other considerations include operational ranges and limits; typical processing times or reactor or unit operation residence times; process control issues; selectivity issues; type of operation (e.g. continuous vs. batch); potential operational problems (e.g. fouling, blockages); ease of technology scale-up and others.

3.3.5. Safety considerations

Potential material and process safety issues are qualitatively identified. Materials with known and reported hazards such as flammability, chemical incompatibilities, explosivity, corrosivity, toxicity, etc.; and or adverse reactions, are identified [14]. Process conditions that may lead to extremes in pressure or temperature should be identified and noted along with any process condition that would tend to make it easier for an uncontrolled process release or reaction to take place or lead to a similarly adverse incident.

3.4. Comparative ranking

Once all the technology options are analyzed, a relative rank is assigned to each category; i.e. environmental, energy, safety and efficiency. The ranking is performed by technical experts and is based on the results of the technology analysis and the values for all the indicators. Technical experts are used because technology comparisons need to be made in their appropriate context and there is a blend of quantitative and qualitative results. In some instances, absolute numbers or ranges (e.g. mass or energy use) may be compared; in other cases, semi-quantitative or qualitative relative comparisons must be employed.

The applicable indicators in each category are assigned a numeric value of 0, 5 or 10. This range of 0–10 is merely used as a relative scale that is assumed to be easily applied and understood. A value of 0 is given if the technology is perceived as having a disadvantage, 10 if it is perceived as having an advantage and 5 if the indicator is not perceived as a significant advantage or disadvantage. Once the indicators are assigned a relative numeric ranking, the arithmetic average of the indicators composing each category is calculated. If the average ranking for the category is lower than 2.5, the category will be assigned a 'red' color. If the average ranking is equal to or higher than 2.5, but lower than 7.5, the category is assigned a 'yellow' color. Finally, if the average ranking is 7.5 or higher, the category will be assigned a 'green' color.

A visually simple presentation of the comparison is employed using the color coding presented in Table 1.

This is also illustrated graphically in Fig. 3.

4. Vortex mixers: an application example

To illustrate the technology assessment methodology, the following sections present a case scenario comparing a batch process, a non-isolation batch

Table 1
Color-code key

Color	Key
🟩	Technologies considered to have significant advantages
🟨	Technologies with no significant advantages or disadvantages.
🟥	Technologies considered to have significant disadvantages

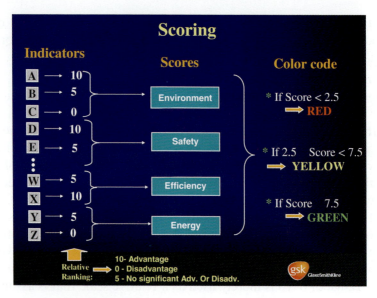

Fig. 3. Graphic representation of scoring process.

process and a continuous process using vortex mixers. The case scenario presented below is based on laboratory results carried at GSK for a Darzen's reaction to produce a pharmaceutical intermediate [20,21].

4.1. Introduction to vortex mixers

Fluid mixing plays a critical role in the success or failure of many industrial processes. Frequently observed effects of poor mixing include an undesirably wide particle size distribution in crystallizations, the production of impurities resulting from competing side reactions or localized heating, among others [22]. The use of a vortex mixer in chemical processing has specific applications where enhanced temperature control and mass transfer is required. In reactions where mass transfer is the rate-limiting step in a reaction sequence, vortex mixing may overcome this limitation [20,21].

A vortex mixer is remarkably simple and compact; it consists of a cylindrical chamber into which the liquid or gaseous reagents are injected tangentially, both in the same direction. The reagents, conserving angular momentum and mass flow, accelerate toward the axial outlet. Because both streams are accelerating, they thin out [23,24], and turbulent energy is dissipated. This gives excellent conditions for diffusion across concentration gradients and mixing at a molecular level takes place [23].

Vortex mixers, as with all fluidic mixers, are completely passive and need no mechanical energy input within the mixer itself. This means there are no rotating shafts, seals or blades. The internal geometry is completely open with no static packing or vanes to foul, erode, corrode or cause cavitation [24]. The small size and rapid mixing offered by the vortex mixer may offer significant advantages for exothermic and fast reactions as compared with a larger reaction vessel where the feed rate for reactants and reagents is limited by the heat and mass transfer capacity of the vessel and its contents [23,25]. In a vortex mixer, the large thermal mass of the mixing block relative to the volume of the mixing chamber may also allow highly exothermic reactions to take place safely. Reagent or reactant rates of addition may also be optimized to enhance yields while reducing solvent volumes normally used to reduce the potential for runaway reactions. Other examples of applications for vortex mixers include precipitation reactions (where the mixing rate can have a profound effect on the morphology and particle size) [25], highly exothermic reactions such as neutralizations using acid or base addition [24], fast reactions such as the iodide–iodate system [26,27], and hydrolysis reactions [28].

In the pharmaceutical industry, another area of potential application for non-isolation processes, including vortex mixers, is in reducing and optimizing solvent use for syntheses. Opportunities to employ non-isolations or semi-continuous processing options have been explored to reduce the amount and number of solvents used. In addition to the dual advantages of lower volume and numbers of solvents used, non-isolation syntheses may reduce cycle times by avoiding difficult crystallizations, filtration, multiple cake washes and drying. Finally, there are likely to be higher overall yields as there are fewer physical product losses during work ups. Several examples of environmental benefits resulting from the successful application of non-isolation processes in pharmaceutical manufacture, have been reported [29,30].

4.2. Description of the batch, non-isolation batch and continuous processes used for the evaluation

4.2.1. Batch process (Benchmark process)
The case scenario involves a reaction to convert pharmaceutical intermediate A to pharmaceutical intermediate B that is limited by the transfer rate (diffusion) of a hydroxide ion from the aqueous phase into the organic phase. While the reaction is an extremely rapid one, the mass transfer exchange between the aqueous and organic phases in a batch reactor configuration is not maximized, so the reaction kinetics are constrained. A phase transfer catalyst (PTC) is used to facilitate the diffusion of the hydroxide ion into the organic phase and increase the reaction rate. Experiments performed without the PTC showed that the conversion to B was extremely low (< 10%) [31]. However, carrying out the Darzen's reaction in a batch reactor with the PTC results in the formation of an undesirable by-product, since intermediate B continues to react with an excess

of the hydroxide ion. The desired product, **B**, is isolated as a wet cake, washed and dried prior to the next reaction. This process was run at manufacturing scale and a flow diagram for this batch process is shown in Fig. 4.

4.2.2. Non-isolation batch process
An alternative to the traditional batch process described above is to eliminate the wet cake work-up (i.e. crystallization, cake wash and drying) and carry the organic layer containing pharmaceutical intermediate **B** directly into the next stage. However, as is the case in the current batch process, the solvent mixture tetrahydrofuran/methylcyclohexane (THF/MCH) used for the extraction of **B** needs to be distilled from the mother liquors in the following stage before the next reaction can take place. This process is presented graphically as Fig. 5.

4.2.3. Continuous process with vortex mixer
A continuous process using a vortex mixer was also investigated as a means of providing better mass transfer rates between the aqueous and organic phases. No MCH is added in this process because of the faster kinetics during the improved phase separation encountered in the vortex mixer. To make the process continuous, a continuous centrifugal separator would be needed for the extraction of product with THF. In addition, as was the case for the non-isolation process, the THF must be removed before the next reaction can take place in the following stage. The block flow diagram for the continuous process using a vortex mixer is shown in Fig. 6.

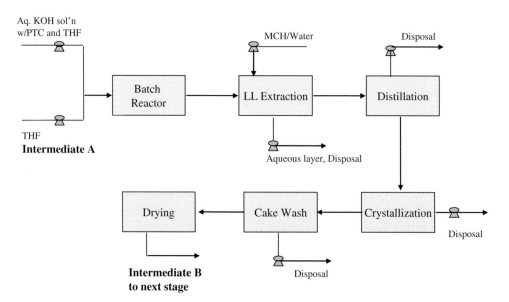

Fig. 4. Flow diagram of the batch process with isolation.

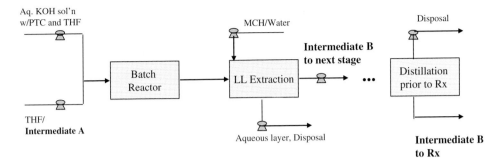

Fig. 5. Non-isolation batch process.

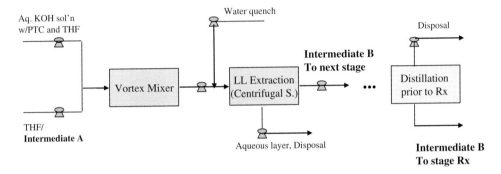

Fig. 6. Vortex mixer process.

The laboratory conditions for the vortex mixer option and the batch process are presented in Table 2. The process parameters in Table 2 used for the traditional batch process are the same for the non-isolation process [20].

4.3. Estimation of indicators

4.3.1. Mass intensity, solvent use and waste production
The production of 1 kg of intermediate B was used as the basis for calculating the mass and energy balances. The non-isolation and vortex mixer processes have better results for mass, waste and solvent intensity. The improved performance is mainly due to lower solvent use, but in the case of the vortex mixers, it is also due to an improvement in the yield. Figs. 7–9 are comparisons of the mass intensity, solvent intensity and waste intensity for each option of the case scenario.

4.3.2. Energy intensity
The energy indicators were calculated from heating requirements for distillations, and drying the product (isolation process), cooling requirements for

Table 2
Laboratory conditions for the vortex mixer and qualification batch conditions

Parameter	Qualification batch	Laboratory vortex mixer
Intermediate A required	125 kg/batch (393.7 mol)	2.5 kg/h (7.9 mol/h)
Weight % of Intermediate A in THF	30.2% w/w	27.3% w/w
Chloroacetonitrile usage	1.12 eq.	1.2 eq.
PTC usage	4.4 kg/batch	90 g/h
Concentration of base	30% w/w KOH	30% w/w KOH
Intermediate B produced	114.7 kg/batch (325.8 mol)	2.4 kg/h (7.1 mol/h)^2
Reaction temperature	0–10°C	Room temperature

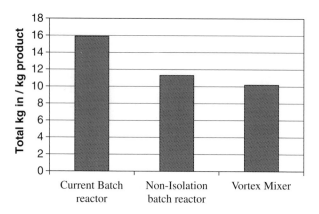

Fig. 7. Comparison of mass intensity for each processing option in the case scenario.

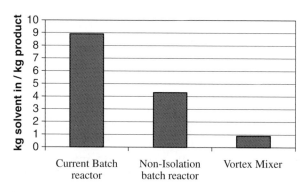

Fig. 8. Comparison of solvent intensity for each processing option in the case scenario.

distillate condensation and refrigeration required for cooling the batch reactor. Cooling is not required for the vortex mixer due to the more efficient heat dissipation and large thermal mass of the vortex mixer relative to the volume of the mixing chamber. Additional energy requirements were calculated from the electricity used to operate pumps, centrifuges and extractors. As can be seen in

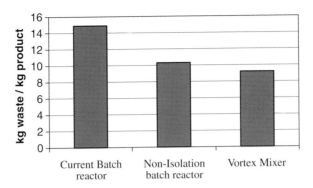

Fig. 9. Comparison of waste produced for each processing option in the case scenario.

Table 3
Summary of energy requirements

Energy requirements	Current batch reactor (MJ/kg product)	Non-isolation batch reactor (MJ/kg product)	Vortex mixer(MJ/kg product)
Heating	3.17	2.58	0.51
Cooling (with cooling water)	−13.23	−2.58	−0.51
Electricity (including refrigeration)	0.53	0.05	0.36

Table 3, the non-isolation process and the vortex mixer have lower energy requirements than the current batch process.

4.3.3. Simplified life cycle indicator

A simplified life cycle inventory for energy and solvent production was included in the comparison of the three processing options. Recent life cycle inventory work [9] has demonstrated that solvent production and use account for the greatest proportion of life cycle impacts in a pharmaceutical manufacturing context, so this approach is not without merit. Using this simplified approach, the use of a vortex mixer results in considerably lower life cycle emissions than the batch and non-isolation processes. For example, as can be seen in Fig. 10, CO_2 emissions associated with vortex mixer use are about 20% of the CO_2 emissions associated with the use of the current batch reactor and 30% of the CO_2 emissions associated with the use of the non-isolation batch reactor process.

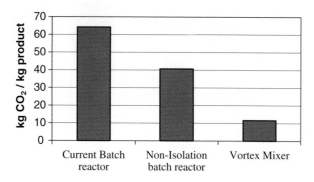

Fig. 10. Life cycle CO_2 emissions for energy and solvent production.

4.3.4. Safety indicators

In general, all processing options described above can be operated safely, and there are only relatively minor safety differences for this case scenario. Both batch processes operate at low temperatures (around $0°C$), whereas the vortex mixer system operates at ambient temperature. It should be noted, however, that vortex mixers represent a potentially safer operating environment due to the small volumes of the mixing chambers and short residence times of the substances in the system at any given time. This reduces the overall risk of accidental fire or explosions. Furthermore, reduced solvent use in the vortex mixer and non-isolation processes has the potential to reduce VOC's emissions during solvent handling and processing, thereby reducing the potential for occupational exposure to solvent vapors.

4.3.5. Operational indicators

In terms of operational considerations, the non-isolation process uses less solvent, has a reduced cycle time (almost eight-fold) and fewer work-ups when compared to traditional batch processes. The vortex mixer process has the following operational advantages over the current batch reactor process:

• Continuous operation of traditional batch processes
• Improved mass- and heat transfer
• Improved control of reactant and reagent residence time and temperature profile
• Reduced solvent use
• The potential for fewer work-ups if continuous separation operations are used
• Generally higher selectivity, yield and quality, especially for exothermic and fast reactions
• Linear scale-up (larger scale vortex mixers will yield the same results given that the residence time and flow velocities using at lab-level are matched) or the ability to easily number-up.

• Faster development of new processes or substances.

4.4. Comparative ranking

Based on the results from an analysis of the three processing options described above, the reaction systems were comparatively ranked. The color-coded summary of the comparative ranking is presented in Table 4.

5. Conclusions

The main purpose of the GTG methodology is to provide comparisons that highlight potentially adverse issues, favorable characteristics and possible trade-offs amongst unit operations or manufacturing process alternatives. Ultimately, the use of this methodology should at the very least facilitate more informed business decisions and at the very best, more sustainable business decisions.

It is recognized that this methodology and the ranking process are limited by the unit operations or manufacturing processing options that are being compared in each case scenario and to a lesser extent, by the subjectivity of the experts ranking the options. These limitations also mean that general conclusions about the technologies should not be drawn from the specific rankings developed for a given case scenario if the technologies are applied using very different conditions or scenarios. It is also recognized that the GTG is a work in progress that needs to be expanded to include more case scenarios, additional indicators and economic factors.

Despite these limitations, it is also recognized that this methodology represents a standard, documented and systematic approach to performing technology comparisons. In addition, the methodology integrates EHS considerations with efficiency and operational aspects that are routinely evaluated when choosing technologies and equipment.

Table 4
Comparative ranking for each technology option in the case scenario

Alternatives	Environment[a]	Safety[b]	Efficiency[c]	Energy[d]
Batch reactor (isolation process)	red	yellow	yellow	red
Batch reactor (non-isolation process)	green	yellow	green	yellow
Vortex mixer	green	yellow	green	green

[a]For the Environment category, a combination of mass and life cycle indicators were evaluated.
[b]For the Safety category, a combination of safety indicators were evaluated.
[c]For the Energy category, energy requirements and life cycle energy were evaluated.
[d]The Efficiency category includes efficiency and operational indicators.

As is demonstrated in the example provided in this Chapter, clear differentiation between technology options is possible, and the results of the analysis may be presented and communicated in a very concise, clear and easy to understand manner. At the same time, the detailed analysis and calculations are transparent and may be easily shared, if so desired.

We are confident that as we and others expand our understanding of what constitutes "green" or "clean" technologies, we will move more rapidly toward more sustainable business practices.

Acknowledgments

The authors wish to acknowledge Robert Herrmann for his outstanding and productive research on vortex mixers.

The authors also wish to recognize the significant contributions and hard work of all our colleagues in GSK's Research and Development and Corporate Environment, Health and Safety Departments that make it possible for GSK to advance toward more sustainable practices.

References

[1] http://www.gsk.com/about/downloads/plan-for-excellence-2003.pdf.

[2] A.D. Curzons, D.J.C. Constable, V.L. Cunningham, Clean Products Processes 1 (1999) 82.

[3] D.J.C. Constable, A.D. Curzons, V.L. Cunningham, Green Chem. 4 (2002) 521.

[4] D.J.C. Constable, A.D. Curzons, L.M. Freitas Dos Santos, G.R. Green, R.E. Hannah, J.D. Hayler, J. Kitteringham, M.A. McGuire, J.E. Richardson, P. Smith, L. Webb, M. Yu, Green Chem. 3 (2001) 7.

[5] A.D. Curzons, D.J.C. Constable, D.N. Mortimer, V.L. Cunningham, Green Chem. 3 (2001) 1.

[6] C. Jiménez-González, A.D. Curzons, D.J.C. Constable, M.R. Overcash, V.L. Cunningham, Clean Products Processes 3 (2001) 35.

[7] C. Jiménez-González, D.J.C. Constable, A.D. Curzons, V.L. Cunningham, Clean Techn. Environ. Policy 4 (2002) 44.

[8] C. Jimenez-Gonzalez, A.D. Curzons, D.J.C. Constable, V.L. Cunningham, Clean Techn. Environ. Policy, published Online First, 8 April 2004.

[9] C. Jimenez-Gonzalez, A.D. Curzons, D.J.C. Constable, V.L. Cunningham, International Journal for Life Cycle Assessment 9 (2) (2004) 114.

[10] C.R. Branan, Rules of Thumb for Chemical Engineers, Gulf Publishing Company, 1994.

[11] R.H. Perry, Chemical Engineer's Handbook, 7th. Ed., McGraw-Hill, New York, NY, 1997.

[12] S.M. Walas, Chemical Engineering 16 (March) (1987) 75–81.

[13] D.R. Woods, Process Design and Engineering Practice, Prentice-Hall, Upper Saddle River, NJ, 1995.

[14] HSE (Health and Safety Executive), Designing and Operating Safe Chemical Reaction Processes. HSE Books, Colegate, Norwich, UK, 78 pp. 2000.

[15] R.D. Dumas, Energy Usage and Emissions Associated with Electric Energy Consumption as Part of a Solid Waste Management Life Cycle Inventory Model. Department of Civil Engineering,, North Carolina State University, Raleigh, NC, 1997.

[16] EMPA, ECOPRO, Life Cycle Analysis Software. EMPA (Swiss Federal Laboratories for Material Testing and Research), St. Gallen, Switzerland, 1996.

[17] PIRA International, PEMS 4, Life Cycle Assessment Software, Randall Road, Leatherhead, Surrey, UK, 1998.

[18] C. Jiménez-González, M.R. Overcash, Clean Products Processes 2(1) (2000a) 57.

[19] C. Jiménez-González, S. Kim, M.R. Overcash, LCA, 5(3) (2000b) 153.

[20] R. Herrmann, R. Hannah, Evaluation of Vortex Mixing: A Comparative Case Study for Production of SB 224773, Stage 8 Intermediate in the SB 207499 Process, Internal Memorandum, Environmental Research Laboratory, 8 pp., 8 June 1998.

[21] R. Herrmann, Applications of Motionless Mixers for Fast Chemical Reactions in the SB-207499 Process. Internal Document, 2 pp., n/d.

[22] D.M. Hobbs, F.J. Muzzio, Optimization of a Static Mixer using a Dynamical Systems Techniques, Chem. Eng. Sci. 53 (18) (1998) 3199–3213.

[23] J. Redman, Getting out of a Mess in Mixing, Chem. Eng., (489) (31 Jan 1991) 13.

[24] S. Brown, The Mixing Revolution, Manuf. Chemist, 65 (1) (1994) 19–21.

[25] M. Knott, Mixers Set to Cause a Stir, Process Eng. (London), 73 (3) (March 1992) 29–30.

[26] P. Guichardon, L. Falk, Characterization of Micromixing Efficiency by the Iodide–Iodate Reaction System, Part I: Experimental Procedure, Chem. Eng. Sci. 55 (19) (2000) 4233–4243.

[27] P. Guichardon, L. Falk, J. Villermaux, Characterization of Micromixing Efficiency by the Iodide–Iodate Reaction System, Part II: Kinetic Study, Chem. Eng. Sci. 55 (19) (2000) 4245–4253.

[28] O.S. Brokers, Power Fluidics or Harnessing Hydrodynamics, Chem. Eng. (London) 521 (1992) 19–20.

[29] D.J. Dale, P.J. Dunn, C. Golightly, M.L. Hughes, P.C. Levett, A.K. Pearce, P.M. Searle, G. Ward, A.S. Wood, The Chemical Development of the Commercial Route to Sildenafil: A Case History, Organic Process Research and Development 4 (2000) 17–22.

[30] C. Jiménez-González, M. Overcash, Energy Optimization During Early Drug Developoment and the Relationship with Environmental Burdens, Journal of Chemical Technology and Biotechnology 75 (983—990) (2000).

[31] R. Herrmann, Internal Communication, 21 March 2001.

Sustainability Science and Engineering: Defining principles
Martin A. Abraham (Editor)
© 2006 Published by Elsevier B.V.
DOI 10.1016/S1871-2711(05)01019-6

Chapter 19

Engineering Sustainable Facilities

J.A. Vanegas

Department of Architecture, College of Engineering, Texas A&M University, 3137 TAMU, College Station, TX 77843–3137, USA

1. Introduction

The *Architecture/Engineering/Construction* (AEC) industry, both in the US. and in every other country in the world, is the main provider and the life cycle custodian of *facilities*, such as residential building, industrial facilities, and of *civil infrastructure systems*, such as transportation, energy, communications, water supply, and waste management systems. Together, facilities and civil infrastructure systems provide the fundamental foundation upon which any society exists, develops, and survives, and the value, quality, performance, integrity, and sustainability of this foundation are functions of the way that the AEC industry plans, designs, procures all resources, builds, commissions, and helps operate and maintain these facilities and civil infrastructure systems over their complete life span.

The results of *what the AEC industry does* (i.e. the facilities delivered), *how it does it* (i.e. the processes followed in delivering them), and *with what* (i.e. the resources used in delivering them), is a significant yearly volume of economic activity. For example, new construction in the United States averaged about 8% of the Gross Domestic Product (GDP) from 1975 to 1995 [1]. This level of economic activity leads to major direct and indirect impacts to the environment, such as natural resource consumption, degradation and depletion, waste generation and accumulation, and environmental impact and degradation, as has been documented extensively in the literature in recent years (e.g. [2–4]). This chapter addresses how the AEC industry can engineer more sustainable facilities toward the mitigation, and hopefully the eventual elimination, of these negative impacts, and also to ensure the highest levels of value, quality, and life cycle performance of facilities.

Building upon the Brundtland Commission's definition of sustainable development as a way to ensure "*...meeting the needs of the present without compromising the ability of future generations to meet their own needs...*" [5], the operational definition for sustainable facilities used in this chapter is: "*A sustainable facility is a facility that incorporates sustainable attributes and characteristics; is delivered following sustainable processes; consumes resources in a sustainable way, both in its delivery and in its operations and maintenance; and meets the needs of its stakeholders without compromising the ability of other stakeholders to fulfill their own facility needs, from an environmental, social, and economic perspectives, in the present and in the future.*"

This chapter presents a road map for engineering sustainable facilities. Its intent is to enable engineers involved in any phase of the life cycle of a facility, i.e. from planning, through design, procurement, construction, and commissioning, to operations and maintenance, to understand (1) the context for engineering sustainable facilities; and (2) how the various elements of the roadmap come together to support the implementation of sustainability in the facility. Since specific engineering guidance is outside the scope of this chapter, a set of implementation resources is presented at the end of the chapter.

2. The context for engineering sustainable facilities

Three cornerstones are important to understand the context for engineering sustainable facilities: (1) an understanding that facilities have unique attributes and characteristics that stem from the specific industry sector of a facility, its specific project type, and the specific context that envelops the facility; (2) an understanding that implementation of sustainability in facilities can be achieved by following a general strategy for engineering sustainable facilities; and (3) an understanding that there are multiple opportunities in the life cycle of the facility for the application of sustainability principles, concepts, strategies, guidelines, specifications, standards, heuristics, best practices, lessons learned, or tools. These three cornerstones are explained next.

2.1. A taxonomy for characterization of sustainable facilities

The first cornerstone for engineering sustainable facilities is an understanding that facilities have unique attributes and characteristics that stem from the specific industry sector of a facility, its specific project type, and the specific context that envelops the facility, as shown in Fig. 1.

Three main axes define the basic structure of this taxonomy. The first axis represents the various types of facilities by industry sector: (1) *residential construction*

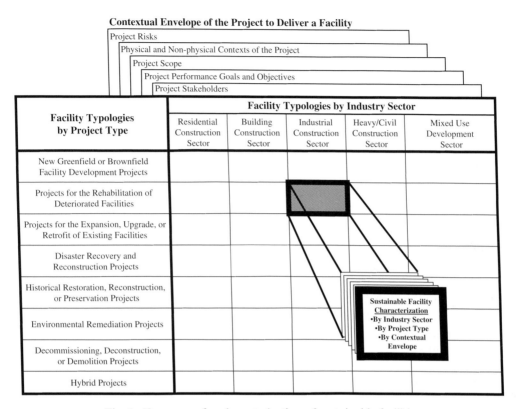

Fig. 1. Taxonomy for characterization of sustainable facilities.

sector, which includes facilities such as detached single family homes, town houses, condominiums, and apartment buildings; (2) *building construction sector*, which includes facilities such as commercial, official, educational, recreational, and institutional buildings; (3) *industrial construction sector*, which includes facilities such as factories, manufacturing, process, and power plants; (4) *heavy/civil construction sector*, which includes facilities such as roads, bridges, tunnels, airports, dams, and wastewater treatment plants; and (5) *mixed use development sector*, which includes facilities resulting from land and real estate developments that have elements from more than one sector. The implementation of sustainability in a facility needs to respond to the unique and specific characteristics of the sector it represents. For example, the attributes and characteristics of a sustainable home may be quite different to the attributes and characteristics of a sustainable manufacturing plant.

The second axis represents the various types of facilities by project type: (1) *development projects* for new facilities, in both greenfield and brownfield sites; (2) *rehabilitation projects* to correct the effects of natural deterioration in a facility; (3) *expansion, upgrade, and retrofit projects* within facilities that are operational; (4) *recovery and reconstruction projects* to correct the damage caused by natural or human-made events to a facility; (5) *restoration, recon-struction, and preservation projects* for facilities that have historical or cultural significance; (6) *remediation projects* to correct negative impacts to the environment caused by the construction and operation of a facility; (7) *deconstruction, decommissioning, and demolition projects* for facilities that have reached the end of their service lives; and (8) *hybrid projects* that combine elements from several of the other project types. The implementation of sustainability in a facility also needs to respond to the unique and specific characteristics of the type of project it represents. For example, the attributes and characteristics of a sustainable brownfield development project may be quite different to the attributes and characteristics of a sustainable rehabilitation project.

Finally, the third axis represents the specific context that envelops the delivery of a project. This *contextual envelope* includes (1) the various *stakeholders* involved in the project; (2) their *performance goals and objectives* for the project; (3) the specific *scope* of the project; (4) the *physical and non-physical contexts* within which the project is executed; and (5) the *risks* associated with the project. The implementation of sustainability in a facility also needs to respond to the unique and specific characteristics stemming from each one of these five elements. For example, the attributes and characteristics of a sustainable facility for an organization in the public sector may be quite different to the attributes and characteristics of a sustainable facility for an organization in the private sector.

2.2. A general strategy for engineering sustainable facilities

The second cornerstone for engineering sustainable facilities is an understanding that implementation of sustainability in facilities can be achieved by following a general strategy for engineering sustainable facilities, as shown in Fig. 2.

This strategy is adapted from research conducted to define an operational framework for sustainability of built environment systems [6], and is discussed in more depth in [7]. Fundamentally, the general strategy is to apply sustainability principles, concepts, specific strategies, guidelines, specifications, standards, heuristics, best practices, lessons learned, or tools as fundamental criteria for making decisions, choosing among various options, or taking actions regarding: (1) *what the AEC industry does*, i.e. the attributes and characteristics of the different types of facilities that it delivers, operates, and maintains; (2) *how the AEC industry does it*, i.e. the processes, practices, and procedures followed at every phase in the life span of the facility, from initial planning to the end-of-service-life of the facility; and (3) *with what*, i.e. the resources required for what they do, and for how they do it — labor, materials, equipment, methods, technology, energy, and money, among many others.

This strategy can be represented conceptually as a tri-axial (X, Y, Z) diagram, with the X-axis representing the *attributes and characteristics* of the facility (as

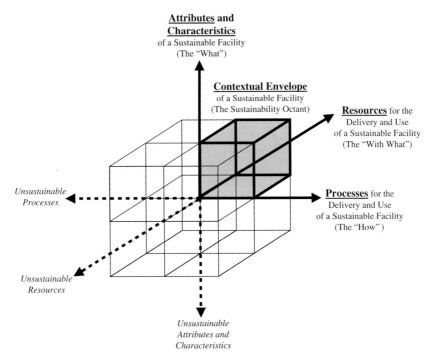

Fig. 2. A general strategy for engineering sustainable facilities (developed from Vanegas [7]).

defined by the type of facility according to the taxonomy discussed in Section 2.1); the Y-axis representing the *processes* followed in the delivery and use of the facility; and the Z-axis representing the *resources* consumed in the delivery and use of the facility. All three axes are expressed on a linear scale that spans from what is unsustainable to what is sustainable in each one, and with a threshold that separates the two extremes within each axis as point (0, 0, 0). The strategy then is to ensure that the outcome of all key project decisions, choices, and actions fall within the *sustainability octant*, i.e. where all three (X, Y, Z) points are on the sustainable side of each of the three axes.

While this general strategy may be conceptually simple, in reality it is quite complex, and much research is yet to be done to provide clear and absolute definitions of what sustainable is, what unsustainable is, and what the threshold is, which separates what is sustainable from what is not.

2.3. Opportunities for engineering sustainable facilities

The third, and final, cornerstone for engineering sustainable facilities is an understanding that there are multiple opportunities in the life cycle of the facility, for the application of sustainability principles, concepts, strategies, guidelines, specifications, standards, heuristics, best practices, lessons learned, or tools, as shown in Fig. 3.

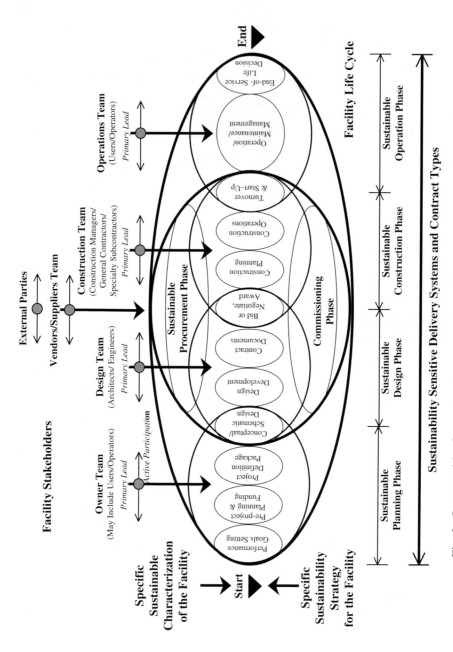

Fig. 3. Opportunities for engineering sustainable facilities (developed from Vanegas [7]).

For a specific facility, two specific points of departure from the current process of delivery of facilities toward a process that leads to more sustainable facilities are: (1) the *sustainable characterization* of the facility, i.e. the definition, from a sustainability perspective, of the unique attributes and characteristics of the facility stemming from the facility's specific industry sector, specific project type, and specific contextual envelope; and (2) the development of a *sustainability strategy* for the facility that defines the processes and resources required to achieve the sustainable attributes and characteristics defined for the facility. These two points of departure are also discussed in more depth in [7].

There are three main categories of opportunities to enhance the sustainability of a facility:

- The first category is composed of the complete set of *stakeholders* involved throughout the complete life cycle of the facility: (1) members of the *Owner Team*; (2) members of the *Operations Team*; (3) members of the *Design Team*; (4) members of the *Construction Team*; (5) *Vendor and Suppliers*; and (6) *External Parties*. Engineering a sustainable facility requires that these stakeholders have a basic understanding of sustainability in any of its manifestations (i.e. principles, concepts, strategies, guidelines, specifications, standards, heuristics, best practices, lessons learned, and tools). In addition, they need to operate as a high-performance, integrated, and aligned team. Furthermore, they also need to use sustainability formally, explicitly, systemically, and systematically as a fundamental criterion when making decisions, choosing among various options, or taking actions for the project at each stage of its life cycle. Thus, every person involved at any stage of the delivery and use of a facility has the potential to support the implementation of sustainability.
- The second category is composed of the complete set of *delivery systems* (e.g. design/bid/build, design/build, build/operate/transfer), and of *contract types* (e.g. cost plus reimbursable fee, stipulated sum, unit cost), used throughout the complete life cycle of the facility. A delivery system can enable or inhibit the implementation of sustainability in a facility by the way it defines the contractual roles, responsibilities, and relationships between or among project team members, and establishes the interrelationships or sequences among the design, procurement, and construction phases of the project. Similarly, a given contract type can be sensitive to, and the implementation of sustainability by the way it defines the primary compensation approach within a specific contractual relationship between two parties, including incentives and disincentives. Thus, every delivery system and every contract used at any stage of the facility life cycle also has the potential to support the implementation of sustainability.
- The third category is composed of the complete set of *processes-, practices-, and operating procedures* followed within the various phases of the complete

life cycle of the facility: (1) the *planning phase*; (2) the *design phase*; (3) the *procurement-, construction-, and commissioning phases*; and (3) the *operations and maintenance phase*.

Each of the phases in the third category of opportunities to enhance sustainability in a facility contributes in different ways to establish the degree, breadth, and depth of sustainability implementation efforts:

- The *Sustainable Planning Phase* for a given facility begins with the characterization of the facility from a sustainability perspective, and with the development of a specific sustainability strategy for the facility, followed by (1) the formal and explicit setting of project performance goals; (2) pre-project planning and funding approval; and (3) the development of the project definition package (PDP). In some cases, the planning phase may extend to also include the conceptual and schematic design for the facility, which marks the transition into design. The planning phase has the greatest potential to influence overall sustainability in a facility at lowest cost. Specific sustainability issues considered in this phase include framing of facility requirements, characteristics, and performance goals within a sustainability perspective; sustainable site selection; and ensuring the compatibility of project objectives and scope with the constraints of its physical and non-physical contexts.
- The *Sustainable Design Phase* includes the (1) development of the conceptual and schematic design for the project; (2) detailed design; and (3) development of contract documents. In some cases, the design phase may extend to also include bidding or negotiation, and award of the construction contract, which marks the transition into construction. The design phase also affords significant opportunities for influencing facility sustainability before any construction operations begin on site. Specific sustainability issues considered in this phase include sustainable site development; integrated building systems design; energy and water efficiency; sustainable material use; and indoor environmental quality.
- The *Sustainable Construction Phase* includes: (1) bidding or negotiation, and award of the construction contract; (2) construction planning; and (3) construction operations. In some cases, the construction phase may extend to also include turnover and start-up, which marks the transition into the operations phase. The construction phase is the bridge between the design solution and the built facility, and offers additional opportunities for increasing the sustainability of the facility. Specific sustainability issues considered in this phase include minimization of site disturbance; indoor environmental quality; increased construction recycling and resource reuse; and construction health and safety.
- The *Sustainable Procurement Phase* parallels the design and construction phases, and provides the interface with the supply chain that provides all the technologies, systems, products, materials, and equipment specified by

designers and procured by constructors to physically realize the project. While the nature, levels of performance, and desired attributes of these resources are fixed by the project design, considerable sustainability benefits can still be realized from the sources selected for specified materials, and from the way they are brought into the project. Specific sustainability issues considered in this phase include reduction or elimination of packaging that cannot be recycled or re-used; increased content of recycled post-consumer and post-industrial waste in materials; minimization of construction debris and demolition waste; and avoidance of manufacturers that use processes with negative environmental impacts.

- The *Commissioning Phase* also parallels the design and construction phases, and ensures that all the building systems and equipment are installed and tested during construction to verify that their performance is within the parameters desired and specified. Lack of proper commissioning leads to higher operations and maintenance costs as a result of inefficient energy and water use. In addition, poor commissioning has a direct negative impact on the productivity of people working in a facility.
- The *Sustainable Operations Phase* includes: (1) turnover and start-up of the facility; and (2) full operation, maintenance, and management of the facility. Sustainable operations, maintenance, and management require effective and efficient planning, allocation, and use of resources over the operational life of the facility. Specific sustainability issues considered in this phase include maintaining indoor environmental quality; thermal comfort; light quality; energy, water, and resource conservation; and waste management.

The decision on what to do with a facility at the end of its service life marks the end of the life cycle of the facility. Since decisions, choices, and actions at any of the previous phases can impact actions at this final point of the facility's life, explicit consideration of what happens to the facility at the end of its useful life of the facility should be considered by all project stakeholders during all the previous phases of the facility life cycle, particularly the planning and design phases. Specific sustainability issues considered at the end of its useful life of a facility include easy disassembly/reuse of components; increase in material recovery and recycling efforts; and site reclamation.

These opportunities for sustainability, together with the characterization of, and the general strategy for, sustainable facilities, establish a firm foundation for the road map for engineering sustainable facilities described in the next section.

3. A road map for engineering sustainable facilities

As shown in Fig. 4, the road map for engineering a sustainable facility contains a set of 15 closely interrelated and interdependent elements that come together, as

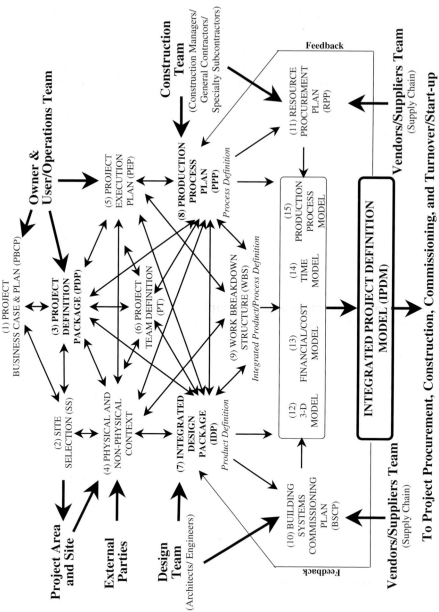

Fig. 4. Road map for engineering sustainable facilities (developed from Vanegas [7]).

a cohesive whole, in an *Integrated Project Definition Model (IPDM)*. The IPDM provides a solid basis for achieving a sustainable facility by defining the parameters and approaches within which the procurement, construction, commissioning, turnover, and start-up processes can be executed in a sustainable way.

Each of these 15 elements contributes in a different way to establish the degree, breadth, and depth of efforts to ensure the sustainability of a specific facility [7]. However, the full potential of the road map is reached when sustainability, in any of its possible manifestations (e.g. principles, concepts, specific strategies, guidelines, specifications, standards, heuristics, best practices, lessons learned, or tools — see Section 2.2 for examples), is incorporated as an integral part of all of its elements in four ways: formally, explicitly, systemically, and systematically. The following section explains in more detail the various elements of the road map, and what is required in each one to achieve sustainability.

3.1. The principal elements of the road map

The first element of the road map is the development of the *Project Business Case and Plan* (PBCP) for the facility. The PBCP establishes, from the owner team and the user-operator team perspectives, both: (1) the justification for the proposed facility, as the basis to compare the proposed project with other projects competing for limited funds, or to assess the options for a project that has already received funding; and (2) the management approach for the project to ensure the delivery of outputs and the realization of outcomes in the project. Engineering a sustainable facility requires that the goal of achieving sustainability be explicitly incorporated as an integral part of the business case and plan for the facility.

The second element is *Site Selection*, which establishes where the facility will be located. Engineering a sustainable facility requires selecting a site for the facility that is optimal from a sustainability perspective, which, among others, can be a site that does not develop on prime agricultural land; does not include habitat for threatened or endangered species; does not lead to wetland destruction of reduction of public parkland; promotes urban or brownfield redevelopment; supports alternative transportation options; and are conducive to minimal site disturbance, natural stormwater management, appropriate landscape and exterior design to reduce heat island effects, and light pollution reduction.

The third element is the *Project Definition Package (PDP)*, which defines, from the owner team and the user-operator team perspectives, what the facility is all about. It provides a cornerstone for the implementation of sustainability in a facility. This element is described in more detail in Section 3.2.

The fourth element is composed of the *Physical and Non-physical Contexts* of the facility, as defined by the external parties involved in the delivery of the facility, by the general area and business environment within which the facility is built, and by the specific site for the facility. The definition of the physical context includes analyses of geographical location, accessibility, transportation options, surface and subsurface conditions, environmental conditions, existing infrastructure, and surrounding activities or assets. The definition of the non-physical context includes analyses of policy issues, legal and regulatory issues, applicable codes, standards, and regulations, economic and financial issues, political issues and public relations, community, social, and cultural issues, industrial, and technological issues. Engineering a sustainable facility requires that the project responds to all the constraints imposed by, and is compatibility with all elements of, its physical and non-physical contexts.

The fifth element is the *Project Execution Plan (PEP)*, which defines, also from the owner team and the user-operator team perspectives, and in a general way, how the project for the facility should be executed. This plan defines the most appropriate delivery system and contract type for the project based on the characterization of unacceptable and prudent reversible risks in the project, and on the assessment of their probability, impact, and allocation. It also establishes the overall strategies in the project required (1) to ensure quality of execution, quality of initial conformance, quality of long-term performance, quality of redesign and improvement, and the prevention of total or partial failures; (2) for risk assessment, risk avoidance, risk mitigation, and risk management; and (3) to ensure the protection of people, property, and the environment from natural and human-caused events and disasters. Engineering a sustainable facility requires that the PEP directly incorporate sustainability as a formal and explicit strategy for project execution.

The sixth element is *Project Team Definition (PTD)*, which defines all the individuals, functional units, and organizations directly and indirectly involved in the delivery of the facility, from the: (1) Owner Team; (2) User/Operator Team; (3) Design Team; (4) Construction Team; (5) Vendors and Suppliers; and (6) any External Parties. Furthermore, it also defines a set of common and well-defined goals and objectives for the project; implements partnering development and team-building processes; implements project alignment and misalignment elimination processes; establishes a set of acceptable tolerances and team norms within which the team will operate; and implements partnering and team maintenance processes, and strong quality leadership on an on-going basis. Engineering a sustainable facility requires that all members of the project team have an awareness and understanding of sustainability.

The seventh and eighth elements define the project from an integrated design/construction point of view: (1) the *Integrated Design Package (IDP)*, which defines, from the design team perspective, the design solution for the project from a product definition point of view: and (2) the *Production Process Plan (PPP)*, which defines, from the construction team perspective, the construction, fabrication, and logistics requirements for the project, from a process definition point of view. These two elements provide two additional cornerstones for the implementation of sustainability in a facility. These two elements are described in more detail in Section 3.3 and in Section 3.4, respectively.

The ninth element is the set of *Work Breakdown Structures (WBS)*, which define various ways in which the work can be packaged in the project. A WBS contains a hierarchy of levels, each representing an increasingly detailed description of project elements. Two main types of WBS define the project, the first one from a product perspective (i.e. Functional, Building Systems/Processes, and Building Components/Elements WBS), and the second one from a process perspective (i.e. Cost, Time, and Cost Control WBS). Engineering a sustainable facility requires that sustainability-related items be explicitly included within the various WBS for the project.

The tenth and eleventh elements include: (1) the *Building Systems Commissioning Plan (BSCP)*, which defines, from a design perspective, the communication, coordination, testing, and verification required to ensure the delivery of a facility in which its building systems and equipment perform as originally intended by the design team; and (2) the *Resource Procurement Plan (RPP)*, which defines, from a construction perspective, how all the resources for the project should be procured. Both these elements are defined with input from vendors and suppliers in the supply chain for the project Engineering a sustainable facility requires extensive commissioning efforts, and also, that sustainable resources be used in the project, e.g. rapidly renewable materials, materials with increased content of recycled post-consumer and post-industrial waste, materials acquired from local or regional sources, or materials that are certified as environmentally friendly.

The remaining four elements include: (1) the *3D Model*, which enables visualization of the facility prior to construction, through the definition of the three-dimensional spatial data and information of the design solution of the project; (2) the *Cost/Financial Model*, which defines the specific parameters required to ensure project performance from a cost point of view, including the financing package; Total Installed Costs (TIC); Operations & Maintenance (O&M) Costs; and Life Cycle Costs (LCC); (3) the *Time Model*, which defines the specific parameters required to ensure project performance from a time point of view, including the cycle times of each of the phases of the life cycle of the project; and (4) the *Production Process Model*, which defines the parameters of the production process to follow in the field during construction of the project. Sustainability needs to be addressed explicitly as an integral component of each of these models.

Although all the elements of the road map provide a means for incorporating sustainability into a project, three of these elements stand out, and are explained later in more detail.

3.2. The project definition package (PDP)

As mentioned in Section 3.1, the *Project Definition Package (PDP)* defines, from the owner team and the user-operator team perspectives, all dimensions and parameters of the facility. The importance of the PDP in a project is explained in more detail in [8].

The general process for the development of the PDP has three phases, as shown in Fig. 5. The first phase is the *Formation* phase, in which, the aligned Owner & User/Operator perspectives provide a point of departure for the project. These perspectives are anchored in the vision, mission, strategic plan, and business plan of the project owner's organization. The second phase is the *Communication*, phase, during which the Design, Construction, Procurement and External Parties perspectives are added. The third and final phase is the *Integration* phase, during which an integrated PDP is developed, within which the perspectives of all stakeholders in the project are aligned.

The specific steps that need to be followed in the development of the PDP after establishing the principal *requirements-, attributes-, and characteristics* of the project by industry sector and project type, include the formal and explicit identification, definition, and documentation of the following items:

- The internal and external *key stakeholders* that are directly and indirectly involved in the project, including organizations, organizational units, functional units, and individuals
- Their *principal project goals and objectives* for 12 parameters of project performance: (1) compatibility and response of the project to its physical and non-physical contexts; (2) functional performance; (3) formal and physical performance; (4) quality and reliability performance; (5) cost performance; (6) time performance; (7) safety and security performance; (8) procurement and constructability performance; (9) commissioning, start-up, and turnover performance; (10) operability, maintainability, and security performance; (11) indoor environmental quality performance, i.e. ensuring the physical health and well-being of the people who will be the project's ultimate users, as a function of indoor air quality, potable water quality, lighting, noise pollution, work environment, comfort, and levels of emissions from materials, such as paints, carpets, adhesives; and (12) sustainability performance, i.e. elimination, reduction, or mitigation of any type of resource base impacts (i.e. resource consumption and waste generation), eco-system impacts (i.e. environmental impacts to air, water, soil, and biota), and human impacts (i.e. current and future impacts to all project stakeholders).
- The *specific elements of scope* of the project associated with each goal and objective
- The *physical and non-physical contexts* of the project

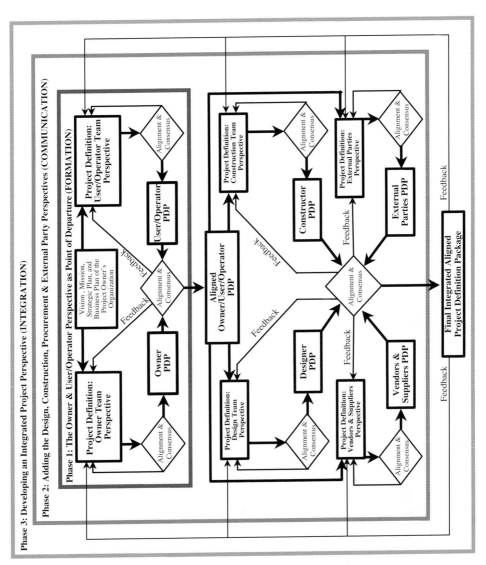

Fig. 5. A general process for project definition.

- The principal *risks* stemming from the project from a product, process, and context perspectives.

Once these items have been identified, defined, and documented, they are used in: (1) the analysis of any *internal and external influences, interrelationships,* and *interdependencies* among the project's requirements, attributes, and characteristics, stakeholders, goals and objectives, scope, physical and non-physical contexts, and risks that affect, or can affect, project performance; and (2) the development of the various work breakdown structures of the project.

The main process within the development of the PDP is a process of alignment and consensus building. This process requires, through on-going strong leadership, the implementation of partnering development, team building, project alignment, and misalignment elimination processes within the project team. It also requires the establishment of a set of acceptable tolerances and team norms within which project the team will operate, as well as the implementation of on-going partnering and team maintenance processes within the project team.

Engineering a sustainable facility requires that explicit consideration of sustainability be an integral part of making decisions, choosing among alternatives, and taking actions within the process of developing the PDP.

3.3. The integrated design package

As mentioned in Section 3.1, the *Integrated Design Package (IDP)* defines, from the design team perspective and from a product definition point of view, the design solution for the project. Hawken et al., in *Natural Capitalism* [9], as well as other sources such as the *Whole Building Design Guide* [10], highlight the importance of an integrated design approach to establish the degree, breadth, and depth of sustainability efforts throughout a project. A general process for integrated project design is shown in Fig. 6.

The principal features of an integrated project design approach include:

- Phased processes for the parallel development, coordination, and integration of design solutions from all the disciplines involved in a given AEC project (e.g. architectural, structural, mechanical, electrical, civil, and others), in increasing levels of detail, from conceptual and schematic design, to design development and contract documents
- Mechanisms to regulate the flow of design data and information from phase to phase, and among stakeholder entities as needed, including processes for analysis, generation, evaluation, selection, and specification; decision-making; and conflict resolution
- Provisions at each phase for performance parameters check before proceeding to the next design phase

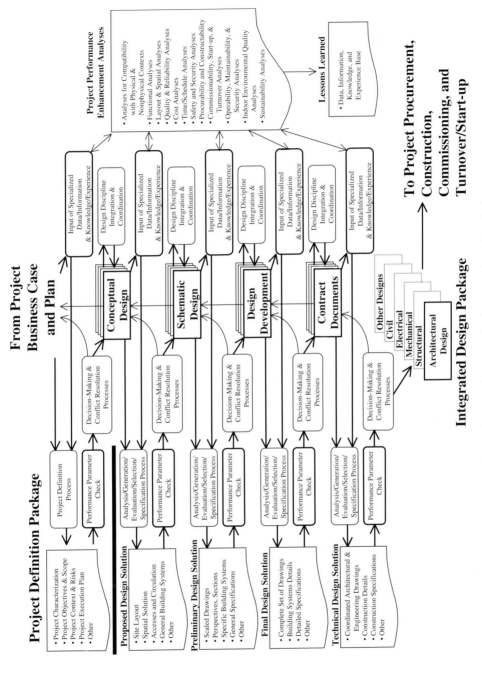

Fig. 6. A general process for integrated project design.

- Mechanisms for formal, explicit, systemic, and systematic input of specialized data, information, knowledge, and experience resulting from project performance enhancement analyses in each of the performance parameter categories
- Capture of lessons learned throughout the process to contribute to the data, information, knowledge, and experience knowledge base.

This process culminates with an IDP, i.e. a set of plans and specifications, which, given the explicit consideration of sustainability as an integral part of the design process, defines the parameters and approaches within which the procurement, construction, commissioning, turnover, and start-up processes can be executed in a sustainable way.

Engineering a sustainable facility requires that explicit consideration of sustainability be an integral part of making decisions, choosing among alternatives, and taking actions within the design process for the facility.

3.4. The production process plan

As mentioned in Section 3.1, the *Production Process Plan (PPP)* defines, from the construction team perspective, the construction, fabrication, and logistics requirements for the project, from a process definition point of view. This plan includes strategies for: (1) the supply and delivery of all required technologies, systems, products, materials, and equipment specified for the project; (2) the construction operations, processes, and tasks required to complete all physical components of the project; and (3) workflow and production unit control. This view of construction as a production process has been promoted in recent years by many researchers and practitioners [11]. The fundamental goals of lean construction of maximizing value and minimizing waste support sustainability efforts for a project.

A general framework for construction as a production process is shown in Fig. 7. The central element of this process is a set of control functions that directly respond to the PDP, the PEP, and the IDP.

Five main functions are critical to this process:

- The first function establishes the work processes required to transform the resources required for the project (e.g. labor, materials, and equipment) into physical components of the project (e.g. building systems and subsystems).
- The second function regulates the flows of the resources required by the work processes established in the first function. These resources serve inputs to the transformation process that will convert them to the desired results.
- If a required resource is not already available on site, the third function triggers a resource acquisition process, which includes: (1) acquiring the resource following the resource procurement plan; (2) transporting the resource

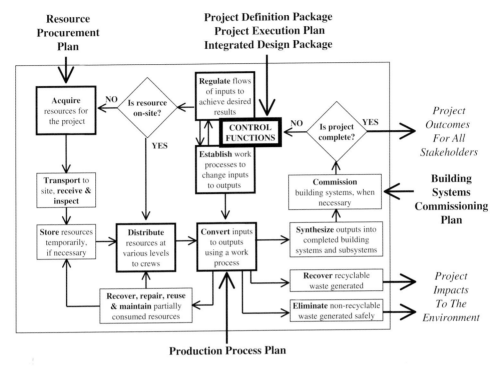

Fig. 7. Construction as a production process (developed from Oglesby et al. [12]).

to the site, receiving it upon arrival, and inspecting it for compliance and quality; and (3) storing it temporarily, if necessary.

• The fourth function distributes the resources required by the various work processes, which are on site, to the various crew at the appropriate levels of supervision of site operations (i.e. superintendent level, foremen level, and crew level).

• The fifth function converts the distributed resources following the production process plan, triggering four additional processes: (1) synthesis of outputs into completed building systems and subsystems, followed by commissioning of building systems and equipment according to the building systems commissioning plan, if necessary; (2) recovery, repair, reuse, and maintenance of partially consumed resources, which then return to storage or to distribution; (3) recovery of recyclable waste generated by the work processes; and (4) safe elimination of non-recyclable waste generated by the work processes.

The process continues iteratively with these five functions until the project is complete. Two main types of outputs result from this process: (1) the project outcomes for all project stakeholders; and (2) the impacts to the environment caused by the recyclable and non-recyclable waste generated by the work processes.

Engineering a sustainable facility requires that explicit consideration of sustainability be an integral part of making decisions, choosing among alternatives, and taking actions within the construction process for the facility.

4. Implementation resources for engineering sustainable facilities

As mentioned throughout the discussion of the roadmap presented in Section 3., effective implementation requires the continuous application of specific sustainability principles, concepts, heuristics, strategies, guidelines, specifications, standards, processes, tools, best practices, lessons learned, or case studies. The challenge is that the body of knowledge on sustainability of the built environment is rich, extensive, and diverse.

Since specific engineering guidance is outside the scope of this chapter, a set of implementation resources is presented in this section, which includes, among others:

- General international information web sites such as: (1) the *International Institute for Sustainable Development* provides an extensive compilation of sustainable development principles from numerous sources that address three major aspects: environment, economy, and community [13]; and (2) the *Sustainability Web Ring*, an Internet tool that allows users to navigate easily between Web sites that address with the principles, policies, and best practices for sustainable development [14].
- Specific guidance web sites such as the: (1) *Sustainable Design and Development Resource* of the Construction Engineering Research Laboratory (CERL) of the Engineer Research and Development Center (ERDC) of the US. Army Corps of Engineers (USACE) [15]; (2) *Sustainable Design Web Site* of the Air Force Center for Environmental Excellence (AFCEE) [16]; (3) *Minnesota Sustainable Design* Guide [17]; and (4) *Green Buildings Center of Excellence* for Sustainable Development of the US Department of Energy [18].
- Organizations that (1) provide technical assistance such as the *Sustainable Facilities & Infrastructure Program* of the Georgia Tech Research Institute (GTRI) at the Georgia Institute of Technology (Georgia Tech) [19]; (2) publish periodicals on various topics related to sustainability such as *Environmental Building News* [20]; and (3) advance the design, affordability, energy performance, and environmental soundness of residential, institutional, and commercial buildings nationwide, such as the *Sustainable Buildings Industry Council* (SBIC) [21].

In addition, a parametric review of existing literature in this area conducted by Pearce and Vanegas [22] identified *general references* (e.g. [23] and [24]); *models and frameworks* (e.g. [25–27]); *heuristics and guidelines* (e.g. [28–30]);

assessment and evaluation tools, (e.g. [31–33]); and *resource guides* (e.g. [34–36]), among many other references.

Any principle, concept, heuristic, strategy, or guideline, among other manifestations of sustainability, from any source such as these, can be integrated within any element of the roadmap for engineering sustainable facilities, and can serve as a metric that can be used in three ways: (1) for data collection and benchmarking on specific sustainability considerations; (2) as a point of reference to assess organizational behaviors and practices of AEC organizations; and (3) as a compass for maintaining an overall vision for built environment sustainability that can be reached incrementally and realistically.

The most important thing to keep in mind is that the journey toward sustainable facilities is a long one, but it is a journey in which sustainability can be achieved in many ways, in a gradual shift to a sustainable future: by making one decision, selecting one choice, or taking one action that is more sustainable than others at a time; by changing one existing paradigm to one that is more sustainable at a time; by making one product or one process more sustainable at a time; by incorporating sustainability one phase at a time in the complete life cycle of a facility; by making one facilities program or one facility project more sustainable at a time; and by making one organization, or one industry sector more sustainable at a time.

References

[1] C. Hendrickson, Project Management for Construction: Fundamental Concepts for Owners, Engineers, Architects and Builders, Second Edition, prepared for World Wide Web publication, 2000; the on-line version of the book can be found at: http://www.ce.cmu.edu/pmbook/.

[2] EBN — Environmental Building News, Buildings and the Environment: The Numbers, Environmental Building News, Brattleboro, VT, USA, Vol. 10, No. 5 May 2001.

[3] J. Birkeland, Design for Sustainability, Earthscan Publications Limited, Sterling, VA, 2002.

[4] C.A. Langston, G.K.C. Ding (Eds), Sustainable Practices: Development and Construction in an Environmental Age, 2 Edition, Butterworth, Heinemann, London, UK, 2001.

[5] WCED — World Commission on Environment and Development, *Our Common Future.* Oxford: Oxford University Press, 1987.

[6] A. Pearce, J. Vanegas, Defining Sustainability for Built Environment Systems: An Operational Framework, International Journal of Environmental Technology and Management, Inderscience Enterprises Ltd., Vol. 2, Nos. 1/2/3 (2002) 94–113.

[7] J. Vanegas, Road Map and Principles for Built Environment Sustainability, Environ. Sci. Technol. 37 (2003) 5363–5372.

[8] J. Vanegas, The Project Definition Package: A Cornerstone for Enhanced Capital Project Performance, Proceedings of the World Congress of the International Council for Research and Innovation in Building and Construction — CIB (2001), Wellington, New Zealand, 2001 (paper in conference CD ROM).

[9] P. Hawken, A. Lovins, H. Lovins, Natural Capitalism: Creating the Next Industrial Revolution, Little, Brown, & Co., Boston, MA, 1999. Available free on the web at: http://www.naturalcapitalism.com.

[10] Whole Building Design Guide, Available only on the web at: http://www.wbdg.org/index.php.

[11] Lean Construction Institute, Available through the web at: http://www.leanconstruction.org.

[12] C. Oglesby, H. Parker, G. Howell, Productivity Improvement in Construction, McGraw-Hill Book Company, New York, NY, 1989.

[13] The International Institute for Sustainable Development Compilation of Sustainable Development Principle; available online at: http://iisd.ca/sd/principle.asp.

[14] Sustainability Web Ring; available online at: http://www.sdgateway.net/webring/default.htm.

[15] Sustainable Design and Development Resource of the Construction Engineering Research Laboratory (CERL) of the Engineer Research and Development Center (ERDC) of the US Army Corps of Engineers (USACE); available online at: http://www.cecer.army.mil/sustdesign/.

[16] Sustainable Design Web Site of the Air Force Center for Environmental Excellence (AFCEE); http://www.afcee.brooks.af.mil/eq/programs/progpage.asp?PID = 27.

[17] Minnesota Sustainable Design Guide; available online at: http://www.sustainabledesignguide.umn.edu/.

[18] Green Buildings Center of Excellence for Sustainable Development of the US Department of Energy; available online at: http://www.sustainable.doe.gov/buildings/gbintro.html.

[19] Sustainable Facilities & Infrastructure Program of the Georgia Tech Research Institute (GTRI) at the Georgia Institute of Technology (Georgia Tech); available online at http://maven.gtri.gatech.edu/sfi.

[20] Environmental Building News; available online at: http://www.buildinggreen.com/index.cfm.

[21] Sustainable Buildings Industry Council (SBIC); available online at: http://www.sbicouncil.org/home/index.html.

[22] A. Pearce, J. Vanegas, A Parametric R of the Built Environment Sustainability Literature, International Journal of Environmental Technology and Management, Inderscience Enterprises Ltd., Vol. 2, Nos. 1/2/3 (2002) 54–93.

[23] D. Barnett, W. Browning, A Primer on Sustainable Building, Rocky Mountain Institute, Snowmass, CO, USA, 1995.

[24] T. Woolley, S. Kimmins, P. Harrison, R. Harrison, Green Building Handbook, E. and F.N. Spon, New York, NY, USA, 1997.

[25] J. Lyle, Regenerative Design for Sustainable Development, Wiley Press, New York, NY, USA, 1994.

[26] CIB — International Council for Building Research Studies and Documentation, Sustainable Development and the Future of Construction: A Comparison of Visions from Various Countries, CIB Publication 225, W82 — Future Studies in Construction, Rotterdam, The Netherlands, 1998.

[27] R. Hill, J. Bergman, Bowen, A Framework for the Attainment of Sustainable Construction, in: C.J. Kibert (Ed.), Proceedings of the First International Conference on Sustainable Construction, CIB TG 16, Tampa, FL, 6–9 November, 1994.

[28] EBN, Checklist for Environmentally Sustainable Design and Construction, Environmental Building News, Building Green, Inc., Brattleboro, VT, USA, Vol. 1, No 2 February 1992 (updated 2001).

[29] PTI — Public Technology, Inc., Sustainable Building Technical Manual: Green Building Design, Construction, and Operations, Public Technology Inc., Washington, DC, USA, 1996.

[30] S. Mendler, W. Odell, The HOK Guidebook to Sustainable Design, Wiley and Sons, New York, NY, USA, 2000.

[31] B. Lippiatt, G. Norris, Selecting Environmentally and Economically Balanced Building Materials, in: Proceedings of the 2nd International Green Building Conference and Exposition, NIST SP 888. 1995.

[32] R. Baldwin, A. Yates, N. Howard, S. Rao, Building Research Establishment Environmental Assessment Method (BREEAM) for Offices, Building Research Establishment, Construction Research Communications, London, UK, 1998.

[33] USGBC — US Green Building Council, Leadership in Energy and Environmental Design (LEED) Green Building Rating System, US Green Building Council, Washington, DC, USA, 2004. Current web site at: http://www.usgbc.org.

[34] St. John, The Sourcebook for Sustainable Design, A Guide for Environmentally Responsible Building Materials and Processes, Boston Society of Architects, Boston, MA, USA, 1992.

[35] J. Hermannsson, Green Building Resource Guide, Taunton Press, Newtown, CT, USA, 1997.

[36] D. Holmes, L. Strain, A. Wilson, S. Leibowitz, GreenSpec, The Environmental Building News Product Directory and Guideline Specifications, E-Build, Inc., Brattleboro, VT, USA, 1999.

Sustainability Science and Engineering: Defining principles
Martin A. Abraham (Editor)
© 2006 Elsevier B.V. All rights reserved
DOI 10.1016/S1871-2711(05)01020-2

Chapter 20

Engineering Sustainable Urban Infrastructure

Anu Ramaswami

Urban Sustainable Infrastructure Engineering Project (USIEP), Department of Civil Engineering, University of Colorado at Denver & Health Sciences Center (UCDHSC), Denver CO 80217, USA.

1. The need for urban sustainability

Urban areas contain a high density of human population and typically function as the engines of commerce and industry in the modern world. Based upon UN statistics [1], approximately 50% of the total world population, and about 75% of the population of the developed world, currently reside in cities. The growth of human population in cities has increased at an unprecedented rate over the past two decades, with most rapid growth occurring in developing nations at an annual rate in excess of 10%. In contrast, population growth within cities of developed nations has stabilized to a less aggressive growth rate of about 3%. However, the spatial extent and human population of *urban corridors* in developed nations is increasing rapidly. Urban corridors are large, contiguous areas that include high-density city centers along with surrounding suburbia, e.g., the Washington, DC–New York City–Boston corridor in the northeast, and, the Colorado Springs–Fort Collins corridor surrounding Denver in the Midwest. Such an urban agglomeration is also termed a megalopolis. Over the next two generations, 4 billion more people will join the 3 billion currently residing in cities worldwide, posing a great burden on nature's carrying capacity [2,3]. It is estimated that by the year 2007, for the first time in human history, more than 50% of the human population on planet earth will reside in cities. Thus urban areas, by sheer density and scale of human population, will have a large impact on the future of our planet. With holistic engineering design of sustainable urban infrastructure and informed urban planning efforts, these high-density urban areas can significantly reduce their ecological footprint, while also reaping the economic benefits of resource-efficient design. Thus, engineering sustainable urban infrastructure represents an economic opportunity as well as a global necessity imposed by natural resource and environmental quality constraints. In this chapter, we describe engineering processes and urban planning strategies that can be applied to improve the sustainability of urban infrastructure, where urban infrastructure refers to the engineered structures that provide energy, water, sanitation, housing and transport (of materials, humans and information) within an urban area.

1.1. Strategies for developing and developed nations

With the new paradigm of sustainable development adopted globally in the 1990s after the Brundtland commission Report [4], developing sustainable infrastructure in urban areas has become an important goal worldwide. However, strategies for urban sustainability are likely to be vastly different in developed and developing nations because the per capita natural resource consumption in developed nations is seven times that in the developing nations [2]. In developed nations, this high per capita resource consumption combined with the fact that more than 75% of the human population in these countries typically reside in cities, creates urban areas that consume the largest proportion of natural resources in the world. Increased resource efficiency in urban infrastructure systems design and engineering is the primary strategy for achieving urban sustainability in developed nations. Public participation in conservation efforts is essential. Perceptions correlating increased resource consumption with an improved quality of life must be addressed. For example, a human behavior assessment of consumption patterns in households in Netherlands examined the underlying psychological basis of natural resource consumption and consumerism related to perceptions of quality-of-life. The study concluded that current human behavior patterns in western/developed nations do not support environmental and ecosystem sustainability [5]. Thus, a high level of public education along with human behavior assessment and modification in certain key sectors, would be needed to attain urban sustainability goals in developed nations. Regulations, including voluntary charters to reduce greenhouse gas emissions, are also expected to be important drivers in the development of sustainable technologies in developed nations.

In contrast, sustainability efforts in developing nations are and will continue to be driven largely by steep population increases. UN population data shows that 17 out of 20 megacities with current populations in excess of 10 million are located in the developing world (see Table 1) [6]. In some of these cities, the

Table 1
Global distribution of megacities with populations in excess of 10 million inhabitants in year 2003
[6]

City and location	Population (millions), year 2003
Tokyo, Japan	35.0
Mexico City, Mexico	18.7
New York, USA	18.3
Sao Paulo, Brazil	17.9
Mumbai, India	17.4
Delhi, India	14.1
Kolkata, India	13.8
Buenos Aires, Argentina	13.0
Shanghai, China	12.8
Jakarta, Indonesia	12.3
Los Angeles, USA	12.0
Dhaka, Bangladesh	11.6
Osaka-Kobe, Japan	11.2
Rio de Janeiro, Brazil	11.2
Karachi, Pakistan	11.1
Beijing, China	10.8
Cairo, Egypt	10.8
Moscow, Russia	10.5
Manila, Philippines	10.4
Lagos, Nigeria	10.1

population is predicted to approach 1 billion by 2050 if current aggressive urban growth rates continue unchecked. In urban areas of the developing world, although the per capita resource consumption is relatively low, the magnitude of the population and the lack of adequate urban infrastructure for water supply, sewage treatment and energy supply causes problems typically associated with such cities: frequent blackouts, water shortage and the resurgence of diseases such as tuberculosis, as well as new diseases associated with close urban proximity, such as SARS and the avian flu. Urban poverty has also been observed to increase with increased urban population growth in developing nations [7,8], as unskilled migrant laborers lacking in economic resources are forced to settle in high-density slum areas and shantytowns with little or no infrastructure. Thus, in developing nations, the lack of adequate urban infrastructure for the vast magnitude of people living in cities causes huge stress on human health and the local environment, as well as the global natural resource support system. The challenge in developing nations is to rapidly install new urban infrastructure to keep pace with the population increases, while also being cognisant of natural resource limitations and the need to protect environmental quality. New and intermediate technologies, as well as lessons learned in developed nations, can help developing nations leapfrog to meet this sustainability challenge, i.e., developing nations can directly utilize newest technologies and well-developed public policy measures tested and evaluated in developed nations, avoiding many of the mistakes and pitfalls associated with these technologies. International collaboration, knowledge sharing and technology transfer hence become important enabling factors for sustainable development of urban areas worldwide.

2. Integrated engineering design of urban infrastructure systems

Given the magnitude of the urban sustainability challenge, new ways of conceptualizing cities as urban ecosystems, and thence designing and engineering urban infrastructure are needed. Traditional boundaries in urban infrastructure planning and management have considered transportation, water, energy, residential structures and industrial units to be compartmentalized sectors. Yet, there are huge opportunities for synergy if a more holistic approach considers these subsystems or sectors to be inter-connected, addressing the combined impact of the total urban infrastructure system on air, water, land and biological resources. For example, smart growth initiatives focus on optimal location of work, residences and amenities to try to minimize daily trips. In the classic example of industrial symbiosis in Kalundborg, Denmark [9], previously wasted heat from industries was used to heat surrounding residential homes. As can be seen from the above examples, developing sustainable infrastructure for an urban system requires an integrated systems approach, which can be conceptualized as shown in Fig. 1.

In Fig. 1, the various subsystems within an urban area — i.e. the residential, industrial, transportation, water supply, sewage, power, health care and information subsystems are considered together, and evaluated in terms of their impact on the total environment addressing water, air, land quality as well as materials cycling through the ecosystem. Along with Fig. 1, measurable goals and well-defined criteria are needed to motivate engineering design, consistent with the integrated systems approach shown in the schematic. In this context, it is useful to specify goals for environmental, ecosystem, economic and social sustainability, where

- environmental sustainability typically refers to preservation and enhancement of air, water and land quality, often in a local context;
- ecosystem sustainability preserves the carrying capacity of the surrounding region and/or the global earth system;
- economic sustainability indicates a need for urban infrastructure to be fiscally viable; and

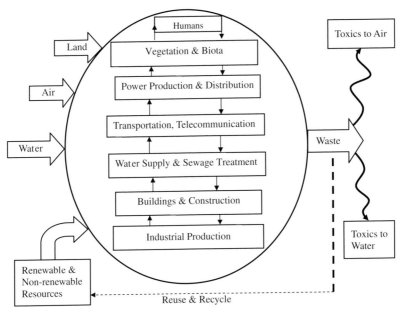

Fig. 1. Schematic showing an integrated urban infrastructure system.

- social sustainability addresses public participation and ownership of sustainability initiatives in a manner that addresses equity considerations.

Fig. 2 shows the inter-relationships between urban infrastructure engineering and the broader goals of environmental, ecosystem and socioeconomic sustainability. In the rest of this chapter, we discuss the current status of three key areas that can have a very large impact on urban sustainable infrastructure engineering:

- urban buildings;
- transportation, energy and air quality; and
- sector integration and symbiosis in urban infrastructure.

We present the current status and background for each area, the engineering strategies and analytical tools that are available, and, regulatory and policy measures that have been or can be applied to promote sustainable urban infrastructure options.

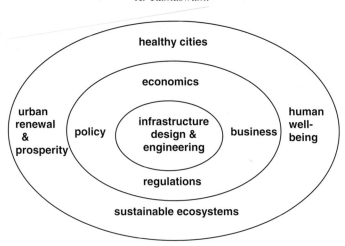

Fig. 2. Linking urban infrastructure design and engineering with environmental, ecosystem and socioeconomic aspects of sustainable development.

3. Urban buildings

3.1. Background

Buildings are the smallest indivisible unit in the urban built environment and play an important role in energy, water and material resource consumption in urban corridors. Buildings, as one of the largest industries in the world, account for a major part of the total energy and materials consumption. More than 30% of America's total energy use, 60% of its electricity and financial resources and 26% of the contents of its landfills are linked to buildings [10]. A total of 54% of the energy used in the US is attributed to the built environment. In addition, 80% of the average American's time is spent indoors. *Hence decreasing the material and energy use associated with buildings, while also using more benign, less-polluting building materials, can simultaneously improve the health of ecosystems as well as the human inhabitants of buildings.* Recent efforts have focused on green building design and construction. However, design innovations have typically targeted individual building subsystems pertaining to energy efficiency, renewable energy resources, water and wastewater and building materials. All the hallmark building projects to-date have featured design innovation in one or at most two subsystems of the building. For example, a high-rise building in Stuttgart features building-integrated wind turbines expected to generate at least 20% of building energy needs [11]. Efforts have been made to showcase Zero Net Energy Buildings (ZNEB) at the scale of a residential home, for example, the 5300-square-foot Ultimate Family Home show-cased in the 2004 International Builders' Show is powered by solar thermal and solar electric energy [12].

The egg-shaped Greater London Authority building optimizes day-lighting while minimizing surface area exposed to direct sunlight; at the same time, solar blinds are placed in response to the building shape in order to collect energy to power the building [13]. In the Farmhouse project in Boulder, Colorado, a 5700-square-foot home and office operates using 70% less energy than a standard wood frame design due to the building material package combined with renewable ground sourcing [14]. Off the shelf green materials are featured in the NRDC building in Santa Monica, CA. Despite the success of the above state-of-the-art buildings, no building project to-date has focused on integrated design, viewing the building as a holistic combination of its water, energy and material subsystems. A holistic systems approach is needed to quantify and minimize the impact of buildings on the *total* environment, encompassing air, water, land and ecosystem resources. A low-rise community center in Commerce City, Colorado, is currently being designed to integrate several building subsystems [15], with the goal of constructing a ZNEB with 75% less embodied energy of its materials, and 50–75% reduction in water usage compared to conventional buildings. Likewise, integrated design of tall high rise buildings is being planned in several large cities across the globe [16].

3.2. Engineering strategies and analysis tools

Building materials: Choice of resource-efficient materials is made based on several criteria such as a high material recycle or reuse content; material synthesis from traditional waste streams; using natural or agri-based materials to *minimize* indoor air pollution due to emissions of harmful volatile chemicals often associated with many synthetic materials, e.g., carpets, flooring products, paints and sealants; minimizing the energy associated with producing, fabricating and constructing with various building materials, termed the embodied energy [17]; addressing greenhouse gas emissions; land disturbance; emission of toxic pollutants during manufacture, e.g., organochlorines from vinyl production; etc. The above impact factors, along with the cost of acquisition of materials, are incorporated into life cycle-based building material selection tools, e.g., BEES [18], enabling the user to choose materials based on cost as well as overall energy and environmental impacts. Since BEES does not yet include a full range of biobased materials, semi-quantitative tools such as the Ecosystems Service Criteria [19] may be applied to assist in material choice. Other green material selection tools include BREEAM/Green Leaf-Green Globes [20].

Building codes: Building materials, whether standard or green (e.g. Green-Spec©) must satisfy building and material codes. Conventional materials will be governed by the American Concrete Institute ACI 318-02 and the American Institute of Steel Construction (AISC). Alternate building materials such as reinforced straw and paper bales, rammed earth and fly ash concrete with recycled aggregates may have to be experimentally tested for structural strength and stability. In addition, distresses such as fire, water proofing, mold and mildew should be in accordance with appropriate standards outlined by the American Standards for Testing Materials (ASTM). It is critical that designers gain the approval of local building code officials as part of the permitting process prior to use of new or uncommon green materials or methods; typically, such approval is granted upon demonstration of accord with ASTM standards.

Energy systems design: Two strategies are used to design energy-efficient buildings:

• The Building Heating and Cooling System is optimized by capitalizing on building orientation (solar access, wind breaks, vegetation), using thermal mass, natural ventilation cooling, solar hot water heating, ground-source heating (EPA, 2002), and other design strategies to reduce energy consumption requirement and maximize energy efficiency, while producing comfortable temperature and humidity, and improved indoor air quality.
• The Renewable Energy System is then designed to generate sufficient energy from renewable, environmentally benign resources (Wind, photovoltaics, anaerobic digester biogas) to support equipment operations and those heating, ventilation and air-conditioning (HVAC) needs not met by thermal systems design.

Design Evaluations for energy efficiency are typically conducted by employing state-of-the-art building simulation tools such as those for energy simulation, air flow modeling, lighting simulation, solar collector design and environmental impact prediction, e.g., DOE-2; TRANSYS [21,22] to compare and optimize the different design solutions. Advanced simulation models, such as Computation Fluid Dynamics (CFD) programs [23] enable prediction of the indoor environmental quality (thermal comfort, humidity, indoor air quality) of buildings, which is directly related to the welfare and productivity of occupants. CFD provides detailed distributions of air velocity, temperature, humidity, and various contaminant concentrations, which are required for building performance evaluation. The National Renewable Energy Laboratories has designed a program to assist in optimal choice of renewable energy resources at a certain site, based on site-specific information pertaining to wind speeds, solar insulation, etc. (HOMER, [24]).

The water and wastewater subsystem within buildings has received by far the least attention in developed countries. The present water and wastewater system is extremely inefficient wherein large quantities of water are treated to high standards of potability in a centralized treatment plant, and then piped across long distances to meet a mix of potable and non-potable water needs in residences and offices. Water exits buildings as a unified wastewater stream, in which a small volume of highly polluting blackwater (carrying human excreta) is co-mingled with large volumes of greywater (non-excrement, non-toilet wastewater, e.g., from laundry, kitchens, hand-washing, etc.) that is once again piped across long distances to wastewater treatment (WWT) plants. The co-mingling of blackwater with greywater doubles the volume of wastewater and more than doubles the energy inputs required for treatment [25]. Source separation of greywater, followed by effective on-site treatment and reuse–recycle of this treated greywater for non-potable needs, e.g., toilets, can significantly conserve water, energy and costs, as has been shown in LCA supported European and Australian case studies [26–28]. On-site greywater and blackwater treatment occurs often in rural areas or in buildings not connected to a centralized sewage system. In fact, on-site WWT systems provide treatment in approximately one of every four housing units in America [29]. Greywater treatment options include reverse osmosis, sand filtration, and the use of wetlands and bioactive filters [30]. Composting toilets [31] and other biobased strategies for blackwater treatment, such as the Living Machine[TM] System, wetlands treatment and anaerobic digestion with methane generation, have also been successfully demonstrated at the site scale both in China and in India as well as in many parts of the United States [32–36]. However, current water and wastewater treatment strategies in urban areas in the US continue along traditional lines.

Opportunities for synergy: Design integration across building subsystems can capitalize on synergies between the systems, e.g., heat recovery from hot water drains can contribute to the energy system [37], biogas produced from WWT can contribute to the building renewable energy subsystem; appropriate siting of a building-integrated water treatment greenhouse can promote passive solar heating–cooling and thermal design benefits. Such design integration is shown schematically in the figure below; the holistic life cycle-informed design approach shown in Fig. 3 is currently being tested in a community building in Commerce City, CO [15]. Various design options will be evaluated using a screening LCA, creating design decision matrices for whole building design.

3.3. Public policy and regulations

The US Green Building Council (USGBC, [38]) has voluntarily developed standards for green building construction, articulating design criteria for many of the building subsystems through the Leadership in Energy and Environmental Design (LEED) certification program. Although the LEED rating does not address holistic building design integration *per se*, innovation credits in the

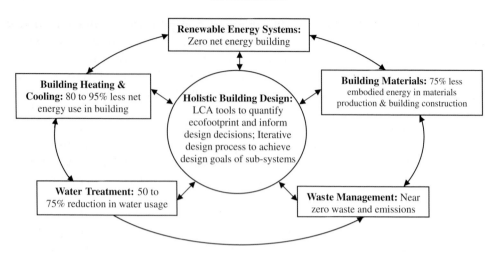

Fig. 3. Schematic showing design integration and illustrative design goals for a building being designed in Commerce City, CO.

program allow for great creativity in building design. USGBCs efforts demonstrate success in voluntarily moving toward sustainable urban infrastructure using programs and criteria developed by industry and professionals, instead of government-initiated regulations. The success of the USGBC efforts may have arisen because of the increasing realization that a resource-efficient building design is economically beneficial in the long term and also presents a marketing advantage. In fact, early life cycle analysis of a residential home conducted in the 1990s indicated that the payback period of resource-efficient upgrades in a home was around 11 years [39], too long for green design to be economically viable as typical homeowners reside in their homes only for an average of 4–5 years. However, steep increases in the costs of residential energy in the US have now decreased the payback period for energy efficient upgrades in homes to 5–6 years, providing a clear economic incentive (see for example the experience of the Big Horn Ace Hardware store in Silverthorne, CO, [40]). Many cities as well have recognized the need for developing green building codes in support of green building construction, e.g., Boulder, CO [41].

To further promote water and wastewater system innovations within a building, state and local regulators must address existing regulations that prohibit or limit rainwater harvesting, as well as on-site water and wastewater treatment. Standardized engineering protocols are also needed to demonstrate the compatibility of the green design with the specific intent of the regulation, such that *implementation of green building design becomes routine and state-of-practice, rather than the state-of-the-art it is today.* Most importantly, pre-construction participatory planning and post-construction training of building inhabitants is essential to ensure their full participation in using green buildings in a sustainable manner. Finally, post-construction building monitoring, tracking energy, water and materials usage, ensures success of the green building design.

Energy-efficient building retrofits: The discussion thus far has focused on new buildings. Many opportunities for energy savings and avoidance of greenhouse gas emissions can be exploited by retrofitting existing buildings for energy efficiency. Typical technological options for building retrofits include improving the energy efficiency of lighting systems and appliances, e.g., replacing incandescent bulbs with compact florescent lamps and use of higher efficiency appliances with E-star rating, improving the mechanical systems that provide heating and cooling, changing the building envelope adding insulation, replacing windows and/or opting for reflecting roofs, and, applying integrated whole systems design for large-scale retrofits [42]. Financing vehicles for building retrofit programs are very important. Several first level retrofits, e.g., replacement of lighting fixtures, provide a 25% cost savings with the up-front capital expenses paying for themselves within a 1–4 year window. More capital intensive retrofits need special financing vehicles facilitated through government policy [43]. A recent review of energy efficiency measures in California [44] shows that the per capita energy use for air conditioning has decreased from approximately 2500 kWh/year in 1972 to 1000 kWh/year in 2003, attributable to higher efficiency standards for appliances. The total energy use per capita in California has increased from approximately 4000 kWh/year in 1962, leveling off at 7000 kWh/person/year in 2003. Looking forward, the most cost-effective strategies to increase energy efficiency in Los Angeles include the use of colored cool roofs, reflective pavements, and, cooling building envelopes such as shade trees. These three strategies alone are expected to cool Los Angeles by 2–3°C and thereby decrease air conditioning costs by $200 million, improve air quality and save an additional $250 million per year in avoided sicknesses [44]. Similar city-scale strategies with appropriate financing vehicles can have a huge impact on urban sustainability worldwide, using existing and available technologies.

Challenges to sustainable building systems design: In summary, there are four main challenges to sustainable building systems design: (1) Design integration across material, energy and water subsystems, capitalizing on synergies; (2) Regulatory acceptance through development of standardized engineering protocols for sustainable building technologies; (3) Participatory planning that involves building users from the pre-design to the post-construction phase, to ensure that green buildings are designed and operated in a sustainable, user-friendly manner, reaping economic benefit through reduced waste generation, and, water and energy consumption, during building opera; (4) Appropriate financing and implementation of building energy efficiency retrofit programs at the city scale.

4. Transportation, energy and air quality

4.1. Background

Approximately 30% (27.2%) of the total energy consumption in the United States is attributed to the transportation sector [45]. Urban air pollution arising from transportation and fuel use includes priority chemicals such as CO (carbon monoxide), nitrogen oxides (NOx, including NO and NO_2), particulate matter (PM), ozone (O_3), SO_2 (sulfur dioxide, most pervasive in developing countries that use high sulfur fuels), lead (Pb), most pervasive in developing nations that use unleaded fuel), and Volatile Organic Compounds (VOCs). PM is reported in different regions of the world as TSP (Total Suspended Particles), PM_{10} and $PM_{2.5}$, i.e., particulate matter with aerodynamic diameters less than 10 and 2.5 μm respectively. Long-lived greenhouse gases such as CO_2 and methane (CH_4) are also associated with transportation and fuels.

Air pollutants have a major impact on human health. The World Health Organization (WHO, [46]) reports that urban air pollution presently contributes to 800,000 deaths and 4 million lost life-years, annually, worldwide. High levels of PM, lead and SO_2 (in developing nations), as well as ozone and air toxics such as polyaromatic hydrocarbons (PAHs), significantly impact human health in many urban areas worldwide. A recent report describes air quality in nine megacities across the world, along with a discussion of contrasting transportation and energy policies applied in these cities [47]. The experience of Los Angeles is described as a success story, demonstrating significant improvements in air quality over the past 20 years despite high increases in population and vehicle ownership. Similar success stories are needed in many of the cities in the developing world where income improvements are contributing both to an explosive increase in urban population as well as vehicle ownership.

4.2. Engineering strategies and analytical tools

Urban infrastructure engineering and planning that integrates transportation and land use planning with environmental and ecosystem impact assessment methodologies can be one of the most effective tools to achieve diminished energy use as well as improved air quality and overall urban sustainability in growing cities. Transportation options for meeting the mobility needs of urban dwellers are largely limited to three options as shown in Fig. 4: non-motorized transport, including walking trips, bicycles and rickshaws; public transportation comprising light rail and buses; and private vehicle ownership, including automobiles as well as two-wheeler scooters and mopeds. Strategies to preserve and enhance non-motorized transport as well as public mass transport, need to be accompanied simultaneously by improvement in vehicle performance and in

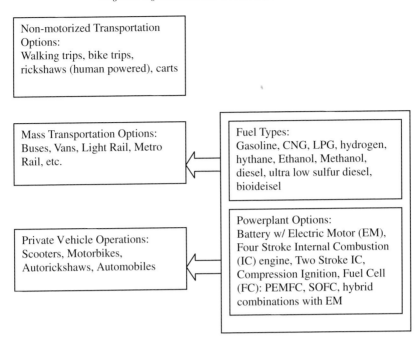

Fig. 4. Schematic showing transportation, vehicle and fuel options.

vehicle fuels. Several new technologies have created many choices for the vehicles of the future — including the traditional gasoline internal combustion engine (ICE); the relatively new gasoline IC-hybrid engine with an onboard electric motor and a battery for storage of energy generated from regenerative braking; a variety of fuel cell (FC) technologies, including the Proton Electrolyte (PEM) and Solid Oxide fuel cells (SOFC), and their potential combinations with hybrid technology; as well as the diesel compression engine that has now seen vast reductions in pollutant emissions due to the new clean fuel technologies [48]. Fuels of the future may include gasoline and blended gasoline with additives such as ethanol, E85 which is 85% ethanol, CNG or compressed natural gas (methane), Liquified Petroleum Gas (LPG), ultra-low sulfur diesel (ULSD), biodiesel, hydrogen as well as hydrogen–methane blends termed hythane.

Environmental-economic life cycle analyses (LCA) are being used as a first step to evaluate the overall impact on the earth system of the new automotive technologies and their accompanying fuels. A complete LCA must consider the environmental costs of (a) manufacturing and processing the fuel, (b) distributing the fuel (often difficult to model or predict), (c) manufacturing the automobile, (e) operating the automobile including on-road emission modeling and testing, and, (f) end-of-life disposal. Few LCAs to date have considered all the five steps above in the life cycle of a vehicle. With the push toward Low and

Zero Emission Vehicles (LEV and ZEV), fuel cells that catalytically combine hydrogen with oxygen in air to form water, with the release of energy, are being touted as the power plants of future vehicles. However, the overall environmental benefit of employing hydrogen in fuel cells may not be significant compared to advanced ICE-hybrid and CIDI-hybrid platforms with improved alternative fuels, when the environmental impacts of our current hydrogen production methodologies are also included in the analysis [49,50].

Researchers at the US National Renewable Energy Laboratories have focused on the first step in the LCA of hydrogen powered vehicles, i.e., an analysis of the overall environmental benefits and costs of manufacturing hydrogen from CNG and from electrolysis of water [51–53]. Pehnt [54] has analyzed the life cycle impacts of manufacturing a PEMFC stack. Lave et al. [55] have examined the hybrid ICE automobile in comparison with a conventional gasoline ICE, while Weiss *et al.* [49] have examined the life cycle impacts of several projected vehicle–fuel combinations of the future addressing fuel production and delivery to the vehicle (wells-to-pump analysis), and on-road fuel combustion during vehicle use (pump-to-wheels analysis), together known at wells-to-wheels or w2w analysis. Models such as the ADVISOR and GREET [56,57] provide useful information on the expected emissions and the w2w impacts from various combinations of fuels and propulsion systems. McAuley couples and compares the w2w impact of a conventional gasoline-powered automobile with other aspects such as vehicle maintenance and end-of-life [58]. Researchers at the CU Denver's USIEP are currently engaged in a project to perform a screening level analysis of various fuel-propulsion systems in India, addressing all five aspects of LCA listed above. Based on the screening level LCA, infrastructure options for alternative fuels and/or alternative vehicles will be compared specific to the region of interest, and political and socioeconomic aspects, also expected to be region specific, will be evaluated [59,60]. Thus, an iterative analysis methodology progressively combining environmental, economic and societal impacts, is being developed.

The above discussion has focused on the private automobile — similar life cycle analysis can be conducted to evaluate the efficacy of mass transit and non-motorized transportation options in the mix of vehicle mode choices available to urban dwellers. Fig. 5 shows that in many growing cities in developing economies, a vast proportion of trips are undertaken through non-motorized modes [61]; such options are rapidly being pushed aside by the increased penetration of individual vehicle ownership. While private vehicle ownership in developing economies has increased exponentially in gross numbers, the proportion of trips conducted using non-motorized modes has not yet been significantly impacted, indicating a window of opportunity that exists in many developing cities during which appropriate transportation infrastructure can

a)

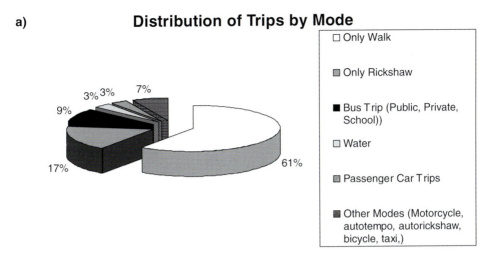

b)

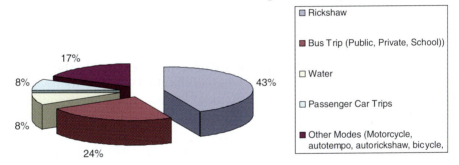

Source: *Dhaka Integrated Transport Study, Final Report Volume 1, 1994.*

Fig. 5. Distribution of transportation modes in Dhaka, Bangladesh [61].

have a positive impact on future sustainable development. Transportation models can be applied to visualize how effectively the mobility needs of the urban population are met through various combinations of private vehicle trips, mass transit and non-motorized transport. In this context, homogeneous traffic models appropriate for use in developed nations must be augmented by traffic simulation algorithms that capture the behavior of more heterogeneous traffic seen in developing nations (see Fig. 5). Khan *et al.* [62] are engaged in characterizing the urban driving cycle and modeling heterogeneous traffic flow in developing nations. Information integrated from fuel-vehicle LCA and from transportation modeling can then be used to inform the public policy deliberations that would control the eventual deployment of a sustainable transportation-energy infrastructure system.

4.3. Public policy and regulations

Public policy options in the transportation and air quality arena include technology-based regulations, economic- or market-based instruments and incentives, and public education programs to ensure participation [47]. Technology-based regulations typically are implemented using a command and control framework, wherein Best Available Control Technologies (BACT) and Best Available Retrofit Technologies (BART) are applied to achieve some specified emission limits or to achieve specified ambient air quality criteria. Emissions monitoring and modeling of vehicles as well as ambient air quality modeling and monitoring are important components of such a program. These models can also be utilized to implement market-based public policy measures that provide economic incentives, such as emission trading, tax incentives, congestion pricing, etc. to achieve the desired outcome. At a micro-level, individual behavior and public education is essential to create effective economic incentive programs.

Many cities are presently exploring a combination of all three public policy options [47]. For example, over the past 30 years, Los Angeles has set up technology-based regulation for fuels, automobiles, as well as household-use engines, e.g. lawnmowers. The technology-based regulations resulted in programs such as the first automotive emission testing in 1961, the first use of catalytic converters in 1975, cars equipped with on-board diagnostics in 1988, standards for cleaner burning fuels, established in 1990, LEV regulations for low-emitting vehicles adopted in 1999, including a requirement for 2% of the vehicles to be ZEVs. Off-road engine regulation for small engines was established in 1998, coupled with economic incentive programs to replace older, more-polluting vehicles. The regional clean air incentives market program was initiated to provide incentives to businesses and facilities to meet emission reduction targets from stationary sources. These measures, together, have improved air quality in Los Angeles despite large increases in urban population and private vehicle ownership over the past 40 years. In London, congestion pricing is being applied in a successful bid to promote mass transport, i.e., a large fee or fine was recently introduced to inhibit external private vehicles from entering the most congested areas of London. Alternate fuel use and vehicle emission testing and phasing out are the two most important strategies being adopted in developing nations, both measures being implemented through a combination of government regulations and economic incentive programs. For example, in Hong Kong, along with emission testing and government regulations to phase out old and diesel-operated buses, economic incentives were set up to make LPG and ultra-low sulfur diesel (ULSD) viable alternative fuels.

While the above discussion has focused largely on technological advancements in fuel-vehicle systems, urban form, i.e., the spatial organization of people in cities, has been shown to strongly influence the demand for motorized transport in cities. Smart growth initiatives and new urbanism measures that reduce the demand for motorized transport can have the greatest impact on sustainable transport. This aspect is discussed further in Section 5. In many cities, urban planning policies are being developed to promote sustainable transport, employing metrics that address system performance, efficiency, environmental impacts and equity in access to transport [63]. Establishing quantitative metrics for transportation system sustainability is a first step in developing policy measures to promote overall system sustainability [64].

5. Sector integration and symbiosis in urban infrastructure

The previous two sections have addressed the building sector and the transportation sector in a holistic manner, evaluating impacts on the local environment as well as the regional and global ecosystem. In this section, we examine sector integration and symbiosis at a local level, and then expand into the broader topic of incorporating biota as an inherent part of sustainable urban infrastructure systems.

5.1. Background

Symbiosis across industries gained international attention after the success of the petrochemical complex in Kalundborg, Denmark, where waste exchanges of more than 2.9 Mt of material per year occurred between four principal industries and surrounding residences, resulting in net water consumption reductions of 25% [65]. The success of Kalundborg gave rise to the concept of eco-industrial parks wherein symbiosis is defined as *"place-based exchanges among at least two or more different entities ... the two unrelated entities exchange materials, energy or information in a mutually beneficial manners"* [66]. A case study of four industrial parks, including Kalundborg and new parks in Styria, Austria, Ruhr, Germany and Sarnia, Canada, attempted to understand the relative importance of key factors that promote industrial symbiosis, namely: technology innovations, geography, planning, market economics and flexible regulations in promoting symbiosis [67]. In industrial symbiosis promoted near the Research Triangle Park area in Virginia, USA, Kincaid and Overcash [68] found that cross-industry education, communication and flexible regulations were also important enablers for symbiosis. Despite the emergence of eco-industrial parks, industrial symbiosis has not gained momentum on a large scale in cities, either in the developed or the developing world. As developing countries industrialize, more work is needed to promote and plan for industrial symbiosis.

As in the case of industrial symbiosis, the heart of sustainable development lies in the ability to circulate material, water and energy between the natural and the urban ecosystem. Urban infrastructure systems that utilize cyclic processes for materials, water and energy use, are expected to more readily achieve the goal of environmental and ecosystem sustainability. However, past references to urban infrastructure have typically encompassed non-living structures exclusively, e.g., pavements, roads, concrete structures, etc. Such references ignore the powerful role that living beings in the natural environment can play in urban infrastructure design. Incorporating natural systems (ecological engineering) into the industrial ecosystem has been considered an important step in sustainability and green engineering [69,70]. Much work is needed in this area of urban sustainable infrastructure development, as symbiotic relationships can be used both to circulate materials within the urban boundaries, as well as between urban areas and the surrounding rural land, forestalling future population migration into cities.

5.2. Engineering strategies and analysis tools

Several strategies can be employed to incorporate biota into urban infrastructure. Section 3.2 described the use of microbiota as well as wetland ecosystems in water and wastewater treatment. Anaerobic digestion of human and animal waste has been used in many rural areas of India and China to produce energy from biogas (methane generation). Although traditional WWT plants in cities also employ anaerobic digestion, the efficiency of energy generation through WWT has not been a design goal. More engineering research is needed in this area. Likewise, while WWT plants in Burlington Vermont and Oberlin College have used whole ecosystems to successfully treat wastewater in Living Machines[TM], design guidelines and the limits of applying ecosystem-based technologies are poorly understood. Furthermore, the economic and environmental benefits of employing biological treatment versus traditional physico-chemical processes has not been quantified for the water-energy sector.

In addition to localized treatment of wastewater, plant systems have also been found to effectively take up and metabolize hazardous and polluting chemicals in air, soil and water. Since the 1990s, phytoremediation, the use of plants and trees for environmental remediation, has emerged as a viable, low-cost green technology for removal of a variety of pollutants from water and soil, including nutrients, pesticides, metals, radionuclides as well as industrial waste chemicals [71–73]. Plants employ many strategies to "process" pollutants including extraction and accumulation of the chemical, enzymatic degradation and volatilization via the transpiration stream. The physicochemical interactions between pollutants and plants has been understood fairly well [71,74], as well as the specificity of specialized plants for certain specific pollutants, e.g., arsenic hyperaccumulation by the Chinese Brake Fern [75], enabling design and implementation of phytoremediation in field applications pertaining to water and soil pollution. Most recently, trees have been found to be capable of capturing air-borne pollutants, both indoors and outdoors. Thus, vegetation buffer strips may be placed at critical locations within an urban system to capture and metabolize diffused non-point pollution in air and in water, for example focusing on benzene, NO_x and SO_x in air, and, runoff of pesticides and deicing chemicals in water. In addition to the environmental benefits listed above, vegetated open areas within urban city centers can mitigate the urban heat island effect. Studies at the University of Chicago have also found correlation between green areas and improved safety and well-being in urban neighborhoods that were located close to playgrounds and trees.

In the area of municipal waste, creative ways of coupling industrial and residential waste streams to yield usable products are needed. For example, shredded automobile tires and other waste materials such as power plant fly ash are incorporated into concrete, often yielding a superior construction material. Building interiors may likewise utilize waste agricultural products, promoting material recycle and reuse. Biodegradability is an important consideration in linking the anthropogenic and the natural ecosystems. For example, plastic bags and tableware (cups, plates, etc.) are now being composed of cornstarch and other agri-based products to promote biodegradation in landfills or in composting facilities. All the above examples represent individual success stories. More fundamental scientific research, engineering design and testing and life cycle environmental-economic analysis is needed to promote material, water and energy circulation in a consistent manner within an integrated urban-natural ecosystem, to achieve the highest goal in urban sustainable infrastructure engineering.

Sustainability and urban form: In recent years, there has been much debate on urban form and sustainable development. It is hypothesized that the degree of compactness versus sprawl is a very important factor controlling the per capita energy and material use in a city [76]. Newman and Kenworthy [77] describe the evolution of cities from "walking cities" developed in the absence of motorized transport, followed by "transit cities" impacted by rail transport, then the "automobile city" currently dominant in developed nations modeled upon the dominance of the automobile, and finally potential future "telecommuting cities" where information technologies could shape the development of urban form. Further, it is postulated that the current transportation infrastructure closely controls the spread and the need for spread of the water distribution infrastructure. Urban infrastructure engineering is closely linked to the evolution of urban form. Urban planners and infrastructure engineers must work together using holistic tools to plan for sustainable cities of the future. The nexus between urban planners and sustainability engineers is just beginning (e.g. [78]), and is expected to be a major factor in urban sustainable development.

6. Conclusions

Urban systems are complex entities, requiring a systems approach to fully understand the spatial and temporal scale of materials, water and energy cycling through densely populated cities that have evolved and continue to evolve into different urban forms. Urban systems studies must integrate the energy, water, sewage, construction, industrial production, information and health care sectors within a city, as well as address impacts on the *whole* environment, including water, air, land and bioresources. Sustainable infrastructure engineering requires addressing technical feasibility of new and innovative technologies in the context of environmental, ecosystem and socioeconomic metrics for sustainability. Engineering evaluations, along with economic analysis and suitable public policy implementation have been shown to lead to success in developing sustainable urban infrastructure in a few cities, worldwide. The challenge lies in understanding these linkages, learning from past experiences and developing a consistent framework for future sustainable urban infrastructure engineering, applying all 10 principles for sustainability described in Section 2.

Acknowledgements

The author acknowledges grants from the US Department of Education GAANN program, the National Science Foundation MUSES program, and the USEPA P3 program for supporting urban sustainability research at USIEP. Dr. Sarosh Khan and USIEP PhD students Mike Whitaker, Mark Pitterle and Tim Hillman provided resources for some sections of this chapter.

References

[1] United Nations (U.N.), Population Division, World Urbanization Prospects, U.N., New York, 1996.
[2] M. Wackernagel, W. Rees, Our Ecological Footprint, New Society Press, Gabriola Island, BC, 1996.
[3] NRC (National Research Council), Our Common Journey: A Tradition Towards Sustainability, Washington D.C., National Academy Press, 1999.
[4] WCED (World Commission on Environment and Development), Our Common Future, Oxford University Press, Oxford, UK, 1987.
[5] K. Noorman, A.S. Uiterkamp (Eds), Green Households? Domestic Consumers, Environment and Sustainability, Earthscan, London, 1998.

[6] United Nations (U.N.) Population Division, World Urbanization Prospect: The 2003 Revision, U.N., New York, 2003.

[7] J.E. Hardoy, S. Caincross, D. Satterthwaite (Eds), The Poor Die Young: Housing and Health in Third World Cities, Earthscan, London, 1990.

[8] J.E. Hardoy, D. Mitlin, D. Satterthwaite, Environmental Problems in Third World Cities, Earthscan, London, 1992.

[9] J. Ehrenfeld, N. Gertler, Industrial Ecology in Practice: The Evolution of Interdependence at Kalundborg, J. Ind. Ecol. 1 (1) (1997) 67–79.

[10] AFCEE (Air Force Center for Environmental Excellence), Sustainable Facilities Guide, John Barrie Associates Architecture, Inc. & US Air Force Combat Command, 2000.

[11] W. Knight, Wind-powered building design revealed, New Scientist 13 (15) (2001) 14.

[12] DOE (US Department of Energy), Energy Efficiency and Renewable Energy website article: Zero Energy Home Displayed at International Builder's Show, 2004.http://www.eere.energy.gov/solar/cfml/news_detail.cfm/news_id = 6607 Accessed February, 2004.

[13] Architectural Record, February 2003.

[14] J. Herdt, Environmental Design: Industrialized Agricultural Architecture, IAA & The Farmhouse Prototype, Proceedings of the National Conference of the American Collegiate Schools of Architecture (ACSA), 2002.

[15] USEPA P3, People, Prosperity and the Planet Design Competition, 2004. http://es.epa.gov/ncer/p3/projects/index.html. Accessed March, 2005.

[16] W. Pank, H. Girardet, G. Cox, Tall Buildings and Sustainability, A report submitted to the Corporation of London, UK, Faber Maunsell, March 2002.

[17] USEPA (US Environmental Protection Agency), Life Cycle Analysis Assessments Inventory Guidelines & Principles, EPA/600/R-92/245, 1993.

[18] NIST (National Institute of Standards and Technology), 2003. Building and Fire Research Laboratory: BEES 3.0 website. http://www.bfrl.nist.gov/oae/software/bees.html. Last updated October, 2003.

[19] V. Olgay, J. Herdt, The Application of Ecosystems Services Criteria for Green Building Assessment, Solar Energy 77 (2004) 389–398.

[20] BREEAM, Building Research Establishment's Environmental Assessment Method, 1998. http://www.breeam.com/ Accessed March, 2005.

[21] DOE2, 2000. http://www.doe2.com/ Accessed March, 2005.

[22] TRNSYS, Transient Energy System Simulator, 2001. http://www.trnsys.com/ Accessed March, 2005.

[23] Z. Zhai, S.D. Hamilton, J.M. Huang, C. Allocca, N. Kobayashi, Q. Chen, Integration of Indoor and Outdoor Airflow Study for Natural Ventilation Design Using CFD. Proceedings of the 21st AIVC Annual Conference on Innovations in Ventilation Technology, The Hague, Netherlands, 2000.

[24] HOMER, The Hybrid Optimization Model for Distributed Power, 2003. http://www.nrel.gov/homer/ Accessed March, 2005.

[25] USEPA (US Environmental Protection Agency) Onsite Wastewater Treatment Systems Manual, Office of Water, EPA/625/R-00/008, 2002.

[26] A. Bjorklund, C. Bjuggren, M. Dalemo, U. Sonesson, Planning Biodegradable Waste Management in Stockholm, J. Ind. Ecol. 3 (4) (2000) 43–58.

[27] M. Hall, H. Lovell, S. White, M. Lenzen, C. Mitchell, Evaluating the Options: Pumpout, Common Effluent Drainage, and Centralised Treatment for High-risk Unsewered Townships in Hornsby Shire: Berowra Creek Case Study — Management of Risk for On-site and Centralised Wastewater Treatment, New South Wales Department of Local Government, Australia, 2001.

[28] A. Balkema, A. Spagni, H. Spanjers, COST 624 Optimal Management of Wastewater Systems, 5th Working Group Meeting, Bologna, Italy, 2001.

[29] National Onsite Demonstration Program. http://www.nesc.wvu.edu/nodp/nodp_index.htm

[30] R.W. Crites, G. Tchobanoglous, Small and Decentralized Wastewater Management Systems, The McGraw-Hill Companies, Inc., Boston, 1998.

[31] D.D. Porto, C. Steinfeld, The Composting Toilet System Book: A Practical Guide to Choosing, Planning, and Maintaining Composting Toilet Systems, and Alternative to Sewer and Septic Systems, The Center for Ecological Pollution Prevention, Concord, MA, 2000.

[32] M. Hessamie, S. Christensen, R. Gani, Anaerobic Digestion of Household Organic Waste to Produce Biogas, Renewable Energy 9 (1–4) (1996) 954–957.

[33] LM (Living Machines, Inc.), Living Machines$^{®}$ website. http://www.livingmachines.com/ Last updated 2004.

[34] OAI (Ocean Arks International), 2004. Natural Water Treatment website. http://www.oceanarks.org/natural/

[35] OEMC (Colorado Governor's Office of Energy Management and Conservation), Colorado Constructed Treatment Wetlands Inventory, 2001.

[36] M. Shyam, Solid-state Anaerobic Digestion of Cattle Dung and Agro-residues in Small-Capacity Field Digesters, Biores. Technol. 48 (1994) 203–207.

[37] WE (Waterfilm Energy, Inc.), GFX Technology website. http://www.gfxtechnology.com/ Last updated 2002.

[38] USGBC (US Green Building Council), 2004. http://www.usgbc.org/ Last updated 2003.

[39] G.A. Keoleian, S. Blanchard, P. Reppe, Life Cycle Energy Costs and Strategies for Improving a Single Family Home, 2000.

[40] Bighorn Ace Hardware, 2004. www.bighornace.com/ Accessed March, 2005.

[41] Boulder Building Code, 2003. http://www.ci.boulder.co.us/cao/brc/index.html. Accessed March, 2005.

[42] US Department of Energy (USDOE): Rebuild America Solutions Center. Potential Energy Efficiency Measures Overview. http://www.rebuild.org/lawson/energyefficiency. Last Accessed February, 2005.

[43] ECEEE (European Council for an Energy efficient Economy), Energy Efficiency as a Commodity: The Emergence of Secondary Markets for Savings in Commercial Buildings, Summer Study Report, 1997.

[44] A. Rosenfeld, Extreme Efficiency: Lessons from California, Presented at the Annual Conference of the American Association for the Advancement of Science, Washington, DC, USA, 2005.

[45] EIA, Energy Information Agency, Annual Energy Review, p. 3, 2002. Available Online: http://www.eia.doe.gov/emeu/aer/pdf/pages/sec1.pdf

[46] WHO (World Health Organization), The World Health Report 2003: Reducing Risks, Promoting Healthy Life, Geneva, Switzerland, 2002.

[47] L.T. Molina, M.J. Molina, R.S. Slott, C.E. Kolb, P.K. Gbor, F. Meng, R.B. Singh, O. Galvez, J.J. Sloan, W.P. Anderson, X.Y. Tang, M. Hu, S. Xie, M. Shao, T. Zhu, Y.H. Zhang, B.R. Gurjar, P.E. Artaxo, P. Oyola, E. Gramsch, D. Hidalgo, A.W. Gertler, Air Quality in Selected Megacities: Online Supplement to the 2004 Critical Review on Megacities and Atmospheric Pollution, AWMA, 2004.

[48] ULSD, 2003. Ultra-Low Sulfur Diesel. http://www.dieselforum.org/factsheet/ulsd.html. Accessed March, 2005.

[49] M.A. Weiss, J.B. Heywood, E.M. Drake, A. Schafer, F.F. AuYeung, On the Road in 2020: A life-cycle analysis of new automobile technologies, Energy Laboratory Report No. MIT EL 00-003, October, 2000.

[50] S. Wright, M. Whitaker, Proton Exchange Membrane Fuels Cells or Internal Combustion Engines for Transportation? Paper for the 2004 World Renewable Energy Conference, April 2004.

[51] M. Mann, P. Spath, Life Cycle Assessment of Renewable Hydrogen Production via Wind/Electrolysis: Milestone Completion Report, NREL Report No. MP-560-35404, 13pp., 2004).

[52] P.L. Spath, M.K. Mann, Life Cycle Assessment of Hydrogen Production via Natural Gas Steam Reforming, NREL Report No. TP-570-27637, 32pp., 2001.

[53] P.L. Spath, M.K. Mann, Life Cycle Assessment — An Environmental Comparison of Hydrogen Production from Steam Methane Reforming and Wind Electrolysis. Hydrogen: The Common Thread, Proceedings of the 12th Annual US Hydrogen Meeting, 6–8 March 2001, National Hydrogen Association, NREL Report No. 31870, Washington, DC, pp. 311–319, 2001.

[54] M. Pehnt, Life Cycle Assessment of Fuel Cell Stacks, Int. J. Hydrogen Energy 26 (2001) 91–101.

[55] L. Lave, H. MacLean, C. Hendrickson, R. Lankey, Life-Cycle Analysis of Alternative Automobile Fuel/Propulsion Technologies, Environ. Sci. Technol. 34 (17) (2000) 3598–3605.

[56] National Renewable Energy Laboratory, ADVISOR, Available Online: http://www.ctts.nrel.gov/analysis/advisor.html

[57] Argonne National Laboratory, GREET, Available Online: http://www.transportation.anl.gov/greet/

[58] J.W. McAuley, Global Sustainability and Key Needs in Automotive Design, Environ. Sci. Technol. 37 (2004) 5414–5416.

[59] M.B. Whitaker, A. Ramaswami, S. Khan, The Future of Diesel in Bus Transport in India, to be presented at StoCon, Annual Conference of the International Society for Industrial Ecology, Stockholm, June, 2005.

[60] J.T. Cohen, J.K. Hammitt, J.I. Levy, Fuels for Urban Transit Buses: A Cost-Effectiveness Analysis Environ. Sci. Technol. 37 (2003) 1477–1484.

[61] Dhaka Integrated Transport Study, Vol. 1, 1994.

[62] S.I. Khan, P. Maini, Heterogeneous traffic flow models. Transportation Research Record, Journal of the Transportation Research Board, No.1678 — Highway Capacity, Quality of Service and Traffic Flow and Characteristics, Transportation Research Board — National Research Council, National Academy Press, Washington, DC, 1999, p. 234–240.

[63] M. Federici, S. Ulgiati, D. Verdesca, R. Basosi, Efficiency and Sustainability Indicators for Passenger and Commodities Transportation Systems. The case of Siena, Italy, Ecol. Indicators 3 (2003) 155–169.

[64] R.K. Bose, Technology and Policy Shifts towards Sustainable Urban Transport System, Pacific Asian J. Energy, 13 (2) (2003) 127–142.

[65] J. Ehrenfeld, N. Gertler, Industrial Ecology in Practice: The Evolution of Interdependence at Kalundborg, J. Ind. Ecol. 1 (1) (1997) 67–79.

[66] M.R. Chertow, Industrial Symbiosis: Literature and Taxonomy, Annu. Rev. Energy Environ. 25 (2000) 313–337.

[67] P. Desrochers, Cities and Industrial Symbiosis, J. Ind. Ecol. 5 (4) 29–44.

[68] J. Kincaid, M. Overcash, Industrial Ecosystem Development at the Metropolitan Level, J. Ind. Ecol. 5 (1) (2001) 117–126.

[69] W. McDonough, M. Braungart, Cradle to Cradle: Remaking the Way We Make Things, North Point Press, New York, 2002.

[70] D.R. Tilley, Industrial Ecology and Ecological Engineering: Opportunities for Symbiosis, J. Ind. Ecol. 7 (2) (2003) 13.

[71] USEPA (US Environmental Protection Agency), Introduction to Phytoremediation, EPA/600/R-99/107, Office of Research and Development, Cincinnati, OH, 2000.

[72] A. Ramaswami, P. Carr, M. Burkhardt, Plant-Uptake of Uranium, Int. J. Phytoremediation 3 (2) (2001) 189–201.

[73] A. Ramaswami, J.B. Milford, M.J. Small, Integrated Environmental Modeling: Modeling Pollutant Transport Fate and Exposure. Graduate-level textbook, Wiley, New York, NY.

[74] A. Ramaswami, E. Rubin, Measuring Phytoremediation Parameters of VOCs: Focus on MTBE, ASCE Practice Periodical of Hazardous and Radioactive Waste: Special Issue on Phytoremediation, July 2001.

[75] L.Q. Ma, K.M. Kumar, C. Tu, W. Zhang, Y. Cai, E.D. Kennelley, A fern that hyperaccumulates arsenic, Nature 409 (2001) 579.

[76] D. Bannister, Reducing the Need to Travel, Environ. Planning B: Planning Design, 24 (1997) 437–449.

[77] P. Newman, J. Kenworthy (1999) Sustainability and Cities: Overcoming Automobile Dependence, Chapter 5: Greening the Automobile-Dependent City, Island Press, Washington, DC, pp. 240–249, 257–281, 1999.

[78] USIEP(Urban Sustainable Infrastructure Engineering Project), University of Colorado Denver-Health Sciences Center, 2005. http://carbon.cudenver.edu/engineering/esi/usiep

Sustainability Science and Engineering: Defining principles
Martin A. Abraham (Editor)
© 2006 Published by Elsevier B.V.
DOI 10.1016/S1871-2711(05)01021-4

Chapter 21

Implementing The San Destin Green Engineering Principles in The Automotive Industry

M. Sibel Bulay Koyluoglu, Stephen L. Landes

Ford Motor Co., Dearborn, MI, USA

Create engineering solutions beyond current or dominant technologies; improve, innovate and invent (technologies) to achieve sustainability. San Destin Green Engineering Principle #7.

On October 16, 2005, with fuel prices above $2.50/gallon, the following story appeared in the Detroit News: "*Hype vs. reality, Will bubble burst once consumers do the math?*" (*The Detroit News*, 10/16/2005, by Jeff Plungis) "What consumers might not realize is that buying a hybrid isn't a big money-saver. Various analysts estimate that it can take up to 10 years for savings at the gas pump to equal the extra cash a hybrid costs.... (N)onbelievers ... think the business case for hybrids is far from proven, and that after the initial excitement dies down, there may not be enough car buyers out there to justify the added cost of what is essentially a second powertrain."

Hybrid technology improves fuel economy which reduces fuel usage, reducing operating cost and emissions. Reduced fuel usage means better air quality and lower greenhouse gas emissions, decreasing the negative impact on the environment.

So what is the math for the consumer? Driving a hybrid vehicle reduces fuel costs, so a hybrid makes sense if the savings are greater than the incremental cost of the hybrid option.

And the math for the company? A hybrid vehicle incorporates two powertrains; one internal combustion-powered system and one electric drive system. Integrating these two systems to operate as one and supplying all the hardware costs more to develop and produce than a conventional powertrain. While a hybrid delivers more benefits than just better fuel economy, the manufacturer can only price for what the buyer is willing to pay: "savings at the gas pump." With vehicle sale price its only source of revenue, costs associated with addressing issues of air quality and climate change are not readily passed through to the consumer. These costs accrue to the company, reducing profit and the rate of return on capital investment.

The challenge to implementing the San Destin Principles in manufacturing lies in integrating them into the business decision-making framework. Current financial, accounting, and tax structures limit the recognition of costs and benefits to those that are tangible to the business. The impact of products on air quality, health costs, biodiversity, resource availability – costs and benefits external to the corporation – are outside the scope of today's corporate financial accounting standards. While these external costs are real, we don't have the means of including in a financial analysis all the social and environmental costs and benefits of a new product or process. Integrating the San Destin Green Engineering Principles into the product development process requires a means of dealing with these external costs. It requires the development of green business principles.

1. The product development process

One of the earliest tasks of a new product program is to define who the customers are and to establish their needs and wants. In the current business model the product development process follows a basic outline:

- Identify the customer base
- Determine customer needs
- Calculate what customers will pay to satisfy those needs

The definition of "customer" includes more than the people who will buy the product. Society, represented by government, is also a customer with needs and wants defined in legislation and regulatory requirements, which the product must meet. For example, public sentiment on environment and safety are distilled in fuel economy standards and safety equipment requirements. And lately, we have seen requirements calling for elimination of certain metals, the recyclability of materials as well as the use of recycled materials, and end of life management. The manufacturer itself is a customer and expresses its needs and wants through corporate standards that relate to performance, manufacturability, material selection, etc. This information, combined with product volume projections and life of the vehicle program, is used to generate the vehicle design specifications and required investment and variable cost targets for the program. The specifications and targets are then cascaded to the engineering community.

Using the target return on investment (ROI) and anticipated revenue, the engineering cost targets for the program are developed. These cost targets, along with the targets for safety, function, quality, and weight are then cascaded to the engineering community.

2. Engineering challenge

The engineering challenge is to develop technology that satisfies the needs at an affordable cost. Each engineering solution requires an investment of business capital to bring it to market. The question of whether or not a particular solution — a product feature — is affordable is a function of variable and fixed cost, the cost of money, and potential revenue (customer acceptability or take rate). Essentially the business question is "will this feature provide at least as good a return on investment as the next best use of capital?"

The vehicle buyer is the sole source of revenue. The cost of the vehicle must be reflective of its value to the customer. Internal marketing studies show that customers have strong positive associations with environmental and social issues, but make little connection between these issues and their purchase decisions. While customers will identify the environment — clean air, water, saving trees — as a high priority they concurrently rate fuel economy as virtually a non-issue in their vehicle purchase decision. The vehicle buyer will pay in the purchase price for the perceived value of each feature. In some cases the perceived value can be measured as a sense of emotional satisfaction: prestige, power, influence, and excitement. In other cases practicality takes over: is there room for the entire family, is the trunk large enough for two sets of golf clubs? Still other attributes, such as operating cost reductions (fewer oil changes, better fuel economy, lower insurance cost), can be measured directly. If the vehicle buyer is unwilling to pay for features or attributes and the manufacturer is unable to pass on the costs to the buyer, the manufacturer loses money, ROI decreases, and ultimately the business will become unsustainable.

Working capital for the company comes from its investors, whose expectation is that the company will provide a return on their investment (ROI) that is at least as good as the return from alternate investments at equivalent risk. A business, which fails to meet this test, will find itself to be unsustainable — capital will flee to alternate investments and the business will be unable to continue. As economic sustainability is one leg of the three-legged definition of sustainability — environment, social, financial — any product solutions must pass the financial hurdle.

Looking at this from a business perspective, the issues are:

- Integrating externalities into the product cost.
- Identify the sources of revenue to cover the incremental costs of a business, which accepts broader responsibility (who pays?).
  - The company
  - End user
  - Investor
  - Society
- Revamping how capital markets value companies.

We must note that niche markets are forming as awareness of the environmental impact of our lifestyle increases. More consumers are considering social and environmental impacts in their purchase decision. And socially responsible investing is growing with the emergence of funds such as the FTSE 4 Good and the Dow Jones Sustainability Index. Some potential buyers have the resources and willingness to pay a premium for environment friendly features in excess of the objective value of those features. Currently, some manufacturers can charge a premium for ultra clean, high fuel economy vehicles that satisfy the desires of this niche market. However, the numbers of these consumers and investors alone are still too small in number to make a real decisive impact on the product decisions of an industry as large as the automotive industry.

3. The hybrid electric vehicle as an example

A hybrid electric vehicle (HEV) incorporates both an internal combustion engine and electric motors and batteries to improve overall vehicle fuel efficiency. An HEV program may require an incremental investment over a program with a traditional powertrain on the order of $200 million. Each vehicle line requires periodic updating to incorporate design changes and technological advances. The number of vehicles over which program investment is allocated is a function of the number of years between major program investments and the number of vehicles produced in each of those years. Assume a program life span of 5 years and an annual vehicle production volume of 10,000 units for total program volume of 50,000 units. HEV program volumes have been about 10,000 vehicles per year (Toyota experience, other companies' plans). On that basis the investment per vehicle is $4000 (Table 1). An HEV effectively adds a second, electric, powertrain to the vehicle. Therefore, assume the incremental variable cost is about the same as a traditional powertrain, or about $2500. The installed cost for an HEV powertrain is estimated to be $6500. (Installed cost equals variable cost plus allocated investment cost per vehicle).

Table 1
Analysis of a hybrid electric vehicle

Element	North America
(a) Program investment	$200,000,000
(b) Program sales volume	50,208
(c) Investment per vehicle	$3983
(d) Equipment variable cost	$2500
(e) Total cost	$6483
(f) Revenue equals fuel savings	$2603
(g) Profit/Loss	($3880)

The primary benefit of the hybrid powertrain is reduced operating fuel cost due to increased powertrain efficiency. Buyers can be expected to pay a higher vehicle purchase price equivalent to a percentage of the present value of the stream of operating savings they realize over some period of ownership. For a best-case scenario, assume customers will be willing to pay the equivalent of the present value of their fuel savings for the HEV feature.

The present value of the fuel economy difference between an ICE (internal combustion engine) and an HEV hybrid is estimated to be $2600 over 5 years. Investment and variable cost for hybrid electric powertrains can only be partially recovered in the vehicle purchase price because more than half the cost of the incremental hardware exceeds the economic benefit accruing to the vehicle buyer.

If the buyer is willing to pay $2600 for the HEV equipment and the cost of that equipment is $6500, there is an unfunded differential of $3900. Who pays this? Who is responsible for this? If the company cannot charge the full cost, they carry the financial cost, which reduces the project ROI. Business will not fund a project it perceives as not financially sustainable. Unwillingness to fund such projects stifles engineering innovation.

The business case is sensitive to changes in two factors: fuel cost and program volume. Higher fuel cost results in larger operating cost savings. Increased production volume spreads investment over more vehicles reducing the investment cost per vehicle.

Compare the case for a hybrid program using North American vs European/Asian assumptions. In the European/Asian case fuel price per gallon has been increased to a more global price of $5 per gallon and program volume has been doubled. There is reason to assume that the volume changes are a rational assumption. Toyota claims to build between 120,000 and 130,000 Prius HEV sedans per year; 47,000 for the North American market and 70,000 for their other markets. Therefore, their investment is amortized over a higher volume than a North America-based program. Changing the operating cost and investment allocation assumptions brings the HEV example into near break-even conditions (Table 2).

It is important to look at the hybrid powertrain and ask what other benefits does this powertrain provide, who are the customers for these benefits, and how can the company derive income from providing these benefits? HEVs provide additional benefits, which the buyer neither receives nor values: because the powertrain is more efficient, so it produces fewer emissions and can be argued to reduce demand for imported (into the US) petroleum. These benefits accrue to society in general and take the form of increased longevity, better health, reduced crop damage, longer lasting structures, and insulation from world energy issues. What are these benefits worth to society and how can a manufacturer generate revenue from society to help make advanced technologies with broad impacts pay for themselves?

Hybrid powertrain-equipped vehicles reduce emissions of VOCs, SO_x, NO_x, CO_2, and CO. The value of abated emissions can be measured directly in emission credit markets. Sample values of emission reduction credits are shown in Table 3.

Emission abatements certified by US EPA can be sold, traded, or retired at market values. With the above assumptions the estimated value of emission reduction credits that could be generated by a hybrid program is $12.5 million after taxes. The after tax value could be applied to the program to reduce total program investment and, ultimately, to reduce the cost of the vehicle to the customer (Table 4).

Table 2
Comparison of the business case for a hybrid electric vehicle in two economies

Element	North America	Europe/Asia
(a) Program investment	$200,000,000	$200,000,000
(b) Program sales volume	50,208	100,083
(c) Investment per vehicle	$3983	$1998
(d) Equipment variable cost	$2500	$2500
(e) Total cost	$6,83	$4498
(f) Revenue equals fuel savings	$2603	$4190
(g) Profit/Loss	($3880)	($308)

Table 3
Cost of vehicle emissions

	NMOG	CO	NOx	HCHO	CO$_2$
$/Ton at current market value	$1324	$7276	$9741	$1324	$5

Table 4
Net cost accounting including full assessment of environmental benefits

Element	Value	Current North America
(a) Program investment	$200,000,000	$200,000,000
(b) ERC after tax value	$12,509,178	
(c) Net program investment	$187,490,822	$200,000,000
(d) Total program sales	50,208	50,208
(e) Investment per vehicle	$3734	$3983
(f) Equipment variable cost	$2500	$2500
(g) Fully accounted cost	$6234	$6483
(h) Lifetime savings	$2603	$2603
(i) Savings B/(W) than cost	($3631)	($3880)

Non-governmental sources estimate that the damage caused by emissions of acid precursors such as NO_x and SO_x may be as much as 16 times the market value of the emission reduction credits. This value is not recoverable from any one source because it is collective damage to buildings, roads, bridges, agriculture, forests, and human health. If the value of emissions reduction credits could be increased to account for the savings to society of these general benefits the investment picture would be different.

Increasing the market value of ERCs by a factor of 10 makes a substantial improvement in affordability of a hybrid powertrain-equipped vehicle. Emission credit markets are a new forum for the valuing and trading of social benefits in measured quantities of abated emissions. A federal policy for certifying mobile source emission reduction credits plus tax credits for purchasers in the amount of value to society of emission reductions would permit manufacturers to recover development and equipment costs of advanced technology powertrains on a par with conventional powertrains.

As an alternative to providing tax credits the Department of Treasury could participate in the emission credit markets as a market maker to buy up and hold credits in an attempt to raise the price to a point where real cost of emissions reductions is recognized. Acting as a market maker would have the effect of strengthening market-based approaches to finding solutions to environmental and social problems.

4. Risk measurement

Several investment advisors, Sustainable Asset Management (SAM) and Innovest, to name but two are beginning to rate businesses on their long-term sustainability. The general model is that businesses must strike a balance among environmental, social, and financial responsibility in order to be viable for the long term. Failing to address any of these areas leaves the business in a riskier position. Two major metrics for business are share price and cost of capital. These metrics are a function of risk: lower risk means more certain returns over time and should mean lower cost of capital. The business that successfully addresses all three areas should have a measurable advantage over its competitors and should be rewarded in the marketplace. The challenge is to be able to measure risk and communicate to investors that taking actions to become sustainable reduces risk. Becoming sustainable effectively changes the yardstick for a business. The SAM techniques of using carbon intensity of profits as a measure of long-term risk would be useful if they can be correlated to real risk to capital. A reduction of a few basis points in borrowing cost for a major company could translate into a substantial competitive advantage and would encourage internalization of external costs and benefits and the adoption of green engineering principles.

5. Conclusion

For green engineering principles to take hold in business, bringing greener products to the marketplace, several things must happen:

1. Public awareness of the linkage between buying decisions and environmental or social consequences must be increased, so that the consumer considers environmental and societal impacts in the purchase decision.
2. The political will must be created to find a way to incorporate the externalities associated with product use into the cost of the product. A value must be placed on the environmental and societal impacts of use and the costs apportioned among the customer, company, capital markets, and society. Will this be in the form of subsidies, tax rebates, emissions credits, or perhaps other creative means? However, there needs to be public discourse to generate political will around the topic. Otherwise, society continues to subsidize the use of non-green products through costly toxic clean-ups, health care costs, lowered productivity, and infrastructure damage.
3. The concept of sustainability must be integrated into the financial accounting system. Parallel principles, similar to the San Destin Green Engineering Principles, should be developed for accounting and finance.
4. Capital markets must reward companies, which innovate, make products that do not pollute, and use renewable resources at a rate that allows their regeneration. This is beginning to happen in the socially responsible investment community, but the volume is not large enough as yet to make a difference. Businesses meeting the definition of sustainability are less risky to the investor. Lower risk must be reflected in lower cost of capital. And that is the reward to the company.

Sustainability Science and Engineering: Defining principles
Martin A. Abraham (Editor)
© 2006 Published by Elsevier B.V.
DOI 10.1016/S1871-2711(05)01022-6

Chapter 22

Infusing Sustainability in Small- and Medium-Sized Enterprises

Bert Bras

George W. Woodruff School of Mechanical Engineering, Georgia Institute of Technology, Atlanta, GA 30332-0405, USA

1. Introduction

Small- and medium-sized enterprises, in short known as "SMEs", form the backbone of virtually every national economy the world over. It is estimated that over 90% of all enterprises are SMEs and they account for more than 70% of goods and services sold worldwide. This number is as high as 97% in Australia, 99% in Canada, 99.8% in the UK, and 99.9% in Spain. [1,2]. This is also the fastest growing sector [3]. The case for SMEs is best represented by the following quotation:

> "Small and medium-sized enterprises (SMEs) are the most important sector of a nation's economy. They provide and create jobs, especially during times of recession; they are a source of innovation and entrepreneurial spirit; they harness individual creative effort; and they create competition that are the seed bed for businesses in the future. In short, small and medium-sized firms are vitally important for a healthy, dynamic market economy." [2]

In addition to their economic importance, SMEs also play a sizeable part in terms of their contributions to pollution. Individually, most of them are too small to show up as large polluters, but collectively, it is estimated that SMEs are responsible for 70% of worldwide industrial pollution levels. Notwithstanding the compounding evidence that corroborates the growing importance of SMEs, this sector remains largely "*under researched*" and therefore deserves our attention [2].

SMEs present a very different set of challenges to sustainable development than the larger companies and organizations do. Owing to their smaller size in virtually all dimensions, SMEs are also heavily constrained in the availability of resources that they can dedicate to new ideas and innovations. Not only do SMEs behave and operate very differently from larger companies, but they differ widely among one another as well. The heavy information processing demands of an initiative such as sustainable development, in what is typically an information impoverished environment, makes infusing sustainability in the SME industry sector and all the more interesting (and challenging) a problem to solve.

In this chapter, we will discuss the challenges in infusing green engineering, environmental management, and sustainability in SMEs. First, we will characterize and define what SMEs are. Second, we will discuss current attempts to introduce environmental management in SMEs and the causes and issues that have caused SMEs to lag behind in the environmental area. Finally, we discuss recommendations for strategies for infusing greater environmental management and sustainability in SMEs. We will start, however, with a general introduction to sustainability in business.

2. Sustainability in business

Given the growing worldwide awareness of environmental issues, the manner in which a company manages its environmental impacts will play an increasing role in the long-term viability of that company. Governments and industry alike are incorporating environment friendly programs into their agendas to appease their respective constituents and clientele. We have seen the emergence of global and domestic legislation in the areas of population control, emission standards, ecological conservation and protection, as well as innumerable in-house company environmental health and safety (EHS) programs. In Fig. 1, the trend of environmental regulations in the US is shown. It is becoming clear to stakeholders everywhere that it is no longer sufficient for companies to simply exhibit economic performance alone; they must now demonstrate how they contribute toward ecological and societal well-being as well — businesses have to find a way of becoming sustainable.

The terms *"sustainability"* and *"sustainable development"* were first coined in the early 1980s. However, ever since the introduction of these terms, a debate has raged over the definition of *"sustainability"* and *"sustainable development,"* and what, precisely, these concepts entail. Consequentially, there are some 70 known definitions of *"sustainability"*. The United Nations World Commission on Environment and Development published a report in 1987 titled *"Our Common Future,"* which presented what is probably the most widely accepted definition of sustainability: *"Sustainability is the development that meets the*

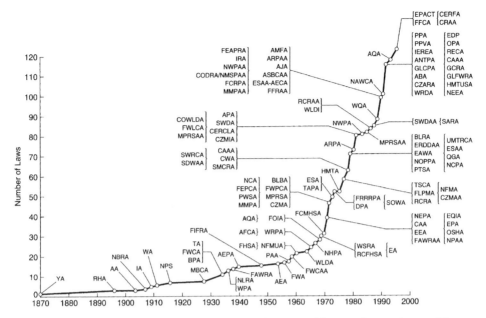

Fig. 1. Cumulative growth in the federal environmental laws and amendments [4].

needs of the present without compromising the ability of future generations to meet their own needs." Another often-cited definition of *"sustainable development"* has been coined by the World Business Council on Sustainable Development:

> "Sustainable development involves the simultaneous pursuit of economic prosperity, environmental quality, and social equity. Companies aiming for sustainability need to perform not against a single financial, bottom line but against the triple bottom line."

Moving toward sustainability involves holistic approaches and systematic thinking. The present consensus, however, is that sustainability leaves business managers with more questions than answers. One of the most fundamental questions being, how do you go about infusing sustainable development into the fabric of a company? In other words, how does one translate the principles of sustainability into actionable steps for business managers to execute? As noted by [5], "Making an environmental monitoring system an integral part of a company's daily operational fabric is often more difficult than actually building the system."

The realization that most proactive business have come to today, is that while sustainable development is a concept many agree to in principle, it can be difficult to translate into practice. Attempting to move a business toward sustainable development requires a change in management philosophy: there needs to be a shift from the traditional responsive, one-dimensional management to one that is proactive, multi-dimensional, and seeks competitive advantage. Managers are discovering the need for new ways to measure and new strategies to improve performance along these new dimensions.

3. Effects on the financial bottom line

There have been several attempts to verify that environmental management and sustainability are in fact good for business from an economic standpoint. However, conflicting opinions on whether this is true or not still exist. Many researchers are of the opinion that corporate value creation has always been affected by three dimensions — environmental, economic, and social. According to, for example, [6] environmental and social issues also affect the bottom line, whether or not they are captured on balance sheets. Consequently, the belief is that addressing sustainability issues will almost certainly translate into increased efficiency and considerable cost-savings in operations, and that companies that adopt a "*triple bottom line*" focus see returns in the following areas:

- *Innovation* — sustainability-oriented companies are constantly positioning themselves to take advantage of competitive pressure and changing markets.
- *Operational efficiency* — sustainability-oriented re-design and re-engineering of products and processes can significantly cut operational costs, including material and energy use.
- *Brand equity* — the enhanced brand equity and reputation that comes with more sustainable business practices attract more customers.
- *Lower risk* — the environmental focus of sustainability-oriented companies decreases the risk of public relations disasters, fines, boycotts, and clean-up costs.
- *Talent* — sustainability-oriented firms find it easier to attract and retain talent owing to their reputation.
- *Shareholder value* — companies pursuing sustainability-oriented business strategy have greater shareholder value than their peers.

Another supporting argument is presented in [5] and captured in Fig. 2 to illustrate the evolution of environmental management in practice, and the corresponding value added to the business in general. Box's contention is that, the more proactive a firm is in managing their business to the "*triple bottom line*", the greater are their opportunities to realize improvements in efficiency and add value to the existing enterprise.

There are many more examples in the literature that suggest that there are several benefits, economic and beyond, in incorporating sustainability into a firm's business model, and a National Science Foundation sponsored study of

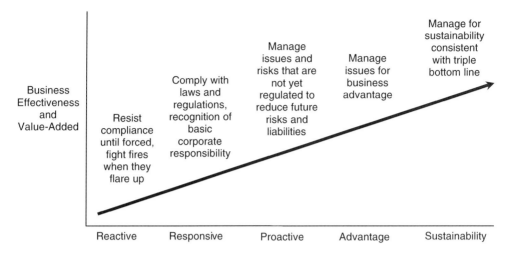

Fig. 2. Evolution of Environmental Management in Practice. Adapted from Box [5]

Environmentally Benign Manufacturing also finds numerous industrial examples to support this claim [7,8]. At the very least, it has been suggested that sustainability does not negatively impact business economics. Citing Germany and Japan as examples of economies with some of the most stringent environmental regulations that are also some of the most productive and innovative, Michael Porter claims that at the micro level, "environmental protection does not harm competitiveness [9]."

However, not everyone is comfortable with the notion of environmental improvement being consistently good for economics. Wagner finds that the relationship between environmental and economic performance is still uncertain. His review of the earlier studies indicates no significant relationship between the two factors, whereas recent studies indicate a significant relationship but are uncertain, whether it is positive or negative. Yet, other analyses have found all types of relationships (i.e. negative, positive, and no relation). He contends that even in stock market performance, there is variability as to what measures provide positive or negative correlation and to what extent of time (long term or short term). Thus, the time period is also important, as the relationship between economics and the environment seems to be changing over time. Ultimately, his conclusion is that one cannot generalize about the relation between economics and environment [10]. This uncertainty about the relationship between environmental and economic performance is one of the reasons that many companies adopt a "wait and see" attitude, especially those with limited resources.

Nevertheless, there are benefits to be gained from environmental improvement, triple bottom line, and sustainability initiatives. While many of these benefits are intangible and may not necessarily result in measurable economic windfall, they nonetheless add value to the overall business operation.

4. Small- and medium-sized enterprises (SMEs)

4.1. Definitions and importance

What are SMEs (sometimes referred to as Small- to Medium-sized Enterprises)? There is an entire range of operational and theoretical definitions for SMEs offered by international bodies, individual scientists, and other organizations with interests in such firms. For instance, companies have been classified into categories by amount of revenue earned, number of employees, amount of capital, sales volume, production capacity, number of computers, extent of Internet use, number of functional areas, age of the firm, amount of government grants and financial assistance, the firm's membership in associations, and so on [1,11]. Some of the more popular classification schemes are those suggested by bodies such as the European Commission DG XXIII, Organization for Economic Cooperation and Development's (OECD) Working Body for SMEs, and the US National Institute of Standards and Technology [2]. However, for our purposes herein we use what is arguably the most common definition of SMEs, i.e. based on their size measured in terms of the number of employees. In Table 1, a classification scheme suggested by Bargmann is shown [12].

As mentioned, SMEs form the backbone of virtually every national economy the world over. Several studies have been conducted, all of which maintain that the vast majority of business are SMEs, not large firms [13]. It is estimated that over 90% of all enterprises are SMEs and they account for more than 70% of goods and services sold worldwide. Over 85% of all manufacturing firms in the continent of Asia are SMEs [11]. Other studies report that, over 97% of all businesses in Australia are SMEs. These numbers are as high as 98% in Asia-Pacific, 99% in Canada, 99.8% in the UK, and 99.9% in Spain. In almost all these nations and more, SMEs provide more than half the employment and contribute somewhere in the vicinity of 60–70% of the Gross National Product [1,2,14]. Additionally, SMEs are also the fastest growing sectors in the business population [3].

Table 1
Classification of companies based on size [12]

No. of Employees	Company Classification
9 and under	Micro enterprise
10–49	Small enterprise
50–499	Medium enterprise
499–999	Large enterprise
1000 and over	Very large enterprise

From a broad global economic perspective, SMEs have been touted as agents in the economy that promote equilibrium between the supply and the demand in the economy and maintain the basis for competition. From an innovation standpoint, new and small firms play a crucial role in experimentation and innovation, which leads to technological change and productivity growth. Moreover, the perpetual turnover of small companies on the market creates a basis for perpetual innovation and development. Clearly, SMEs form an important sector of national economies and society in general.

4.2. Differences between SMEs and larger companies

How do SMEs differ from larger companies? A common misconception is the assumption that SMEs behave and operate in a similar manner as large companies but on a smaller scale. Research that focuses specifically on smaller firms can be summed up by the notion that "*a small business is not a little big business* [15]." Accordingly, findings, generalizations, and approaches for large companies cannot necessarily be applied to smaller companies [2,16,17]. First and foremost, SMEs suffer from "resource poverty," which is a lack of various resources in an organization — time, financial, and expertise [15,18–20].

- *Financial resources* — SMEs lack of financial resources is widely corroborated [15,20,21]. As such, SMEs are financially ill-equipped to absorb the costs of a mistake [15,22,23].
- *Expertise resources* — SMEs are often unfamiliar with changing technology, techniques, and management practices (lack of "awareness knowledge"). Also, SMEs often lack "how-to knowledge" (training and experience) with technologies and practices [15,23–25].
- *Time resources* — Personnel in smaller firms do not have "extra" time on their hands, and more importantly, there are no financial resources to hire more personnel [20,21,23].

Smaller firms also have a management focus that is more "day-to-day" and short-range than strategic and forward looking [15,17,21,26–28]. Coupled with resource poverty, this short-term focus results in a general lack of formal management practices. Formal use of management practices such as written procedures, problem-identification tools, basic quality-improvement tools, and performance measurement have been found to be infrequent and generally poorly supported in SMEs [17,26,29]. SME management invests in new innovations only when they feel it is necessary to carry out activities, particularly administrative and transaction-based activities [28]. The emphasis on such cost reduction projects is due to the focus on short-term financial benefits [30]. Other organizational factors in SMEs include a heavy dependency on the top manager

[23] and the fact that SMEs tend to employ generalists who typically wear many hats, rather than specialists.

It would appear that SMEs are severely disadvantaged compared to larger companies. However, there are certain advantages SMEs possess, over their larger counterparts. For example, due to the generally lower degree of bureaucracy, shorter internal lines of communication and greater levels of informal communication, SMEs are able to make decisions faster [26]. Magnusson also noted these shorter decision cycles in SMEs in addition to a more flexible organization, and greater payback potential from innovations compared to the amount invested [23]. SMEs are valued in industry because of their ability to respond quickly with shorter production runs and turnaround times compared to larger firms [28].

In addition to the differences with larger companies, SMEs are an incredibly diverse group in and of themselves as well. SMEs' concerns and issues have been found to vary by several factors such as the market sector they operate in (e.g. high-tech or metalworking), maturity (young or old), geographic location (rural or urban), and where along the supply chain they are situated (e.g. supplier or final producer). For example, SME usage of information technology and systems varies tremendously by industry sector [31]. This high degree of variance precludes the effective use of general, boilerplate solutions. Since all companies have their own special features such as size, industry, method of production, employee structure, and development strategy, custom solutions are needed [32]. This conclusion has implications not only for SMEs, but also for those seeking to assist or do business with them in any way.

4.3. Growing pressures on SMEs

What are some of the challenges that SMEs are coping with? SMEs are increasingly affected by global competitive forces partly due to the emergence of a tiered supply chain that ties the future success of suppliers to manufacturers as they seek to respond to the changing demands of globalization. Most Tier-1 manufacturers and suppliers are returning to their core competencies in design and production: "non-core activities are being pushed down the supply chain. This means the core competencies on which smaller firms had built their businesses are no longer sufficient as a basis for growth or even survival [19]." These new competencies are requiring a longer-term planning horizon from these small suppliers, which traditionally have been reactive and adaptive.

Generally, SMEs operate under conditions involving short production runs, changing product characteristics, and defect-free, ontime production at decreasing prices [21]. Currently, market pressures for manufacturers to deliver products with increasing shorter development cycles and better quality are driving the needs for new systems and innovations [32]. In the case of such

SMEs, it is generally agreed upon that the most important drivers for change have been customer pressures for cost reduction and increasing global competition [30]. Typically, in a tiered supply chain, one is liable to find that SMEs are dependent upon customers who purchase large quantities, and these customers wield power over the SMEs (are able to influence prices, for example). In such cases the consensus is that for SMEs, customer numbers and customer power tend to be inversely proportional: a small number of customers with considerable power dominate most SMEs [28].

Magnusson explains this through, the growing pressure on companies of all sizes to adopt e-commerce and cites for example the Swedish governmental objective to have 95% of all purchases in the public sector be made electronically. However, she reports that despite all the pressure, the diffusion rate of innovations such as e-commerce in SMEs has been slow thus far [23].

5. SMEs and the environment

5.1. Environmental pressures

What are the environmental impacts and pressures that SMEs deal with? From a regulatory and compliance perspective as well, SMEs recognize that there are pressures to improve their environmental performance. They are experiencing unprecedented legislative, business-to-business and other stakeholder (local community, NGOs) pressure to meet, comply, and participate in a host of environmental programs and initiatives [2]. These pressures are growing because, next to their economic impact, SMEs also play a sizable part in terms of their contributions to environmental impact and global pollution. It is estimated that collectively, SMEs are responsible for 70% of worldwide industrial pollution levels [2]. Reasons for the individually obscure, yet cumulatively high levels of SME pollution are [11]:

- Use of old and inefficient technology;
- Lack of information on newer, cleaner technology;
- Lack of waste disposal and treatment systems;
- Poor infrastructure; and
- Organizational and social barriers for adoption and implementation of environmentally sound technologies.

Individually, most SMEs are too small to show up as large polluters, and due to the large number of SMEs, it is virtually impossible to check their environmental performance. As a result, the focus of pollution control remains for the most part on the larger corporations. Nevertheless, many governments and policy makers have realized the importance and impact of the SME sector and are seeking ways to have SMEs comply better with existing environmental regulation and become more proactive in terms of environmental management if they ever want to achieve sustainability. For that reason, a large number of environmental management initiatives exist focused on SMEs.

5.2. Environmental management and SMEs — incentives and barriers

What are the issues in environmental management for SMEs? Contrary to the once-popular approach of focusing on "cleaning up" the wastes after they are produced, the focus of today's environmental management methodologies is on identifying the sources of the impacts within an organization, and reducing (or eliminating) the impact at these sources. Arguably, if a company is continuously working to become more aware of and reduce their impacts, then new developments in legal or market requirements will likely not be a surprise. Meeting these requirements, then, will likely require less of an investment than in a company not in tune with their impacts.

To accomplish a more efficient environmental feedback loop within an organization and help companies focus on their environmental bottom line, a variety of methodologies are available. These methodologies go by many different names, including Environmental Management (EM), Environmental Assessment (EA), and Pollution Prevention (P2). Standards such as ISO 14001, the European Union's Eco-Management and Audit Scheme (EMAS), the EPA Performance Track, the New Mexico Environment Department (NMED), and the Green Zia Environmental Excellence Program, can also serve as methodologies. Other methodologies include approaches such as the Natural Step, Factor Four, and Six Sigma approaches. Often, these methodologies are not used independently of one another. As one example, Green Zia is a performance-based environmental management system that can be used as an EMS on its own, or combined with the ISO 14001 standard [33].

While the names and exact definitions of these and other similar methodologies vary depending on the source and even the application, the fundamental concepts of the modern environmental management methodologies remain uniform throughout. The primary focus of companies is to become a more proactive and less reactive organization, to consider environmental effects and eliminate wastes before they arise through a more thorough understanding of their processes.

The US EPA lists the following general benefits on its website for environmental management systems (EMS):

• Improve environmental performance;
• Enhance compliance;

- Prevent pollution and conserve resources;
- Reduce/mitigate risks;
- Attract new customers and markets (or at least retain access to customers and markets with EMS requirements);
- Increase efficiency/reduced costs;
- Enhance employee morale, also possibility of enhanced recruitment of new employees;
- Enhance image with public, regulators, lenders, investors;
- Achieve/improve employee awareness of environmental issues and responsibilities; and
- Qualify for recognition/incentive programs such as the EPA Performance Track Program.

It is also noted that developing and implementing an EMS may have some costs, including:

- An investment of internal resources, including staff/employee time;
- Costs for training of personnel;
- Costs associated with hiring consulting assistance, if needed; and
- Costs for technical resources to analyze environmental impacts and improvement options, if needed.

Many SMEs are aware of the benefits of improved environmental performance in terms of better customer relations, cost savings, and competitive advantage. The key benefits of improved environmental performance in SMEs are seen in the form of improved organizational and managerial efficiency, continuous monitoring of compliance, improvement of the company's image, the possibility of economic benefits through streamlined processes and optimization of resources (reduced raw materials, energy, water, etc.), and better waste management, and re-use programs [34]. Benefits of programs such as ISO14000 include better-trained employees, process improvements, increased recycling, reduced pollution, improved safety and working conditions, costs savings from lower insurance premiums, defensible legal positions, positive public image, and improved credibility and ratings among customer base can lead to new business opportunities [2].

Despite the benefits, the implementation of environmental assessment and management programs is still a daunting task, and not just for SMEs. Some particular issues that can cause problems for implementation are:

- *Company knowing the value of a program — "Programs do not sell themselves* [35]." Despite the values that many environmental programs can have to a company, the awareness of the value many times is not enough to provoke a company to either decide to adopt the particular program or even successfully implement a program. There is a need on the part of external sources to reinforce the value that a good environmental assessment and management program can have for a company.

- *Fear of the unknown* — Many times companies are unsure what type of environmental assessment and management program to adopt into their company. Selection of the "correct" system is very critical in the successful adoption of new technologies [17]. The inability for many small companies to properly select the "correct" system due to uncertainty or lack of experience led many companies to choose not to implement environmental assessment. The uncertainty associated with what the new program may bring outweighs the possible value that the assessment program may have for the company.
- *Follow the leader* — Many companies are unable to select the "correct" system for their particular application because of their lack of experience. This uncertainty caused by the inability to select a system leads some companies to "follow the leader." Some companies follow in the footsteps of others that may have a similar process or produce the same products when it comes to selecting and implementing an environmental assessment and management program. This selection can have negative effect on the company if significant time and effort is spent to implement a program that is not suited for the company's specific application.
- *Quick fix programs* — Many environmental assessment and management programs that are based in the area of pollution prevention have developed easy guidelines to identify sources of pollution. These programs provide a great approach for eliminating waste at its source, however for the most part they do not systematically track waste throughout the entire process. They mainly look for solutions for a specific compliance or regulation issue and are seen as a quick fix to monitor and reduce environmental impact instead of systematically tracking waste throughout the entire process.
- *Proper training* — The proper training in the development of any program is essential to have the necessary support of employees documented earlier. The increase in training of company personnel leads to a reduction in uncertainty and a better understanding of individual responsibilities to the program [36].

Perkins also alludes to problems involved in implementing a P2 program and reasons why many companies may be sluggish in the shift from traditional pollution control to pollution prevention:

> "in spite of numerous success stories, obstacles continue to impede the adoption of aggressive P2 programs and to hinder investments in pollution prevention equipment and processes. Lack of proven technology or technical information, existing command-and-control regulation and infrastructure, insufficient financing, and underestimation of financial and less-tangible benefits all stand in the way of more rapid shift from pollution control to prevention" [37].

According to [2], compliance with regulation is not driving environmental performance, even in the minority of proactive SMEs that are taking action. Some SMEs do not view compliance with environmental programs as a potential source for competitive advantage, a marketing issue, or of importance to their customers. There is a need to derive effective and acceptable criteria for measuring the benefits of environmental action if SMEs are to meet sustainability goals. Another related observation is that SMEs tend to consider environmental aspects a delicate and confidential matter. While they like to have good relations with authorities, they are concerned about negative reaction from local patrons and the community. This is the reason why SMEs are anxious about publishing such information through an environmental statement. Many SMEs think it wiser not to have any formal environmental policy in place at all [34]. Clearly, this disposition and attitude can be a significant deterrent in promoting sustainability within SMEs.

In [38], the following barriers to EMS adoption in terms of motivation, resource issues, and implementation are given:

Motivation:
- Lack of customer requirements or demand for having an EMS;
- Misconceptions that environmental issues are a low organizational priority or are believed to be under control;
- Beliefs that an EMS is not important or relevant to the businesses or capable of adding to the bottom line;
- Lack of public pressure or NGO pressure to implement an EMS;
- Beliefs that an EMS is the current management "flavor of the month"; and
- Beliefs that an EMS is not widely accepted or used in an industrial sector or geographical area.

Resource issues:
- Concern about the cost and time necessary for establishing an EMS;
- Concern about operational management costs following implementation; and
- Perceptions that an EMS is complicated and unattainable.

Implementation concerns:
- Fear of discovering non-compliance with regulations or permits; and
- Fear of discovering or uncovering internal problems within the organization (staff issues, process issues, company policies, etc.).

5.3. Resource poverty

What is the root cause of the SMEs lagging adoption of environmental management systems, and other methods, tools and technologies that promote sustainability? The general lack of resources appears to be at the root of most barriers SMEs face on the path to environmental management and ultimately sustainability. As mentioned before, small businesses are not just smaller big businesses but suffer from a condition known as "resource poverty [15]". The shortfall of resources can be in the form of a lack of commitment from top management, a lack of training and technical knowledge, the lack of specialist staff, and lack of financial capital [2]. Four main categories of constraints that SMEs face when implementing new technologies can be identified:

- Organizational constraints — there is often resistance within the entire organization to change from the normal procedure or operation.
- Expertise constraints — there is typically a narrow focus of expertise of the employees.
- Financial constraints — there will likely be less capital available for projects different from standard operational procedures.
- Time constraints — with a smaller workforce, each employee will have more responsibilities, leaving less time for out of the ordinary project implementation.

These constraints are not restricted to environmental and sustainability issues, but are applicable to other technology-related areas (e.g. information technology and advanced manufacturing) as well. Sustainability and environmental management bring some unique challenges. For example, there are so many different established environmental assessment and management methodologies that, it is a challenge (even for large organizations) to understand them all and choose one among them that is appropriate for the specific organization, see, e.g. [39]. While an approach may be considered the *best* by one group at one particular time, the field is evolving at such a rapid pace that there is no single accepted methodology that can be agreed upon by all organizations. Even the ISO 14000 Standard, which is undoubtedly accepted and used by many companies, has many criticisms against it and is sometimes claimed that it is being oversold [40]. Noting this, many times companies choose to take an approach that is a hybrid of two or more methodologies, playing off the strengths and weaknesses of each. Impediments indigenous to SMEs make adoption of environmental management all the more challenging compared to a large company, including:

- less capital,
- a smaller workforce,
- a less formal management structure,
- a focus on the satisfaction of short-term goals verses long-term goals, and
- a lack of understanding of or experience with sophisticated software tools when compared to larger firms in similar industries.

The lack of resources also causes researchers to note that ISO 14000 and EMAS standards are ill-suited to SMEs. These programs were designed to be exhaustive in their requirements. The consensus is that the overwhelming amount of documentation may be too detailed and complex for SMEs [34]. Even a positive attitude and "concern for the environment will not necessarily mean adoption of environmental standards [2]."

Not only are SMEs unfamiliar with extensive environmental auditing tools and procedures, but the absence of existing environmental management systems in most SMEs means that they typically have to start from scratch, and this is costly. Managers of smaller companies often have to bridge a cultural gap regarding awareness of environmental matters in their organizations. The sheer amount of time that management must devote to sustainability initiatives such as an EMS implementation is considerable, e.g. [34,38].

A disturbing picture is painted by, e.g. the Georgia Pollution Prevention Assistance Division (P2AD) that very few firms allocate waste-related costs directly to products and processes. This means that companies are unable to make the business case for environmental management. Furthermore, in those cases where these costs are allocated, the bases of allocation used are often inappropriate. This has led to situations where managers have targeted the wrong process or products for prevention efforts. In essence, this implies that SMEs are likely to pick the wrong starting point on their road to sustainability, potentially resulting in disappointing or less than expected environmental and economic benefits.

The selection of the "correct" system is very critical in the successful adoption of new technologies [17]. The implementation of the wrong program for whatever reason — be it that the company did not have financial resources to afford the "correct" program, or did not have the experience to select the "correct" program, or simply followed another company that was similar to theirs — reduces the changes that the program will be successfully adopted into the company.

The SMEs' unfamiliarity with existing systems is a reflection of their general lack of knowledge about formalized systems and their benefits. This seems to manifest itself primarily in the implementation stages where implementation is often interrupted, due to the SMEs' inability to see relevance of all stages, or perhaps the lack of a champion. There is a predominant uncertainty over how to maintain continuous improvement within the SME community [2].

This lack of continuous improvement has also been observed in Dutch and Swedish studies on implementation of EcoDesign in SMEs. In a Dutch study [41], it was stated that SMEs were many times eager to be helped, but that as soon as the consultants left, the motivator seemed to have gone as well. The reasons cited were that SMEs tend to think short-term and do not have many resources to spare. A later Swedish study confirmed this [7]. An unresolved issue, therefore, is how to create the motivation for self-sustaining efforts after an external party has performed initial analyses.

6. Improving environmental management in SMEs

What are strategies that would improve environmental management and sustainability in SMEs? The literature is full of suggestions and recommendations on how to promulgate environmentally conscious thinking and other aspects of sustainability in small-to-medium enterprises. We will present some general strategies, followed for particular strategies related to tools, but start with the key strategy: clear message and communication.

6.1. Communication

The two most important considerations in any strategy to promote or disseminate an idea (or solution) such as sustainability are:

• the characteristics of the message itself, and
• the manner in which it is communicated to the target audience.

It is encouraging to note the widespread consensus about the ideal structure of the message to SMEs. Most experts agree that any message must present a solution in a manner that is simple, clear, highlights economic benefits, presents easy-to-do actions, and indicates where to go for further assistance [11]. The message about the problem, its solution, and where to go for more information should be clear and simple. It should use examples and success stories in other companies as motivating examples, and the information should be perceived as "*reliable*," and not from a "*faceless*" government organization [2].

Another method is to design the message so as to allow the solution to market itself. SMEs need a solution that concentrates more on the day-to-day running of small firms, increases the availability of advice and reduces the cost of consultancy through more detailed developments for each industrial sector based on best practices, not on strategic management. Thus, a solution needs to be inexpensive, sensitive to the limitations of SMEs, locally based, user-friendly, and flexible.

The latter issue of how the message is communicated is also relatively well agreed upon. It appears that the most effective way to reach and change the behavior of SMEs is through personal contacts, made through existing, trusted routes. Additionally, targeted materials are likely to be more effective than blanket marketing messages [11]. Contacts should also be made through existing information dissemination pathways [38]. Palmer and France found that direct mailings and flyers were the most successful ways to bring a new service to the attention of SMEs. However, they too identified using local networks as the next best alternative in getting their attention [3]. This is in line with our experiences.

6.2. Beyond communication

Clearly, communication in itself is not sufficient. Numerous organizations have experienced and written about "lessons learned" that form excellent sources for new strategies, see for example [42], and it would be impossible to list them all. Some frequently mentioned strategies are listed below. We have used a number of these strategies too in our work, although with mixed results [12,39,43]. This indicates that no strategy is a "silver bullet" that solves the problem of disseminating environmental management and sustainability in SMEs. Nevertheless, some frequently cited (and used) strategies are as follows:

- Because SMEs are very short-term-bottom-line focus, sell the idea that EMS can reduce costs and increase efficiency
- SMEs are compliance-driven and reactive — refer to EMS as a way to make compliance easier.
- Go for the low hanging fruit — promote "*easy*" changes that can be quickly implemented and show results.
- Facilitators should target top management.
- Charge the "*appropriate*" price for information, but not for general information.
- Businesses should consider turning products into services and fundamentally transform their business models from a product-maker to service-provider.
- Cooperation is the best way of diffusing information and facilitating EMS use in SMEs. This cooperation can come from:
  o Customers, Vendors, Supplier, other enterprises.
  o Public institutions (universities), local communities.
  o Peer networks and trade associations, peer-auditing programs.
  o Local and Federal government training programs.
  o Environmental NGOs.
- Develop measures for promoting networking and cooperation among SMEs.
- Provide training and technical support to SME personnel.
- Provide financial support and/or incentives for SMEs.
- Develop simplified EMAS/ISO14001 requirement/guidelines targeted at SMEs. SMEs should be permitted and encouraged to take "*baby-steps*", i.e. an incremental approach to implement such EMS.
- Promote development of environmental performance indicators and environmental benchmarking and comparisons.

- External pressures are good motivators:
  - o Pressure from supply chain, big supplier/buyers demanding improvement in environmental performance.
  - o Many foreign governments and companies are using environmental performance criteria in their purchasing.
- Creation of a telephone Helpline and free half-day site visits (such as employed by Manufacturing Extension Programs in the US) are also attractive schemes since they help reduce the cost of environmental management, and these services are compatible with the ad hoc, informal approaches that are typical of SMEs.

Numerous additional recommendations can be identified in order to facilitate the dissemination of environmental management (and sustainability) in SMEs. A recent report [38], however, provides a good summary of recommendations that is indicative of the current state-of-art and practice. The following recommendations were derived from a variety of workshops and stakeholder meetings:

- Support sector-specific EMS implementation tools;
- Build the business case for EMSs;
- Work with Trade Associations and other industry groups to promote EMSs;
- Integrate EMS into other government programs [than EPA regulatory programs].
- Develop a Web-based EMS technical assistance providers resource center.
- Create a standard "EMS 101" training program for SMEs;
- Develop tools to assist with aspect analysis and target setting for EMSs; and
- Inventory EMS assistance programs and providers.

With respect to EMS tools for SMEs, a number of other recommendations can be made, in part based on general technology transfer lessons from infusing Information Technology in SMEs.

6.3. Implementing computer-based tools

From our experience [12,39,43] and others, we can also provide recommendations regarding the implementation of computer-based tools supporting environmental management and sustainability with SMEs. Given that SMEs differ greatly from larger companies as well as from each other, blanket solutions will generally not work in SMEs. Furthermore, highly customized tools are a bad idea due to the fact that specialized knowledge is required and generally not available in SMEs. Based on our experiences and those found in the literature, the following general requirements for computer-based tools for improved environmental performance in SMEs can be postulated.

First and foremost, the tool should support decision-making for improved environmental performance. This implies that the tool:

- Allows decision-makers to target and prioritize environmental and operational improvements;
- Allows managers to compare environmentally benign product and process alternatives, including "what if" scenario analysis;
- Provides managers with a range of metrics to measure impact of their decisions, allowing them to select the right ones for their project/company;
- Supports and/or automates environmental reporting needs for compliance and/or EMS; and
- Has the ability to perform integrated analysis of environmental and financial impacts.

Second, given the lack of resources in SMEs, the tool needs to be affordable and simple. In other words, the tool:

- Requires little or no outside expertise (consultants) to install or maintain;
- Must not be overly complicated — it must meet the needs of the problem with minimum information intensity or complexity;
- Requires minimal implementation time; and
- Is inexpensive *up-front* as well as has a short payback;

Third, to avoid "re-creating the wheel", the tool should leverage existing systems where possible. The ideal tool:

- Is based on established environmental accounting techniques/methods;
- Is based on a widely used, familiar information technology platform;
- Is compatible with existing information technology systems and platforms; and
- Supports data from management and engineering estimates, operational logs, and existing accounting systems, whether the information management is formal or informal.

In addition to the design of the actual tool, there are some strong suggestions from the literature on increasing the chance of implementation within an SME.

- A senior manager must champion the entire implementation initiative as project manager.

- Upper management support must exist prior to and during the implementation effort.
- The tool should be implemented using either a phased or pilot implementation scheme. In certain cases, when extremely mature systems with plenty of support and proven benefits are available, a more aggressive implementation can be pursued.

Ideally, a sector approach should be used so that many firms within an industrial sector can use the tool. This means identifying the common elements within a sector and enabling the user to assemble the ones needed. Also, to achieve another level of economy of scale, the tool may be generalized at the conceptual level for use by firms across multiple industrial sectors. This implies that the structure of the tool needs to be repeatable across industrial sectors. Developing fundamental building blocks (such as unit processes, material, energy, and dollar flows, metrics, etc.) for the tool structure will enable this.

Despite all good intentions, tools, and recommendations, some company owners and SME personnel can (still) be very averse to changes in (environmental) management and/or operating practices due to perceived risk and added burden. One strategy to overcome such (initial) skepticism and pilot sustainability in a company is to focus on capital investment decisions — and their potential for environmental savings — rather than "*just*" operating practices. This approach seems to bear more fruit because a major capital investment requires significant attention and focus of upper management and provides a unique window for introducing and cascading process and even organizational changes through the company (see, e.g. [43]).

7. In closing

In summary, SMEs form an important sector of the global economy that has a major impact on the environment. Improved environmental management within SMEs is required, but major issues exists that affect the adoption and implementation of environmentally conscious tools and practices in SMEs.

The vast majority of impediments that affect an SME's ability to adopt innovations stems from the fact that they suffer from a severe lack of resources (manpower, money, expertise, information, and time), which in turn makes them apprehensive toward adopting new innovations. The implication is that environmental considerations are not a priority for an SME that is struggling to meet its financial goals and obligations.

In addition, SMEs are rather unique. They are very different from larger companies in that they typically exhibit slower technical development, lower level of awareness, heavy dependency on top managers, short-term and ad hoc decision-making processes, fewer external knowledge sources, greater dependency on external agents, etc. But they are more flexible and responsive to changes than larger firms. This makes it inappropriate to assume that a SME is "*a little big company*," and researchers caution against attempting to translate findings and solutions that may have been effective in large companies, to smaller firms.

SMEs vary tremendously among themselves as well. Their concerns and issues vary by industry sector, maturity, geographic location, and position on supply chain. Blanket solutions will generally not work, and that custom solutions may be needed for each SME on a case-by-case basis.

Furthermore, SMEs tend to consider environmental aspects a delicate and confidential matter. While they like to have good relations with authorities, they are concerned about negative reaction from local community and patrons. This is why SMEs are anxious about publishing such information through an environmental statement. Many SMEs think it wiser not to have any formal environmental policy in place at all. This disposition and attitude can be a significant deterrent in promoting sustainability within SMEs.

Channels of communication have an influence over how successfully an innovation is diffused in an SME. Use of pre-existing networks through established contacts is preferred in dealing with SMEs. Thus, typically cold-calling an SME will not be fruitful, and researchers must seek out ways of delivering their message through established (personal) contacts.

Moreover, any message to SMEs about sustainability or otherwise must present a solution in a manner that is simple, clear, highlights economic benefits, presents easy-to-do actions, and indicates where to go for further assistance. A solution needs to be inexpensive, sensitive to the limitations of SMEs, locally based, user-friendly, and flexible.

It should be clear that achieving sustainability in SMEs is critical to achieving sustainability in general and, given their challenges, significantly more resources and attention may have to be directed toward SMEs to accomplish this.

Acknowledgments

We gratefully acknowledge the support from the National Science Foundation (grant no. DMI-0086762) and Georgia Tech's Manufacturing Research Center. We would like to acknowledge Melissa Bargmann, Chris Robb, and Sharad Rathnam (our students), P.J. Newcomb and Colin Kiefert at the Georgia Pollution Prevention Assistance Division (P2AD), our colleagues Carol Carmichael, Leon McGinnis, and Chen Zhou for their valuable input, and last but not least Chris Moore for volunteering his time for project management.

References

[1] E. Dans, IT Investment in Small and Medium Enterprises: Paradoxically Productive? Electron. J. Inf. Systems Eval. 4 (1) (2001).

[2] R. Hillary, Small and Medium-Sized Enterprises and the Environment: Business Imperatives Sheffield, United Kingdom, Green Leaf Publishing, 2000.

[3] J. Palmer, C. France, Informing Smaller Organizations about Environmental Management: An Assessment of Government Schemes, J. Environ. Plann. Manage. 41 (3) (1998).

[4] D.T. Allen, D. Shonnard, P. Anastas, Green Engineering: Environmentally Conscious Design of Chemical Processes, Prentice-Hall, Englewood Cliffs, NJ, 2000.

[5] W.J. Box, Sustainability is IT, Pollut. Eng. 1 (2002) 13–17.

[6] S. Girshick, R. Shah, S. Waage, Information Technology and Sustainability: Enabling the Future, The Natural Step Working Paper Series, San Francisco, 2002.

[7] T.G. Gutowski, C.F. Murphy, D.T. Allen, D.J. Bauer, B. Bras, T.S. Piwonka, P.S. Sheng, J.W. Sutherland, D.L. Thurston, E.E. Wolff, Environmentally Benign Manufacturing, International Technology Research Institute, World Technology (WTEC) Division, Baltimore, MD, 2001.

[8] D.T. Allen, D.J. Bauer, B. Bras, T.G. Gutowski, C.F. Murphy, T.S. Piwonka, P.S. Sheng, J.W. Sutherland, D.L. Thurston, E.E. Wolff, Environmentally Benign Manufacturing: Trends in Europe, Japan and the USA, ASME J. Manufacturing Sci. 124 (4) (2002) 908–920.

[9] D. Hitchens, J. Clausen, K. Fichter, International Environmental Management Benchmarks: Best Practices from America, Japan and Europe, Berlin, New York, Springer, 1999.

[10] M. Wagner, The Relationship between the Environmental and Economic Performance of Firms: What does Theory propose and What does Empirical Evidence tell us? Paper for the Sustainability, Technological Innovation and Competitiveness of the Firm 2nd POSTI meeting, 2000.

[11] A. Harrer, Strategy for the Dissemination of Good Environmental Practices to SMEs in the People's Republic of China, the Republic of Korea, Malaysia, Philippines and Thailand, Institute of Sustainable Techniques and Systems, University of Technology Graz, Austria, 2001.

[12] M. Bargmann, Development of a Tool to Support Environmental Management within Small and Medium Sized Enterprises, MS Thesis, George W. Woodruff School of Mechanical Engineering, Georgia Institute of Technology, Atlanta, GA, 2002.

[13] G. Gable, G. Stewart, SAP R/3 Implementation Issues for Small to Medium Enterprises. Information Systems Management Research Center, Queensland University of Technology, Australia, 2000.

[14] C. Hall, Entrepreneurs, E Commerce, and SMEs in APEC. in Pacific Economic Cooperation Council, Session 5 PECC XIV, Hong Kong, 2001.

[15] J. Welsh, J. White, A Small Business is not a Little Big Business, Harvard Business Review 59 (4) (1991) 18–32.

[16] S. Blili, L. Raymond, Information Technology: Threats and Opportunities for Small and Medium-Sized Enterprises, Int. J. Inf. Manage. 13 (1993) 439–448.

[17] J.Y.L. Thong, Resource Constraints and Information System Implementation in Singaporean Small Businesses, Int. J. Manage. Sci. 29 (2001) 143–156.

[18] L. Raymond, Information Systems in Small Business: Are they Used in Managerial Decisions, Am. J. Small Bus. 6 (4) (1982) 20–26.

[19] G. Lee, D. Bennett, I. Oakes, Technological and Organizational Change in Small-to-Medium-Sized Manufacturing Companies: A Learning Organization Perspective, Int. J. Prod. Manage. 20 (5) (2000) 549–572.

[20] H. Rantanen, Internal Obstacles Restraining Productivity Improvement in Small Finnish Industrial Enterprises, Int. J. Prod. Econ. 69 (2001) 85–91.

[21] NRC, Learning to Change: Opportunities to Improve the Performance of Smaller Manufacturers, Washington, DC, National Research Council, National Academy Press, 1993.

[22] W.H. DeLone, Determinants of Success for Computer Usage in Small Business, MIS Quarterly 12 (1) (1988) 51–61.

[23] M. Magnusson, E-Commerce in Small Businesses: Focusing on Adoption and Implementation, in: 1st Nordic Workshop on Electronic Commerce, Halmstad, Sweden, 2001.

[24] P. Attewell, Technology Diffusion and Organizational Learning: The Case of Business Computing, Org. Sci. 3 (1) (1992) 1–19.

[25] P. Cragg, M. King, Information Systems Sophistication and Financial Performance of Small Engineering Firms, Eur. J. Inf. Systems 1 (6) (1992) 417–426.

[26] P. Kueng, A. Meier, T. Wettstein, Computer-based Performance Measurement in SMEs: Is there any option? Institute of Informatics, International Working Paper, 2000, 00–11.

[27] M. Levy, P. Powell, B. Galliers, Assessing Information Systems Strategy Development Frameworks in SMEs, Inf. Manage. 36 (5) (1999) 247–261.

[28] M. Levy, P. Powell, P. Yetton, SMEs and the Gains from IS: From Cost Reduction to Value Added, in: Information Systems: Current Issues and Future Changes, IFIP, Helsinki, Finland, 1998.

[29] R. Chapman, T. Sloan, Large Firms Versus Small Firms — Do they Implement CI in the Same Way?, The TQM Mag 11 (2) (1999) 105–110.

[30] K. Soderquist, J. Chanaron, J. Motwani, Managing Innovation in French Small and Medium-sized Enterprises: An Empirical Study, Benchmarking Quality Manage. Technol. 4 (4) (1997).

[31] A. Kagan, K. Lau, K. Nusgart, Information System Usage Within Small Business Firms, Entrepreneurship Theory Pract 2 (1990) 25–37.

[32] J. Tavcar, J. Duhovnik, Typical Models of Product Data Integration in Small and Medium Companies, Int. J. Adv. Manuf. Technol. 16 (2000) 748–758.

[33] R.B. Pojasek, How Do You Measure Environmental Performance?, Environmental Qual. Manage. 10 (4) (2001) 79–88.

[34] V. Biondi, M. Frey, F. Iraldo, Environ. Management Systems and SMEs: Motivations, Opportunities and Barriers Related to EMAS and ISO14001 Implementation, TGMI, 29 (2000) 55–69.

[35] T. Lindsey, Key Factors for Promoting P2 Technology Adoption, Pollut. Prev. Rev. 10 (1) (2000) 1–12.

[36] P.J.G. Stapleton, M.A. Glover, S.P. Davis, Environmental Management Systems: An Implementation Guide for Small- and Medium-Sized Organizations, NSF International, Ann Arbor, MI, 2001.

[37] S. Perkins, Improving Your Competitive Position: Strategic -and Financial Assessment of Pollution Prevention Investments — Training Manual, Third Edition, The Northeast Waste Management Officials' Association, Boston, MA, 1998.

[38] NEETF, Standardizing Excellence: Working with Smaller Businesses to Implement Environmental Management Systems, Green Business Network and The National Environmental Education & Training Foundation, Washington, DC, 2001.

[39] C. Robb, An Approach for Systematically Developing Environmental Assessment Information for Small to Medium Enterprises, MS Thesis, George W. Woodruff School of Mechanical Engineering, Georgia Institute of Technology, Atlanta, GA, 2002.

[40] J. Pringle, K.J. Leuteritz, M. Fitzgerald, S. Butner, R. Gupta, B. Kelley, N. Roy, ISO 1400 Workgroup White Paper — ISO 14001: A Discussion of Implications for Pollution Prevention, National Pollution Prevention Roundtable, 1998.

[41] C. Van Hemel, EcoDesign Empirically Explored — Design for Environment in Dutch Small and Medium Sized Enterprises, Ph.D. Dissertation, Design for Sustainability Research Programme, Delft University of Technology, Delft, The Netherlands, 1998.

[42] C.A. Branson, S.P. Davis, Environmental Management Systems: A Guide for Metal Finishers, NSF International, Ann Arbor, MI, 1998.

[43] S. Rathnam, Designing an Environmentally Conscious Decision Support Tool for Capital Investments in Small and Medium Enterprises, MS Thesis, George W. Woodruff School of Mechanical Engineering, Georgia Institute of Technology, Atlanta, GA, 2003.

Sustainability Science and Engineering: Defining principles
Martin A. Abraham (Editor)
DOI 10.1016/S1871-2711(05)01023-8

Chapter 23

Sustainable Design Engineering and Science: Selected Challenges and Case Studies

S.J. Skerlos[a], W.R. Morrow[a], J.J. Michalek[b]

[a]*Environmental and Sustainable Technologies Laboratory (EAST), Department of Mechanical Engineering, The University of Michigan at Ann Arbor, Ann Arbor, MI, USA*
[b]*Design Decisions Laboratory, Department of Mechanical Engineering, Carnegie Mellon University, Pittsburgh, PA, USA*

1. Introduction

As an instrument of sustainable development, *sustainable design* intends to conceive of products, processes, and services that meet the needs of society while striking a balance between economic and environmental interests [1]. By definition, the benefits of sustainable design are publicly shared, and to achieve them individual designers must place their decisions into a context larger than any single company, and even larger than the society or generation within which the design functions. It is therefore difficult to define sustainable design in an operational sense, and thus sustainable design is easy to ignore, especially in the fast paced and competitive process of bringing design artifacts to market. Complicating sustainable design further is the fact that environmental impacts depend on the consequences of specific stressors, rather than on which product or process causes the stressor (e.g. the atmosphere is indifferent to a kg of CO_2 saved by changing the design of a refrigerator versus changing the design of a television). Owing to these characteristics, sustainable design requires consistent and well-coordinated implementation to be achieved in a meaningful way.

Given the challenge of coordinating the complex trade-offs between economic, societal, and environmental factors influenced by design, it can be expected that governments interested in operationalizing sustainable development will begin to directly legislate the feasible space of options available to designers. This has been the approach in the EU, where the last few years alone have seen the proliferation of Directives on Waste Electric and Electronic Equipment (WEEE) [2], Restrictions on Hazardous Substances (RoHS) [3], and End of Life Vehicles (ELVs) [4]. For instance, according to RoHS, the new electrical and electronic equipment cannot contain lead, mercury, or cadmium after July 1, 2006, *except* for listed applications (e.g. leaded glass in CRTs) where substitution via design changes or materials is technically or scientifically impracticable, or where their substitution would cause environmental, health, and/or consumer safety impacts larger than their use [3]. Such regulations attempt to level the competitive playing field for environmental improvement, and to reduce the need for companies to make subjective and isolated judgments regarding the sustainability of design decisions.

While prescriptive environmental directives such as RoHS and WEEE intend to simplify sustainable design, they do not necessarily achieve its objectives. For example, eliminating a toxic substance from a product, such as mercury from fluorescent lamps, might lead to greater use of incandescent lamps that consume more energy, which on balance could have a negative impact on the environment [4]. In industrial cleaning machines, reduced use of detergents might typically lead to increased water temperature and hence higher energy consumption, which on balance could have a negative impact on the environment [4]. In the design of fuel cell vehicles, selecting materials on the basis of recyclability could ultimately lead to vehicles of larger mass, and consequently increased emissions associated with hydrogen production, which on balance could have a negative impact on the environment [5].

The need to coherently resolve such trade-offs among environmental attributes, and between environmental attributes and product performance, provides the rationale for the European Commission's (EC) recent proposal for a framework Directive to set *eco-design* requirements for energy-using products. Eco-design is the focus as it is estimated that over 80% of all product-related environmental impacts are determined during the product design stage [4]. With government entities now targeting the design process, it is becoming imperative that companies and their designers understand the environmental and economic implications of their design options. Moreover, the impending consideration of such eco-design legislation will require companies to become actively engaged in the broader development of environmental product policy, not as a matter of environmental altruism, but as a matter of maintaining competitive position.

Against this backdrop, it is an interesting and perhaps ironic observation to note that those who apply knowledge of science toward fulfilling society's needs through technological invention and selection (e.g. engineers and designers) rarely have a quantitative understanding about society's preferences, business decisions, economics, and the environmental impact of technological decisions. In other words, it is rare for engineers and designers to have the ability to systematically address the trade-offs inherent to sustainability. Unfortunately, this is more than just an educational shortcoming. At present, there is a clear need for a comprehensive body of knowledge and quantitative approaches that integrate engineering, economic, societal, and environmental science models toward a holistic definition of sustainable design.

For the purpose of this text, we define *design as a creative decision-making process that aims to find an optimal balance of trade-offs in the production of an artifact that best satisfies customer and other stakeholder preferences.* The artifact can be a product, manufacturing process, or service, with typical trade-offs including those between performance characteristics (e.g. light weight versus high strength), manufacturing capability, cost, safety, time-to-market, degree of customization, and the often contradictory preferences of different stakeholders. In our view, *sustainable design* only adds specific focus to design: *design, with particular attention paid to life-cycle trade-offs between functional performance, economic success, and the establishment of healthy social and environmental systems.* In other words, sustainable design is a consideration of the balance between public and private interests in the course of satisfying customer and other direct stakeholder interests.

In this chapter, we focus on the following challenges to sustainable design:

1. Understanding Incentives and Inhibitors to Sustainable Design (Section 2)
2. Establishing Targets, Metrics, and Strategies for Sustainable Design (Section 3)
3. Accounting for Variability in Product-User Interactions (Sections 3 and 4)
4. Evaluating Alternative Technologies for Sustainability Characteristics (Section 4)
5. Estimating the Market Value of Sustainable Design Attributes (Section 5)
6. Developing Market-Conscious Policies to Encourage Sustainable Design (Section 5).

Figure 1 serves as a framework for organizing these challenges in a manner that suggests a flow of abstract societal values regarding sustainability into products and services with economic, environmental, and societal consequences. Various influences are listed in one of the many possible progressions from values to artifact, including the designer's perceptions of technical and environmental alternatives (Challenges 2–4) and the implementation of societal values as regulatory and market variables (Challenges 5 and 6). Influenced by the designer's perceptions, and against the backdrop of current market conditions, the company will optimize its design decisions and set them into action, thus affecting the balance of factors in the sustainability triangle.

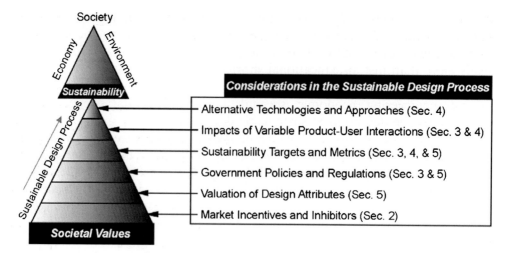

Fig. 1. Framework for conceptualizing sustainable design challenges described in this chapter.

In this chapter, we begin by providing an overview of business incentives and inhibitors to sustainable design (Section 2). This is followed by a brief review of sustainable design processes and metrics (Section 3). In these introductory sections, we focus primarily on environmental aspects of sustainable design, although issues of corporate social responsibility and trade-offs between societal and environmental variables are mentioned. The introductory sections are followed by two case studies, which highlight specific trade-offs that arise in sustainable design applications. The first case study (Section 4) provides an overview of economic, environmental, and societal aspects of mobile telephone production, use, and remanufacturing. The second case study (Section 5) provides a quantitative methodology for the evaluation of sustainable design policies related to automotive fuel efficiency. The two case studies are starkly different in approach. While the first takes a high-level and empirical view of existing mobile phone design and remanufacturing activities, the second takes a mathematical approach toward modeling the impacts of environmental policy options on engineering design. By presenting both case studies in this chapter, we contrast the strengths and weaknesses of these approaches as they apply to sustainable design.

2. Selected incentives and inhibitors to sustainable design

2.1. Incentives for sustainable design

In the ideal situation, sustainable design decisions would spontaneously self-assemble in the marketplace. For this to happen, sustainable design would need to create more business value than could be captured by designs not considered sustainable. But how can sustainable design add value for companies? Here we define three categories of value created by sustainable design: *adding positive value*, *eliminating negative value*, and *creating negative value for competitor firms*. Each of these categories is discussed below.

2.1.1. Adding positive market value

Inspiring innovation. Sustainable design need not be considered an additional constraint for producers, especially if the sustainability perspective can encourage the designer to search a previously unexplored region of the design space, leading to a breakthrough design. Examples of environmentally inspired breakthrough innovations include hybrid power train systems for automobiles, novel production facilities and methods (e.g. [6]), and advanced renewable electricity generation systems (e.g. [7]).

Increasing market share or consumer willingness to pay. According to [8], only about 15% of US consumers will consistently pay more (up to approximately 22% more) for products perceived as being environmentally friendly. These customers tend to exist in niche markets, such as the organic food market, which has recently been growing by 25% per year in the US [9]. Similar examples are currently difficult to find in North America.

Development of new markets for environmentally conscious products. This route to capturing environmental market value is exemplified by the discipline of industrial ecology [10], where resource cycling is investigated with the aim of converting waste from one product or process into an input for another industrial activity. This simultaneously creates market opportunities while addressing significant environmental problems. Toward this end, economically successful examples of recycling and remanufacturing are on the rise. In fact, one report has estimated that the US remanufacturing industry exceeds $53 billion per year in annual revenue and employs almost a half million individuals spanning 46 major product categories [11]. However, due care must be taken in evaluating the environmental characteristics of reused or remanufactured products, since such products need not be environmentally superior to manufacturing new products (see Section 4).

2.1.2. Removing negative market value

Reducing production costs. The pollution prevention literature is replete with examples describing how the redesign of manufacturing processes has inspired simultaneous reductions in production costs and pollution. Some of the most common examples exist in the Green Chemistry literature, where large cost savings in chemical and pharmaceutical manufacturing have been observed [12]. As one example, Dow Chemical claims to have reduced emissions of targeted substances by 43% and the amount of targeted wastes by 37%, primarily through green chemistry innovations. In this case alone, a one-time investment of $3.1 million is now saving the company $5.4 million per year [13]. Other profitable pollution prevention examples come from diverse areas such as membrane filtration recycling of industrial fluids [14,15], novel metal finishing technologies [16,17], and alternative integrated circuit production methods [18,19].

Minimizing regulatory losses and avoiding litigation. Pollution prevention investments by US companies are small relative to investments made toward compliance with EPA regulations, which amounted to 2.1% of GDP in 1990 (approx. $241 billion in 2003 dollars) [20]. While it has been estimated that $1 invested in complying with EPA regulations returns $10 to $100 in terms of ecological and health benefits [21], it is widely accepted that the current US regulations fail to address pressing sustainable design issues such as excessive resource consumption (e.g. petroleum), the proliferation of toxics in the environment (e.g. the disposal of electronic waste), and the accumulation of greenhouse gases (e.g. CO_2) in the atmosphere. With respect to each of these issues, the US is lagging in sustainable design policy drivers relative to Europe and Japan, both of which have been more progressive in eco-design-oriented legislation.

Minimizing damage to public image. Since the development of the Toxic Release Inventory, public reporting of environmental emissions has driven many companies to reduce the amount of pollution they produce. Moreover, companies such as Exxon, Union Carbide, and Nike learned the hard way that public image related to environmental and corporate social responsibility (CSR) issues can directly affect profitability. Now such issues are a key component of public image management for large companies across a wide range of industries ranging from oil and chemical production, to consumer electronics, to the automotive industry [8]. In fact, the need for accountability and visibility with respect to CSR issues has been an influential driving force behind corporate backing for initiatives such as the United Nations Global Compact program [22].

2.1.3. Increasing negative market value for competitors
Strategic utilization of legislation for competitive advantage. Sustainable design can create negative value for competitor organizations when it facilitates the development of government policies that favor organizations in a relatively strong sustainable design position. For example, at the time of debate over the Montreal Protocol, DuPont and ICI were major producers of ozone-destroying chlorofluorocarbons (CFCs) and held patents on costly CFC substitutes. While initially resistant, DuPont and ICI eventually supported the Montreal Protocol, which served to increase the value of the companies' proprietary technologies [23]. For similar reasons, it has been occasionally observed that larger companies, with a greater capacity to manage sustainability issues, are more

supportive of stringent health and environmental protection than smaller and/ or environmentally weaker companies.

Strategic utilization of product attributes for competitive advantage. Changing the system of societal valuation by altering consumer perception and education regarding the sustainability attributes of products can create opportunities for profit. For instance, between January 2003 and January 2004, US sales of the Toyota Prius increased by 82% as consumers became more comfortable with the technology. Toyota not only profited from the increased sales, but also from the sales of hybrid technology patent rights to Ford and Nissan [24,25]. More generally, this concept is beginning to take hold as indicated by growing attention being paid to programs, such as the Eco-Label program in the EU [26] and the Swedish Environmental Products Declaration program [27], that are predicated on the notion that eco-friendly attributes can be used strategically by corporations to gain competitive advantage.

2.2. Inhibitors to sustainable design

Numerous factors can serve to overcome the incentives listed above, precluding the manifestation of sustainable design. While some barriers are technological, many of the greatest challenges are products of the economic system itself [28]. Perhaps most importantly, sustainable design characteristically requires one firm or entity to pay its costs, while the benefits are widely shared. Since the designer's traditional stakeholders receive only a small fraction, and in some cases none, of the benefits of sustainable design, deciding who and how much to pay for sustainable design is a complex endeavor.

For example, private preferences that individual US consumers have for larger vehicle size and faster acceleration are well captured in the market, while public preferences that the same individuals may have for greater environmental protection, human health, and sustainability are not as easily captured. Since any individual is both a private player in the market and a member of society, inherent conflicts of interest exist that must be resolved in a fair and equitable manner. Incorporation of public value in the marketplace is usually achieved by direct incentives or regulations imposed by elected government officials (e.g. through tightened corporate average fuel economy (CAFE) standards), with some government policies being more economically efficient than others (see Section 5). Naturally, such decisions extend beyond trade-offs between environmental protection and performance into issues of vehicle safety, production cost, dependence on foreign oil, and consumer preference. A committee of the National Academy of Sciences recently concluded that such trade-offs can only rightly reside with elected officials, and that the trade-offs themselves are inherently difficult to quantify [29].

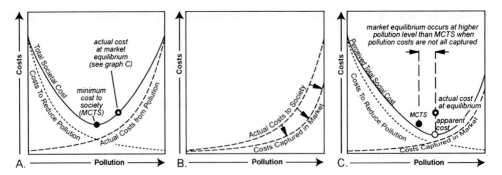

Fig. 2. (a) Pollution level at minimum total cost to society; (b) Undervaluation of pollution costs to society; (c) Resultant level of pollution observed at equilibrium when pollution costs are undervalued.

The example of automobile costs and benefits also makes clear that at points of optimum economic efficiency, where total costs to society are minimized [28], some pollution and resource consumption still exists. As suggested by Fig. 2a, the benefits associated with sustainable design (e.g. reduced pollution) have diminishing returns and increasing costs, such that when the point of minimum total social costs is reached, dollars invested in pollution prevention are best spent in other arenas where the marginal "benefit" to the environment (as valued by society) exceeds the marginal costs to society. While quantifying the costs of sustainable design is relatively straight forward, it is extremely difficult to quantify the benefits. The undervaluation of benefits skews the optimum point in Fig. 2 toward excess pollution. Quantifying the benefits of sustainability has been a growing topic of interest in the field of natural resource and environmental economics. While much progress has been made toward this end in the field of contingent valuation and behavior methods, the limitations of the methods are also now well established [28]. Moreover, even if the benefits could be quantified precisely, the fact remains that while individuals pay the costs to achieve sustainable design, they only receive a small fraction of the benefit [30].

Beyond trade-offs between public and private value, a number of other inhibitors to sustainable design are inherent to the US economic system. Such inhibitors that have been discussed in the literature include: technology and infrastructure cycle times that are either too fast (e.g. electronic equipment) or too slow (e.g. manufacturing facilities) [15,31], emphasis on short-term profits driven by quarterly reporting cycles [32], financial structures biased against prevention-based investments [33], difficulties valuing non-financial assets [28], financial discounting [34], and lost opportunity costs related to sustainability investments [21]. While these are significant inhibitors to sustainable design, they are not insurmountable. As recent EU directives are demonstrating, barriers to sustainable design can be removed through government actions requiring businesses to adhere to design targets.

3. Targets, metrics, and strategies for sustainable design

The basic challenge of sustainable design can be summarized by the old business management adage: "if you do not measure it, you do not manage it." Ultimately, governments bear the bulk of responsibility for managing sustainable design, and recent EU directives on RoHS, WEEE, and ELVs are a reflection of this responsibility. The EU approach to the management of sustainable design is conceptually similar to Fig. 3.

Figure 3 is an idealized approach to establishing quantitative sustainable design targets. The approach has a scientific component in that life cycle impact assessment is utilized for quantifying environmental impact magnitudes and uncertainties as inputs to a political decision-making process. Government then facilitates a discussion among stakeholders, industry, and the general public toward establishing sustainability objectives for society as a whole. It is these sustainability objectives for society that are to be met by establishing tangible design targets for specific products. Partitioning society's overall objectives into sustainability targets for specific product categories is a total cost minimization problem (Fig. 2), which must account for performance, societal, economic, and environmental aspects of the products to be regulated. In the words of the

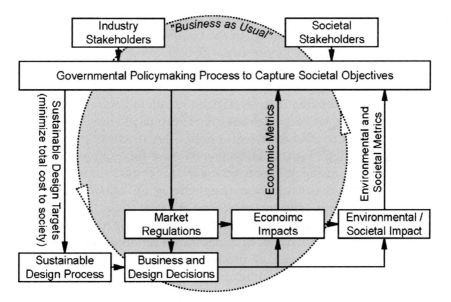

Fig. 3. Overview of high-level considerations in the development of sustainable design targets.

European Commission [4], simultaneous consideration of these factors is needed to assure that proposed sustainability targets do not result in "unacceptable loss of performance or utilities to customers". Once established, a competitive environment must be created where companies can pursue these sustainable design targets without fear of economic loss, as discussed in Section 5.

3.1. Targets, metrics, and processes for sustainable design

After product targets for sustainability are developed, specialized tools are needed at the product design level to predict the environmental stressor profile associated with different design options and to compare them with established targets. Such tools are particularly important since designers would suffer in their work if taxed by the need to generate stressor profiles from scratch for each design option. Existing sustainable design tools are used for the following purposes: (1) to create awareness about potential environmental impacts and possible mitigating design strategies (e.g. checklists, guidelines, and case studies), (2) to provide the ability to rank or score the environmental performance of a product with respect to a limited number of environmental aspects (e.g. toolboxes or advisor software tools), or (3) to perform a life cycle assessment (LCA).

The 2003 Sandestin Conference on Green Engineering, in addition to several other initiatives, has led to the development of useful principles that serve as a starting point for sustainable design [35,36]. From here, experience-based checklists and guidelines are often developed by companies, in most cases pointing out what not to do or suggesting how sustainability principles can be specifically utilized in a given application. Sustainable design guidelines and checklists are currently in widespread use throughout the consumer electronics, appliance, and automotive sectors of the economy (e.g. [37–39]). Some examples of guideline-based and case study resources tailored to specific life cycle stages include: material selection (e.g. [9,40]), assembly and disassembly (e.g. [41–43]), packaging and transport (e.g. [44]), recycling (e.g. [41,45]), and remanufacturing (e.g. [46–48]).

With the large number of guidelines found in typical checklists, it is almost certain that they will conflict, either with each other or with other performance attributes of the design. Typical conflicts may arise for example between mass and recyclability (e.g. using polymers versus metals in automotive applications), reusability and energy consumption (e.g. reusing an old refrigerator versus producing a new energy-efficient one), and between toxic chemical use and energy consumption (e.g. using mercury-containing compact fluorescent lamps versus incandescent lamps). Without a significant amount of experience or investigation, and in the absence of product-specific sustainable design targets established by government, it is difficult to know, which guideline is the most applicable to the current situation? For instance, it has been suggested that the EU directive on ELVs is currently biasing design options away from high-strength, low-weight composite materials, even though this may not be optimal from the life cycle design perspective.

To resolve conflicts between different sustainable design guidelines and to support innovation, a number of application-specific software tools have been developed. For instance, Motorola has developed a Green Design Advisor that stores information regarding component recyclability along with disassembly information in order to calculate the maximum degree to which products can be recycled [49]. A similar, but more general End-of-Life Design Advisor has also been developed at Stanford University [50]. Such software tools are now widely reported in the consumer electronics sector, where further developments have extended beyond end-of-life (EoL) considerations into the assessment of product and process materials toxicity and energy intensity (e.g. [51]).

Application-specific software tools such as these have both the advantage and disadvantage of requiring less information than a full life cycle assessment (for details on LCA methodology, see [52]). These tools allow design options to be quickly ranked and have demonstrated the ability to inspire respectable eco-design solutions [53]. On the other hand, they tend to lack the transparency of full LCAs and do not normally capture the environmental characteristics of the supply chain, which can be rather significant (e.g. in the case of integrated circuits). Application-specific software tools are also unlikely to account for situational factors in production, use, and disposal.

LCA-based methods are generally considered to provide the most comprehensive and reliable product evaluations, although they are intended for existing activities and are therefore difficult to use in the creative design process. Since a properly conducted LCA can take several months to perform and cost tens of thousands of dollars even for a relatively simple product, a number of software tools have been developed that contain representative environmental emission and resource consumption quantities for typical engineering materials and manufacturing processes. Some of the most commonly used tools in the design of consumer products include EDIP LCV [54], Umberto [55], Simapro [56], TEAM [57], and GaBi [58]. These software packages generally contain three components: (1) open frameworks for life cycle inventory development, (2) a database of representative materials and process inventories, and (3) impact assessment frameworks for comparing design options.

While the inventory methods and data presentations are fairly similar across existing software packages, the impact assessment methodologies can vary significantly. As an example of these differences, the Eco-Indicator 99 (hierarchist) and EDIP methods were compared in the production of vegetable- versus petroleum-based metalworking fluids (MWFs) [59]. Resource consumption and emissions associated with the production of both MWFs (2000 kg each) were assembled into an inventory, which is provided in Fig. 4 in terms of aggregated equivalent inventory categories. Figure 4 shows that the vegetable-based MWF

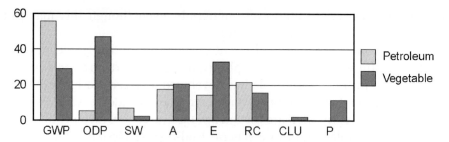

Fig. 4. Comparison of life cycle inventories for the production of 2000 kg of vegetable versus petroleum metalworking fluid. GWP: Global Warming Potential (10 kg CO_2); ODP: Ozone Depletion Potential (mg CFC11); A: Acidification (kg SO_2); E: Energy Consumption (gigajoules); SW: Solid Waste (kg); RC: Resource Consumption (100 kg); CLU: Cultivated Land Use (1000 m2); P: Pesticides (g).

is superior in some categories, while the petroleum-based MWF is superior in others. Since the goal of impact analysis is to resolve such differences, Fig. 5 shows the conversion of the inventory into single score environmental impact results using the Eco-Indicator 99 and EDIP assessment methodologies. According to the Eco-Indicator 99 methodology, the bio-based MWF is superior to the petroleum-based MWF, resulting in a score 60% lower, while the EDIP analysis indicates that the petroleum-based MWF is superior, with a score 57% lower. Several categories comprise the key differences in the single score results from these two methods. In the EDIP analysis, pesticides used in the production of the vegetable-based MWF account for a significant portion of the final score due to their chronic and acute toxicity in water. It is the weighting of pesticide impacts (relative to the weighting of petroleum consumption) that shifts the final outcome from favoring bio-based MWFs to favoring petroleum-based MWFs when using the EDIP methodology. Utilization of impact scoring methods is therefore inconclusive in this application, and a decision based on any single scoring metric taken in isolation will only serve to propagate the assumptions used for characterization, normalization, and valuation in that method.

Such issues of interpretation, situationality, and appropriateness associated with environmental impact metrics complicate their use in design applications and run counter to their intention to allow the designer to utilize such metrics comfortably without developing expertize in environmental science. Such complications have also led to a provision in ISO 14042, which discourages the use of weighted impact scores for comparative assertions [60]. Therefore, there is a growing interest in utilizing the results of life cycle inventory data more directly in sustainable design activities. For instance, the Swedish Environmental Management Council has promoted the development and distribution of standardized Environmental Product Declarations (EPDs) [27]. The EPD approach establishes product-specific requirements for selected product groups, as

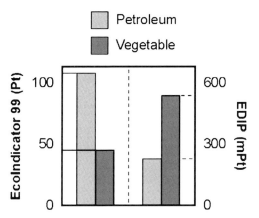

Fig. 5. A comparison of life cycle impact scoring results for the production of 2000 kg of vegetable versus petroleum metalworking fluid using the EDIP and Eco-Indicator 99 methodologies.

well as harmonized rules for LCA data collection, calculation, and presentation of the results. The EPD metrics are typically expressed within equivalent emission categories, similar to Fig. 4. EPD metrics typically include greenhouse gas emissions, ozone depletion potential, acid rain forming potential, etc. Taking such product declarations one step further, the EPA has suggested displaying such metrics in the form of a "nutrition label" (Fig. 6), which provides a familiar aesthetic for consumers [61,62]. Figure 6 also illustrates how such an environmental inventory database could be used during design to evaluate evolving product concepts.

With respect to the establishment of quantitative sustainable design targets, the proposed Energy using Products (EuP) framework similarly distinguishes between actual product *environmental impacts* (e.g. climate change, forest degradation due to acid rain, ozone depletion, eutrophication, etc.) and product *environmental aspects*, which are stressors leading to those impacts (e.g. emissions of greenhouse gases, emissions of acid substances, emissions of substances disturbing the oxygen balance, emission of substances affecting stratospheric ozone, etc.) [4]. The proposal, which intends to harmonize environmental regulation impacting the eco-design of energy using products across the EU, has stated a strong preference for the regulation of environmental aspects rather than impacts. This is because the environmental aspects are more easily measured and controlled by the producer through design (whereas impacts depend on additional factors such as locality, time, and user choices), they can be measured consistently, and they are more transparent in interpretation. Also, for small and medium enterprises with fewer resources, the prediction of environmental impacts may not be feasible, while the measurement of environmental aspects is relatively straightforward.

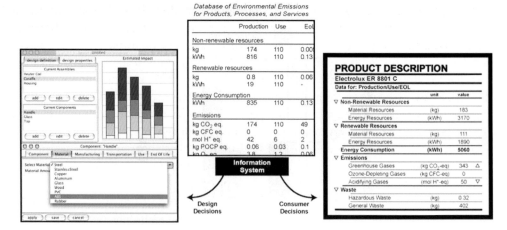

Fig. 6. Conceptual use of environmental inventory databases to support design and consumer decisions.

3.2. Research opportunities related to establishing sustainable design targets

The goal of setting product-level targets and metrics for sustainable design through a process such as Fig. 3 presents a number of opportunities for quantitative research, especially in the areas of industrial ecology, LCA, economic impact analysis, and product performance modeling. With respect to LCA, inventory and impact profiles for different product categories are required, including their supply chains. For example, the inventory profile of the supply chain is particularly important for the case of integrated circuits that are utilized in consumer electronics [63]; however little product-specific information is available in the public domain regarding their environmental profile, as discussed in Section 4.

With respect to environmental impact assessment, it can be assumed that the *selection* of product-level environmental aspects to be targeted by EuP will be based on life cycle impact analyses performed across product sectors, as suggested by Fig. 3. Ongoing research towards establishing cause-and-effect relationships between environmental stressors and impacts, including geographic, temporal, and statistical uncertainty information, will be particularly helpful in establishing these targets. Such issues have recently been raised in the context of developing the TRACI life cycle impact assessment method [64]. Quantitatively modeling the relationship between eco-design options, performance, cost, environmental emissions, and resource consumption is an issue for engineering research. This begins with the quantitative prediction of environmental emissions and resource consumption as a function of design variables. As a simple example, consider the case of modeling the electricity consumption of a refrigerator/freezer. Design variables include the volume of the refrigerator/freezer, its configuration (e.g. side-by-side, top over bottom, etc.), insulation type and thickness, compressor characteristics, evaporator/condenser characteristics, etc. Using basic heat transfer equations, material data (e.g. for insulation), and a limited number of calibration experiments to estimate the efficiency of heat rejection systems, it is relatively straightforward to predict the steady-state electricity consumption of different design options as they would be reported (for instance) on the EnergyGuide label utilized in the US. A modeling approach is therefore useful for sustainable design, as it can permit the calculation of eco-efficiency (e.g. cost per unit of environmental emission reduction) associated with different design options.

Typically, the ability to model steady-state or standardized operational performance (e.g. EnergyGuide ratings) is sufficient for comparison of the relative environmental impacts of two designs. However, there are instances where the ability to model subtle and/or dynamic behavior of the product is also useful in the sustainable design process. For instance, the electricity consumption of a refrigerator/freezer may actually be up to 30% higher than predicted from the EnergyGuide label due to factors in use that would not be captured from steady-state engineering models [65]. In the case of the refrigerator/freezer, losses associated with opening and closing the door usually account for 5–10% of the total life cycle energy consumption of the refrigerator [65]. The ability to predict the effectiveness of design measures intending to reduce losses from the door opening and shutting would require models and assumptions related to the convective replacement of cold air in the refrigerator with warm, humid air from the kitchen, using non-steady-state calculations. While such advanced design modeling intending to reduce a 5–10% loss may not seem worthwhile, depending on the product, such design efforts can reduce the overall environmental impact of the industry significantly. In fact, reducing energy consumption of all US refrigerators by just 1% would save approximately $140 million dollars in energy costs and 1.5 million tons of carbon released to the atmosphere each year [66]. This carbon savings is greater than the emissions from most nations on the African continent [67]. The point is that modeling subtle, second-order impacts of design decisions on environmental performance can be important for products with a relatively large environmental impact, and that are in widespread use.

The refrigerator case also demonstrates that while basic engineering modeling can be useful to reveal the *relative* environmental impact of one product versus another, rather advanced modeling may be needed to reveal the *absolute* impact of eco-design changes on the environment. These absolute impacts may have particular relevance to policymaking. For example, the EPA is currently considering for the first time in two decades changing the way it reports standard automotive fuel efficiency to better reflect real-world performance [68]. In filing a petition to the EPA, the Bluewater Network (San Francisco, CA) argued that real-world gas mileage could be up to 1/3 lower than calculated using EPA's current test methods, even though these EPA estimates are already adjusted downward 22% for highway and 10% for city. They believe that "more accurate estimates of fuel economy would benefit both consumers and those involved in setting national energy policy" [69]. In short, while current fuel economy estimates provided by EPA are useful in selecting one vehicle over another on a relative scale, they may understate the actual magnitude of fuel consumption, and by consequence, they may also understate the benefits of sustainable design strategies for the automobile.

As discussed above, quantitative modeling of technological performance and emissions can allow design options to be compared with sustainable design targets at the product concept level. Once this capability is achieved, it is necessary for sustainable design to be seamlessly integrated into traditional design processes. For example, research in [70] describes the integration of environmental variables and targets into an engineering design process, using a quality function deployment approach. Within such a framework, it becomes possible to evaluate trade-offs between cost, functional performance, and environmental emissions. While it has been shown that such trade-offs can be established on a quantitative basis, a quantitative prediction of how sustainable design attributes might impact market performance is much harder to achieve in the analysis, and has not traditionally been considered as part of the design process.

Section 5 describes how mathematical models of consumer preference can be utilized within a decision-making framework to understand the relationship between sustainable design attributes and market performance. As a lead-up to this theoretical treatment, the next section describes empirical observations of sustainability attributes for the case of mobile phone production, use, and remanufacturing. The case study emphasizes the complexity associated with simultaneously balancing the economic, environmental, and societal implications of technological decisions, as well as the challenge of developing metrics for sustainable design.

4. Case study: sustainability characteristics of mobile phones[1]

In 2002, Original Equipment Manufacturers (OEMs) sold over 420 million mobile telephones worldwide [71]. By 2005, it has been estimated that the number of discarded mobile phones will grow to more than 500 million [72],

---

[1]The results described in this section are derived from research conducted between the Technical University Berlin (TUB) and The University of Michigan (UM). The TUB participants included Professor Guenther Seliger, Ph.D. Candidates Bahadir Basdere and Marco Zettl, and M.S. graduate Aviroot Prasitnarit. The UM participants included Professor Steven J. Skerlos, Ph.D. pre-candidate W. Ross Morrow, and M.S. graduate Aaron Hula.

providing the stockpile necessary for the continued acceleration of mobile phone reuse and remanufacturing (or "re-marketing") activites. Currently, third party re-marketers of mobile phones are making significant profits from reselling mobile phones in emerging markets. *But is remanufacturing of mobile phones consistent with the goals of sustainable design?*

This section describes the synthesis of empirical research related to the economic, environmental, and societal aspects of mobile phone production, use, reuse, and remanufacturing. The research on economic and societal aspects is largely literature based, while also including significant input from personal communications with parties currently engaged in remarketing mobile phones. The research on environmental aspects is largely LCA based, drawing from direct observation of mobile phone production and remanufacturing activities, as well as the literature.

4.1. Economic characteristics of mobile phone reuse and remanufacturing

Mobile phone reuse and remanufacturing is currently economically attractive for various reasons. First, mobile phones in advanced markets are not purely technological objects but trendy or stylistic objects, leading to rapid disposal rates and a large supply pool of functionally reusable phones. Currently, only third-party "remarketers" are involved with the reuse-oriented treatment of obsolete phones, serving a market estimated to represent less than 1% of the annual OEM market share [74,75]. A scan of clearinghouse websites such as Ebay also indicates a large but informal activity in discarded mobile phone resale. According to a major mobile phone re-marketer in the US, 2003 sales of discarded mobile phones were expected to reach 4 million. For that company, processing of discarded mobile phones follows Fig. 7, with about a 90/10 distribution between direct phone resale and remanufacturing operations for the over 300 phone models claimed to be profitable to resell.

At first, handset OEMs may consider third party remarketing as a threat to their market share. However, taking into account that the majority of remarketed handset users are first-time customers, originally not able to purchase mobile telephony, but tending to change to new handsets later on, market shares could be expected to increase in medium-term. In fact it has been shown in [76] that flourishing second-hand sales can lead to accelerated sales of virgin product. Obviously, remanufacturing conducted by handset OEMs themselves carries the potential for increased process efficiency relative to the operations of third parties, due to reduced technical and logistical barriers. Especially for the European market, where WEEE makes handset OEMs responsible for takeback and EoL treatment of phones by 2006, reuse and remanufacturing with OEM participation would have economic and technological advantages.

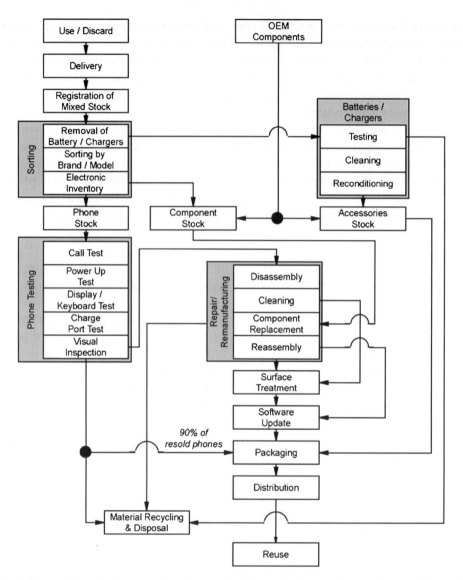

Fig. 7. Remanufacturing process flow diagram for the organizational re-marketing of discarded phones.

Driving the growth of remanufacturing operations is an increasing demand for mobile communication, especially in emerging markets (EMs). Despite their low purchasing power, sales of mobile phones (both new and reused) are growing rapidly. At present, the majority of remarketed phones are distributed to EMs in Africa and South America. Distributing these mobile phones in developed markets (DMs) would offer some potential for profits also, although the bulk of sales currently exist in EMs. Interestingly, it has been found that the attractiveness of the second-hand mobile phone market is rising in DMs, especially in European countries such as Germany. Supported by recent changes in legislation, such as the new warranty law, which grants customers a 1- to 2-year warranty for used products purchased, there is impending competition of remarketed mobile phones with new ones, creating real, albeit slight, competition for OEM market share.

4.2. Environmental characteristics of mobile phone production and reuse

It is widely known that mobile phones have a potentially hazardous EoL profile: land filled or incinerated mobile phones create the potential for environmental release of heavy metals or halocarbon materials from batteries, printed wiring boards (PWBs), liquid crystal displays, plastic housings, wiring, etc. Over the past few years, OEMs have been particularly active in pursuing environmental improvements, which has resulted in a number of life cycle investigations related to mobile phones. For example, the Ericsson 2001 Sustainability Report claims that mobile phone production accounted for 10% of CO_2 releases for the company that year [77]. In [78], an LCA of the Phillips Fizz and Genie phones suggests that the manufacturing stage accounts for 77–79% of the phones' life cycle environmental impact, as assessed using the Eco-Indicator 95 method.

Useful cross comparisons of publicly reported mobile phone LCAs such as these are not possible, not only due to different reporting units, but also due to the use of differing LCA boundary scopes, inventory categories, or use of aggregated impact metrics. For instance, while the Ericsson life cycle assessment included overhead activities such as travel and commuting, they did not include integrated circuit (IC) manufacturing [77]. IC manufacturing was explicitly included in the Philips study [78]. For mobile phones, such scope variations can be of particular importance, especially with respect to the inclusion of IC components. This is due to the large quantity of energy and emissions necessary to produce ICs [63], as well as the number of ICs utilized per phone (which can exceed 40).

4.2.1. Emissions inventory: mobile phone production
Prasitnarit (2003) describes an LCA of a mobile phone with the scope definition shown in Fig. 8. Total emissions and energy use over material acquisition, manufacturing, use, and EoL stages were estimated using a mixture of database information (primarily for material acquisition and EoL phases) and direct process measurement (primarily for manufacturing and use phases). The life cycle inventory of over 400 materials was included in the material acquisition phase. Only a small number of low concentration metals and chemicals in the phone were not included [79]. In the investigation, the manufacturing process energy and emissions were directly measured. This included manufacturing of

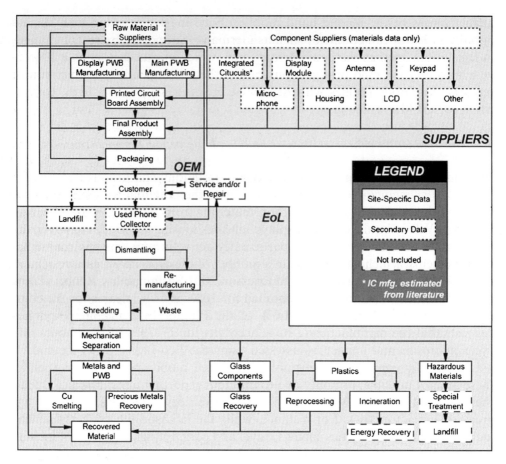

Fig. 8. Boundaries considered in mobile phone life cycle inventory of [79].

the display PWB, the main PWB, chip shooting and placing, reflow, screen printing, assembly, and testing. IC-related energy and emissions were not directly measured, but were taken from the literature.

The results of the investigation showed that mobile phone production accounts for almost all of the non-energy-related emissions in the life cycle. It was also found that the ICs, display module, and main PWB accounted for nearly three-quarters of the energy consumed in the production phase. Not including IC manufacturing, the production stage itself consumed approximately 250 MJ of energy, which was over two times the amount of energy consumed by the normal use of the mobile phone estimated over 2 years.

4.2.2. Emissions inventory: mobile phone use and remanufacturing

Use Phase. In [80], a model was developed to help understand the effect of mobile phone user habits on energy consumption. The model considered efficiency losses during charging, as well as in call and standby power consumption. Table 1 lists the three representative use scenarios that were considered. As observed in Fig. 9, significant variation in energy consumption (expressed as CO_2 emissions in different electricity grid situations) arises due to variation in user behavior. Although this variation is large, even in the worst case the use phase energy consumption per year is below 20% of the energy consumed during phone production (without including ICs). Further, a "typical" charger profile (e.g. Profile #2) over 1 year has the same energy consumption as the production of only six of the "typical" ICs investigated in [73]. For reference, the phone considered in [78] had 40 ICs.

Remanufacturing and redistribution. Apart from a relatively small quantity of emissions from cleaning operations and packaging, emissions from remanufacturing processes and distribution are almost entirely associated with energy consumption associated with the use of electricity. In the remanufacturing operations listed in Fig. 7, the top three energy consuming activities are: sorting (driving a conveyor belt), battery testing and reconditioning (e.g. using a Cadex® C7000 series analyzer), and software updating (standard PC usage). The total amount of process energy consumption per remanufactured phone has

Table 1
Profiles of charger use as modeled in [80]

Profile	Charger Use Behavior
1	Charger always left in wall socket
2	Charger left in the socket while phone is charging overnight; removed during day
3	Charger is left in socket only for amount of time needed to recharge the battery

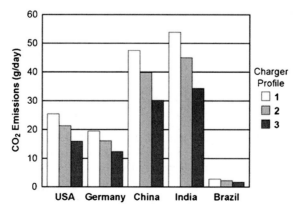

Fig. 9. CO_2 emissions per year for reused phone with different charging and electricity grid profiles.

been estimated to be between 0.8 and 1.6 MJ, with the variation almost completely dependent on the method and amount of battery testing and reconditioning, as over 90% of energy consumption during remanufacturing occurs in the testing, charging, and reconditioning of batteries [31].

After remanufacturing, shipping the restored mobile phones to emerging markets is typically accomplished by air transport owing to large distances (ranging from 5000 to 13,000 km), relatively small volumes, and the urgency of transactions (due to volatility in the second-hand market). Especially for remanufacturing, this air transportation represents a dominant energy consumption and emissions activity. For example, the estimates presented in Table 2 are based on an assumed mobile phone mass of 100 g, and a CO_2 release of 1110 g/ton*km for air travel according to [81]. It is seen that distribution to EMs can release an amount of CO_2 that is at least an order of magnitude higher than the remanufacturing process, but still insignificant relative to the production phase.

Once a mobile phone is resold, the environmental impact of its "second life" is likely to be greater than the impact of its "first life". In EMs such as those in South America, Central Asia, and Africa, there is not typically an infrastructure to properly handle toxic battery and circuit board materials that remain after the phones are discarded. Moreover, since remanufactured batteries generally hold less charge than new batteries (80–100% of original capacity), higher energy consumption per unit service will occur in the second life. The associated environmental emissions may be compounded further where power generation and distribution systems of EMs are relatively inefficient and/or more dependent on polluting energy-generation technologies than in DMs.

Consideration of electricity grid technology leads to Fig. 9, which highlights such situational factors among different use-profiles of remanufactured mobile phones (transmission line losses not included in the analysis). For instance, Fig. 9 shows that India and China are likely to have among the highest CO_2 emissions for remanufactured mobile phones on a per day basis. Brazil, on the other hand, has the lowest proportion of CO_2 emissions since most of its electricity is generated from hydroelectric sources. For Brazil, reduced environmental impacts due to less CO_2 release are traded off against the environmental impacts associated with the use of large amounts of hydroelectric power.

Table 2
CO_2 release for air transportation between New York and target markets overseas [12]

EM	Distance	CO_2 Released	Percent of Remfg.	Equivalent Use Duration
	(km)	(g)	(%)	(days)
Bombay	12,536	1400	1140	31
Rio de Janeiro	7757	900	740	360

Another situational issue to be considered with the diffusion of remanufactured mobile phones to EMs is the heightened pressure that this creates for base stations and a supply chain for both auxiliary and replacement components. Compared to a remanufactured mobile phone sold in a market closer to purchase saturation, a mobile phone sold in an EM would create a disproportionately higher demand for new base stations and supply chains. In other words, a mobile phone in a DM generally creates less "infrastructure demand" than one in an EM.

4.2.3. Emissions inventory: summary

Fig. 10a illustrates a summary inventory profile for the mobile phone of [79], highlighting relative contributions of each life cycle stage. Fig. 10b compares the production, use, and EoL energy consumption for two of these phones under the following three scenarios: (1) both phones are manufactured and disposed at landfill without recycling or remanufacturing, (2) both phones are manufactured and completely recycled (even though 100% recycling is neither economically nor technically feasible), and (3) one phone is manufactured as new, and the other identical phone is restored as new from a discarded phone of the same model. Perhaps unsurprisingly, it is evident from Fig. 10b that a remanufacturing pathway has by far the least energy consumption. This is because the remanufacturing pathway, unlike recycling, avoids repeating manufacturing steps with characteristically high-energy consumption and environmental emissions.

While these results are encouraging for remanufacturing, it should be noted that reduced environmental impact is only achieved if the remanufactured phone replaces the production of a new one. However, the vast majority of

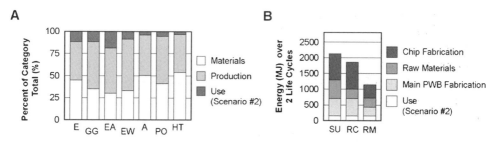

Fig. 10. (a) Relative contribution of life cycle stages for the mobile phone in [78]. E: energy consumption (MJ); GG: greenhouse gas emissions (kg CO_2 equivalent); EA: emissions to air (kg dichlorobenzene [DCB] equivalent); EW: emissions to water (kg DCB equivalent); A: total acidification potential (kg SO_2 equivalent); PO: total photochemical oxidant creation potential (kg ethane equivalent); HT: human toxicity potential (kg DCB equivalent). (b) Energy consumption comparison for two mobile phones under single use and disposal (SU), recycling (RC), and remanufacturing (RM) scenarios.

remanufactured mobile phone customers are first-time users in EMs, for whom the low cost of the remanufactured mobile phones serves as a conduit for entry into the market. Consequently, new use-phase, transportation, and EoL environmental impacts are created by remanufacturing where they did not exist before, adding to the overall environmental impact of the mobile phone industry. *The mobile phone example therefore highlights a disconnect between realizing the narrow goal of remanufacturing, and achieving the broader goal of lowering global environmental impact in the context of sustainability.* Currently, in the case of mobile phones, remanufacturing is creating new users and is increasing environmental impact without significantly reducing the production of new phones. Since the environmental impact of the cell phone industry is currently increasing due to the cell phone remanufacturing activity, it must be asked whether cell phone remanufacturing is consistent with the goals of sustainable design. For this reason, the societal dimension of cell phone remanufacturing is explored in Section 4.3.

4.3. Societal characteristics of mobile phone use in emerging markets

The importance of telephony as a requisite for economic development is well established. In fact, telephony has been described as a basic human need, which is implied by the fact that the function of the telephone (two-way conversation over distance) has not changed over the past 100 years [82]. In addition, it is widely recognized that modern economic development can only occur if there is a communications infrastructure to support it. Telecommunication is a critical part of a modern economy, along with a steady supply of electricity to power factories, good roads, rail systems, ports, and a steady financial system that can support the supply chain [83]. For this reason, and due to the close relationship between telecommunications, information systems, democracy, education, and job creation, the United Nations has placed a high priority on expanding communications systems within the poorest countries, such as those in Africa [84]. Although 80% of mobile phones are currently found in the more developed nations, the 1990s saw the number of subscribers in EMs grow faster than anywhere else [84]. The rapid expansion of mobile phone use in EMs is largely due to the fact that a mobile phone network can be up and running much more quickly and inexpensively than a fixed one.

4.3.1. Anecdotal evidence of mobile phone benefits in emerging markets

There exist a number of examples which highlight the role of mobile phones in improving lives for individuals living in EMs. For example, groups of small farmers in remote areas of Côte d'Ivoire share mobile phones so that they can follow hourly fluctuations in coffee and cocoa prices. This means that they can choose the moment to sell their crops when world prices are most advantageous to them. A few years ago, they could only have found out about market trends by applying to an office in the capital, Abidjan. Even then their deal making was based on information from buyers, which was not always reliable [85].

A study conducted by Bayes (2001) observed the effects of mobile phones on rural villages in Bangladesh [86]. Bayes' study found that the introduction of mobile phone services led to improved law enforcement, communication during natural disasters, and the ability to call doctors for health-related information. In addition, the phones helped families keep in touch with relatives living far away, strengthening family bonds. The study also described positive effects of mobile phones with respect to the empowerment of women, and suggested that the services from mobile phones can most greatly benefit poor members of the community. These examples, while not discussing the potential negative impacts of mobile phones on developing societies, provide some of the context and justification for their rapid diffusion into developing countries.

4.3.2. Quantitative metrics of mobile phone societal impacts

Although the incorporation of economic and environmental metrics with metrics for societal development has been recognized as a critical need in sustainability evaluation and life cycle assessment [87,88], quantifying the benefits of expanding mobile phone utilization in EMs remains difficult. Thus far, societal indicators have not been incorporated into decision-making frameworks because, even more so than environmental metrics, societal impact metrics are subjective, confounded with other causal variables, and situation-dependent [88]. However, it is also generally agreed that subjective indicators are needed in societal policymaking because objective indicators only provide part of the information needed to understand the decision context [89].

To quantify if and to what extent expanded mobile phone use might foster accelerated societal development, one can begin by analyzing the statistics of the United Nations Development Programme (UNDP). The annual UNDP Human Development Report provides measurements of various indicators of progress in specific categories under six main areas. Under the category of "Technology: Diffusion and Creation" there are estimates of the number of fixed telephone lines and mobile subscribers per 1000 people for numerous countries that can be cross-compared with the human development index (HDI) reported by the UNDP. Such an analysis shows that countries rated with a "high" HDI (e.g. Sweden, USA, Singapore) had an average of 556 fixed telephone lines and 487 mobile users per 1000 people, while countries with a "low" HDI (e.g. Nigeria, Ethiopia, Bangladesh) had an average of eight telephone lines and three mobile users per 1000 people. The effect of telephony on development is not explored in the report, but the correlation between telephone access and development can be clearly seen [90].

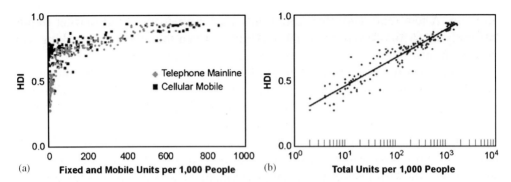

Fig. 11. (a) Human Development Index (HDI) vs. number of mobile/fixed line phones per 1000 people based on data from [90]. (b) Semi-log plot of HDI versus total teledensity based on [89].

A plot of HDI versus number of fixed and mobile phone users per 1000 people is shown in Fig. 11 [90]. A logarithmic-type relationship can be seen: lower teledensity exists in countries with lower HDIs, and higher teledensities in countries with higher HDIs. The slope of this curve decreases significantly as teledensity increases. Put simply, expanding phone access in less developed countries has a higher positive correlation with HDI than expanding phone access in more developed countries, which has little correlation with HDI.

4.3.3. Situational differences in the ethics of pollution

For cases such as cell telephone remanufacturing where net environmental impact is increasing, but where societal development benefits exist, it seems appropriate to factor-in the potential for increased HDI (or other similar-intending metrics) in the context evaluating sustainability. For instance, the global warming potential (GWP) associated with providing 50 MJ of electricity to power a mobile phone in an EM for a year might be compared with the equivalent GWP of an activity that might be less correlated with increasing HDI (e.g. watching a high-end, 190 W flat screen television for 73 h). Is one GWP emission more appropriate or acceptable than the other? If so, how can such ethical metrics be built into LCA frameworks? Moreover, what are the ethical implications of making discarded mobile phones available to countries not able to handle the waste, and who have not been offered technical assistance or financial aid in this regard?

As yet, the state of the art is unprepared to discuss such questions quantitatively in the context of sustainable engineering. Discussion of such issues has recently begun to appear in the literature (e.g. [91]), and will be important to consider with respect to decision-making for sustainable development. Correlative analysis between HDI and expanded use and consumption can help in such analyses, but naturally these analyses need to be accompanied by research aimed at understanding if such correlations between expanded access to telephony (e.g. through expansion of mobile phone use) and HDI are truly causal in nature.

In summary, with respect to the balance of sustainability factors, the recent growth of mobile phone reuse and remanufacturing is a case where the economic and societal benefits are positive, while the short-term environmental impact is negative due to increased energy consumption and the potential for toxics release at EoL associated with the currently observed flux of second-hand mobile phones toward developing countries. Although growing concern over the latter issue is being voiced in the press [72], it remains to be seen whether developed countries exporting discarded electronics will take steps to limit the EoL impact of electronic waste in developing countries. Even if such action were to be taken, a new discussion would begin regarding how to minimize toxics release from electronic waste at minimum cost to contributing governments and international organizations. In short, the question would turn to one of maximizing the "eco-efficiency" of the environmentally targeted intervention.

5. Case study: eco-efficiency, public policy, and vehicle design[2]

The core concept of eco-efficiency is to maximize the societal and environmental benefits of a design decision or policy, while minimizing its economic cost and negative impact on individual consumer preferences. The need for developing eco-efficient government policies arises when (1) the economic drivers described in Section 2 are not strong enough to achieve the self-assembly of environmentally conscious actions in the marketplace (e.g. environmentally conscious disposal of electronic waste), or (2) if a societal decision-making process such as described in Section 3 concludes that the environmental or societal consequences of a particular activity are too large to be considered acceptable (e.g. the excessive consumption of gasoline by automobiles). In this section, we consider two basic questions related to the eco-efficiency of government policies: (1) What are the impacts of environmentally conscious policy alternatives on engineering design and business decisions, and (2) how can the relative eco-efficiencies of sustainable design policies be quantified? To highlight how such questions can be addressed from a mathematical modeling perspective, we consider the case of automotive fuel economy and emissions.

---

[2]This section summarizes research conducted as part of the Antilium project (http://antilum.umich.edu) at The University of Michigan. The research was performed by Assistant Professor Jeremy Michalek and Professors Panos Y. Papalambros and Steven J. Skerlos. The research is described in detail in Michalek et al., 2004 [93].

In recent years, the environmental burden created by automotive emissions has been increasing in the US due to falling average fuel economy and an increase in total vehicle miles traveled [92]. Reversing this trend will require a balance between reducing vehicle emissions, meeting consumer mobility demands and preferences, and minimizing added vehicle costs (since alternatives to gasoline engines, such as diesel, hybrid, fuel cell, and electric systems are currently more expensive to manufacture than traditional gasoline systems). Government policies can provide incentives to bring these alternative choices into the market, but the problem of quantifying the impact of specific government policies on engineering design and business decisions is as yet not well studied.

In the development by Michalek et al. [93] that is summarized in this section, the paradigm of Fig. 12 is applied to quantify the impact of fuel economy and emission policies on design decisions of competing automotive companies. The links between engineering and business decisions, including models of cost and demand, are at the core of the investigation. Each of these considerations is represented by a separate analysis model, and their interactions are captured within an integrated design decision model. By performing a series of optimization routines with respect to the local perspective of each producer, one can explore the effects that vehicle emission policies have on consumers, manufacturers, and the design decisions that a particular policy encourages.

Section 5.1 provides a conceptual overview of a basic modeling paradigm, which intends to capture all of these factors within the context of a market simulation. Section 5.2 discusses specific mathematical models that were developed to analyze the impact of environmental policies on design decisions, with particular attention paid to the strengths and weaknesses of individual models used in the market simulation. Section 5.3 reviews the results of the modeling approach and provides a discussion regarding the calculation of eco-efficiency.

5.1. *Quantitative models of economic and environmental design characteristics*

Figure 12 illustrates the interplay of engineering design decisions, cost drivers, demand forces, and government policy in determining the environmental impact of a consumer product. Ultimately, the engineering design determines the overall cost and environmental characteristics of the product, as well as the extent to which consumers demand the product. In addition, business decisions such as price and production volume have a major influence on both the costs and revenue generated by a specific design. Government policy, as mentioned in Section 3, also plays a role in influencing design by restricting the feasible space of options and by changing producer cost structures through penalties and incentives.

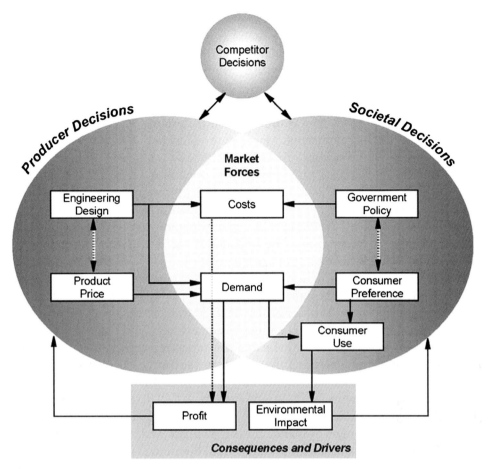

Fig. 12. Conceptual model of the interaction between producer, consumer, and public policy decisions in the market, along with their relationship to engineering design and environmental impact.

Consumer preference is the key driver for the revenue generated by a specific design, while also playing a major role in determining environmental impact through the way in which products are used. In the automobile case, variability in environmental impact caused by consumers arises mostly due to differences in fuel economy during highway versus city driving, as well as distances traveled. Auxiliary functions such as the use of air conditioning also have a substantial, but situational, influence on automotive fuel consumption and emissions.

Although Fig. 12 is intuitive and conceptually simple, the development of a meaningful model for the system at an appropriate level of detail is complex. For some of the sub-models, such as the calculation of manufacturing cost, the model forms can be simple, but necessary data to support them may be difficult to obtain. In other cases, such as the modeling of demand, commonly utilized model forms may be straightforward, yet lacking of a cause-and-effect meaning, particularly when based on observed choice data from the market. Moreover, preference for a specific product is not only a function of performance characteristics, which vary in relative importance to consumers over time, but also of price, which is perceived differently as a function of time-dependent economic conditions. Also, data for demand models as a function of the product characteristics relevant for a specific study can be difficult to find and expensive to collect, and due to the dynamics of consumer preference as a social phenomenon, such models are difficult to validate in the traditional scientific sense, since the market cannot be manipulated in the context of a controlled experiment. In the case of environmental and health impact, inventory data may be lacking, and the impacts of specific factors may be so confounded by other variables that casual relationships may require decades to establish.

Predicting vehicle performance characteristics such as acceleration, fuel economy, and emissions as a function of detailed vehicle design decisions is also challenging. Vehicle performance must be considered during sustainable design, as it is a key parameter in influencing consumer preference that, in turn, determines which products are bought and how they are used. While the complex chemistry of combustion can be modeled from fundamental scientific principles, such models are too complicated to be run over a full driving cycle and may not be able to capture heterogeneities in temperature, pressure, and airflow in the engine that have a major influence on the quantity and nature of vehicle emissions. Empirical measurements of automotive emissions are widely available, but are only useful to engineering design when put into a model context as a function of design variables.

The next section describes how the separate analysis models of Fig. 12 can be developed and integrated within the context of game theory toward an analysis of how government fuel economy policies impact engineering design decisions in a competitive market. We consider that each producing firm chooses engineering design decisions, production volume, and selling price for each vehicle in its product line with the aim of maximizing profit. We also consider that the engineering design decisions of each producer not only impact the decisions of competitors, but the decisions of competitors also impact the decisions of the producer. Furthermore, government penalties are imposed on specific vehicles proportional to the quantity of emissions produced or fuel economy attained. Individual consumers in the market choose among alternatives by maximizing benefit (utility) to themselves, considering their own preferences as captured in the model. By considering the self-interested decisions of producers and consumers in the market, based upon real observations of the marketplace, the potential success of environmentally conscious policymaking can be evaluated.

5.2. Overview of specific analysis models and optimization framework

5.2.1. Profit model

To start, each producer k decides on a set of products $\mathcal{J}_k$ to produce including design decisions, prices, and production volumes. Specification of the design topology τ_j (engine type, diesel or gasoline) and design variables x_j (engine size and final drive ratio) determine the product characteristics z_j (fuel consumption and acceleration) that are observed by the consumer. Vehicle topology τ_j, design variables x_j, and production volume V_j of each product in $\mathcal{J}_k$ together determine the total cost c_k to producer k. Consumers make purchasing choices among the set of all products $\mathcal{J} = \cup_k \mathcal{J}_k$ based on the product characteristics z_j and price p_j of each product, resulting in an overall demand q_j for each product j calculated by the demand model. Each producer k attempts to maximize its profit Π_k (defined as revenue minus cost) by making the best possible design, pricing, and production decisions according to Eq. 1:

$$\text{maximize } \Pi_k = \left(\sum_{j \in \mathcal{J}_k} q_j p_j \right) - c_k$$

$$\text{with respect to } \left\{ \tau_j, x_j, p_j \right\} \forall j \in \mathcal{J}_k \tag{1}$$

subject to engineering constraints.

5.2.2. Engineering performance model

In the Michalek et al. (2004) study, the scope is limited to the small vehicle market segment and the following variables: (1) engine type τ, either a gasoline or diesel, (2) engine size x_1, taken as a scaling of the baseline engine size ranging from 0.75 to 1.50, and (3) the final drive ratio x_2, taken in the range of 0.2–1.3. Each producer in the market selects the engine type, the engine size, and the final drive ratio on the basis of profit maximization. Using the engineering model ADVISOR [94], these producer decisions are mapped to product characteristics z, upon which consumer purchasing decisions are based. In this case, it is assumed that the relevant performance criteria z consist of the vehicle gas mileage z_1 (in mpg) and the time for the vehicle to accelerate from 0 to 60 mph z_2 (in seconds). It is also assumed that vehicles only differ by engine and transmission design: specifically, the default small car vehicle parameters in ADVISOR are used in all simulations (based on the 1994 Saturn SL1), and only the engine variables $\{\tau, x_1, x_2\}$ are changed. The EPA Federal Test Procedure

(FTP-75) driving cycle is used to compute the performance and fuel economy characteristics for all vehicle simulations.

5.2.3. Consumer demand model

The consumer demand model is based on discrete choice analysis (DCA), which presumes that consumers make purchasing decisions on the basis of the *utility* value of each product option. Utility u is measured in terms of an observable deterministic component v, which is taken to be a function of the product characteristics $\{z_1, z_2\}$, and a stochastic error component ε. The probability P_j of choosing a particular product j from the set $\mathscr{J}$ is calculated as the probability that product j has a higher utility value than all alternatives,

$$p_j = Pr(v_j + \varepsilon_j \geq v_{j'} + \varepsilon_{j'}; \forall j' \in \mathscr{J}). \tag{2}$$

Various probabilistic choice models follow the DCA approach, including the widely used logit [95] and probit [96] models. The logit model, which was originally developed by McFadden to study transportation choices, is utilized here and has been used extensively in the marketing literature. Only recently have logit models begun to be applied to engineering design problems [97].

The logit model assumes that the unobserved error component of utility ε is independently and identically distributed for each alternative, and that ε follows the double exponential distribution (i.e. $Pr[\varepsilon < \alpha] = \exp[-\exp(-\alpha)]$). Assuming the double exponential distribution for ε terms in Eq. (2), the probability P_j of choosing alternative j from the set $\mathscr{J}$ takes the form,

$$P_j = \frac{e^{v_j}}{\sum\limits_{j' \in \mathscr{J}} e^{v_{j'}}}, \tag{3}$$

where each utility function v_j depends on the characteristics $\mathbf{z}_j$ and the price p_j of design j. Given a functional form for $v_j(\mathbf{z}_j, p_j)$ and observed choice data, a model fitting procedure is performed to arrive at the parameters defining the function $v_j(\mathbf{z}_j, p_j)$. Given the empirical nature of $v_j(\mathbf{z}_j, p_j)$, the model must be developed and interpreted carefully.

In the Michalek et al. (2004) investigation, the utility model developed by Boyd and Mellman is used [98]. This model, originally developed using vehicle purchase data from 1977, was found to be the best logit model available in the public literature that included engineering design variables and an appropriate level of detail for the study. Although several other variables are included in the demand model (e.g. vehicle style, noise, and reliability), these variables are assumed equal across vehicles in the Michalek et al. (2004) investigation [93].

The utility equation developed by Boyd and Mellman is

$$v_j = \beta_1 p_j + \beta_2 \left(\frac{100}{z_{1j}} \right) + \beta \left(\frac{60}{z_{2j}} \right), \tag{4}$$

where $\beta_1 = -2.86 \times 10^{-4}$, $\beta_2 = -0.339$, $\beta_3 = 0.375$, p_j is the price of vehicle j, z_{1j} is the gas mileage of vehicle j in mpg, and z_{2j} is the 0–60 mph acceleration time of vehicle j in seconds [98].

In [92], Eq. (4) is applied to the small car sub-market (assumed population s to be 1.57 million people based on [99]), with the recognition that this could introduce error since the equation was developed based on the entire car market. Using the logit model, the demand q_j for product j is

$$q_j = sP_j = s \frac{e^{v_j}}{\sum\limits_{j' \in \mathscr{J}} e^{v_{j'}}}, \tag{5}$$

where v_j is defined by Eq. (4).

While the Boyd and Mellman demand model is adequate for a preliminary analysis, it does introduce several sources of error that highlight the need for additional research:

- The model is fit to purchase data from 1977–1978.
- The model utilizes purchase data only: Consumers who choose not to purchase vehicles were not studied. Thus, the model can only predict *which* vehicles consumers will purchase, not *whether* they will purchase. The size of the purchasing population is treated as fixed, independent of vehicle prices (i.e. there is no outside good).
- The model is an aggregate model, and therefore it does not account for different segments or consumer groups.
- The use of the logit model carries with it a property called *independence from irrelevant alternatives* (IIA), which implies that as one product's market share increases, the shares of all competitors are reduced in equal proportion [100]. For example, a model with the IIA property might predict that BMW competes as equally with Mercedes as with Chevrolet. In reality, different vehicles attract different kinds of consumers, and competition is not equal. In this investigation, predictive limitations of the IIA property were mitigated since the model is applied only to the small car market (a relatively homogeneous market) rather than to the entire spectrum of vehicles.

5.2.4. Cost model

The total cost to manufacture a vehicle c^P is considered to be the sum of two parts: the investment cost to set up the production line c^I and the variable cost per vehicle produced c^V. The variable cost is composed of the engine cost c^E and the cost to manufacture the rest of the vehicle c^B, so that $c^V = c^B + c^E$. The cost to manufacture q units of a vehicle with engine type τ and design variables $\mathbf{x}$ is then calculated as

$$c^P(\tau, \mathbf{x}) = c^I + qc^V(\tau, \mathbf{x}) = c^I + q(c^B + c^E(\tau, \mathbf{x})). \tag{6}$$

In Eq. (6), it is assumed that $c^B = \$7500$ and $c^I = \$550$ million per vehicle design. c^B is estimated based on data for the Ford Taurus [101], and c^I is based on an average of two new product lines described in the literature [102]. c^E is determined based on regression of established engine cost data for diesel (compression ignition) and gasoline (spark ignition) engines. Finally, the total cost to producer k is calculated as the sum of the production cost of each vehicle in k's product line and the regulatory cost c^R:

$$c_k = \left(\sum_{j \in \mathscr{I}_k} c_j^P \right) + c_k^R. \tag{7}$$

5.2.5. Environmental policy models

Three specific producer penalty scenarios are considered here: CAFE standards, a hypothetical use-phase CO_2 emission tax, and a hypothetical quota system for producing a minimum percentage of diesel vehicles. To start, the current CAFE standard for cars ($z_{CAFE} = 27.5$ mpg) is used, and two different penalty charges are explored. The first penalty charge is the current standard: $\rho = \$55$ per vehicle per mpg under the limit, and the second is a hypothetical double-penalty scenario. The total regulation cost c^R incurred by design j is therefore $\rho q_j (z_{CAFE} - z_{1j})$, where ρ is the penalty, q_j is the number of vehicles of type j that are sold, and z_{1j} is the fuel economy of vehicle j. In this investigation only a single market segment is utilized even though CAFE applies to all passenger vehicle markets in which the producer operates.

A CO_2 valuation study from the literature [103] is utilized to estimate the economic cost to society associated with environmental damage caused by the release of a ton of carbon dioxide in the use-phase. A CO_2 tax per vehicle sold is calculated as $\upsilon d\alpha_M / z_1$, where υ is the dollar valuation of a ton of CO_2, d the number of miles traveled in the vehicle's lifetime, α_M the number of tons of CO_2 produced by combusting a gallon of fuel, and z_1 the fuel economy of the vehicle. For this investigation, it is assumed that $d = 150{,}000$ miles, α_M is 9.94×10^{-3} tons CO_2 per gallon for gasoline or 9.21×10^{-3} tons CO_2 per gallon for diesel fuel, and υ is taken from [103] to range from $\$2/\text{ton}$ to $\$23/\text{ton}$, with a median estimation of $\$14/\text{ton}$.

A quota regulation was also modeled to force alternative fuel vehicles into the market, as was attempted for electric vehicles in the California market [104]. Here, the quota policy is to levy a large penalty cost for violation of a minimum diesel to gasoline engine ratio quota. For the quota case, the regulation cost is modeled as

$$c_k^R = \max(0, \ \rho(q_k^{SI} - (1 - \phi)(q_k^{SI} + q_k^{CI}))) \tag{8}$$

where ρ is the penalty per vehicle over quota (\$1000), ϕ the minimum diesel vehicle percentage required by the quota (here, $\phi = 0.40$), q^{SI}_k the total number of gasoline engines sold by producer k, and q^{CI}_k the total number of diesel engines sold by producer k.

5.2.6. Simulated oligopoly competition

Substituting Eqs. (2)–(8) into Eq. (1) yields the following profit objective for each producer:

$$\Pi_k = \left(\sum_{j \in \mathcal{I}_k} q_j p_j\right) - c_k = \left(\sum_{j \in \mathcal{I}_k} q_j(p_j - c_j^V) - c^I\right) - c_k^R. \tag{9}$$

To account for competition in the design of vehicles subject to government regulations, game theory is used to find the market (Nash) equilibrium among competing producers. In game theory, a set of actions is in Nash equilibrium if, for each producer $k = 1, 2, ..., K$, given the actions of its rivals, the producer cannot increase its own profit by choosing an action other than its equilibrium action [105]. It is assumed that this market equilibrium point can provide a reasonable prediction of which designs manufacturers are driven to produce under various regulation scenarios, even though Nash equilibrium does not model preemptive competitive strategies by producers. In order to find the Nash equilibrium point for a set of K producers, the decision variables of each producer k are optimized to maximize the profit of that producer Π_k while holding the decisions of all other producers constant. This process is then iterated, optimizing all producers $k = 1, 2, ..., K$ in sequence until convergence, yielding the Nash equilibrium for K producers, where K is set to the largest value that yields positive profit for the producers. Additional details can be found in [93].

5.3. Results and discussion

The results of the investigation are summarized in Table 3. For each regulation scenario, Table 3 lists the maximum number of producers K that yield a positive profit Nash equilibrium and the market share per producer. Owing to the use of an aggregate demand model, each producer makes the same decisions (i.e. produces the same designs) at market equilibrium, so Table 3 summarizes the decision variables, product characteristics, costs, and profits for a typical producer in each scenario.

It is found at equilibrium that each producer manufactures only a single design rather than a product line (except in the quota case) due to competition

Table 3
Model predictions yielded by market simulation under various policy scenarios

		None	CO$_2$ tax			CAFE	SCAFE	Quota	
			Low	Med.	High				
No. Producers (-)	K	10	10	10	10	10	10	5	
Market share (%)	q/s	10	10	10	10	10	10	11.9	8.1
Engine type (SI/CI)	M	SI	SI	SI	SI	SI	SI	SI	CI
Engine size (-)	$b_M x_I$	127.9	127.7	114.3	110.3	113.3	88.4	127.9	98.0
FD ratio (-)	x_2	1.28	1.28	1.28	1.27	1.28	1.29	1.28	0.88
Price ($)	p	12,886	13,031	13,719	14,259	13,058	12,772	13,372	16,083
Gas mileage (mpg)	z_I	20.2	20.3	21.8	22.4	22.0	25.5	20.2	29.8
Accel. Time (s)	z_2	7.46	7.46	7.93	8.10	7.97	9.29	7.46	7.84
Investment cost ($)[a]	c_I	550	550	550	550	550	550	550	550
Var. cost/vehicle ($)	c_2	9,001	8,999	8,878	8,844	8, 869	8,670	9,001	11,713
Reg. cost/vehicle ($)	c_R/q	0	147	956	1,530	304	217	0	0
Profit ($)[a]	Π	60.5	60.5	60.5	60.5	60.5	60.5	276	6.5

[a] In millions of dollars.

and the existence of substantial investment cost. This result may have been caused by factors such as neglecting the possibility of commonality among designs and the use of an aggregate model for demand that ignores consumer heterogeneity. Table 3 also indicates that producers accrue equal profits in all regulation scenarios (except the quota case), and all incurred costs are passed to the consumers at equilibrium. This is because the demand model assumes a fixed car buying population (there is no option not to buy) and does not consider the utility of outside goods.

The Michalek et al. (2004) study lists a number of important caveats that are useful to consider when using observation-based demand models as the basis for sustainable design analysis. For example, the demand model used here (Eq. (5)) predicts a preference for vehicles with faster acceleration. Therefore, a vehicle that dramatically sacrifices unmeasured characteristics such as maximum speed for a slight improvement of acceleration time will be preferred according to the model. However, in practice a consumer would observe the unmeasured limitations during a road test, especially if the limitations are extreme. To account for this issue, each optimum vehicle design was tested to ensure the vehicle's ability to follow the standard FTP driving cycle and achieve a speed of at least 110 mph on a flat road. All vehicle designs in the study passed this test [93]. The example highlights the importance of thoughtful modeling and of remaining cognizant of the limitations inherent to quantitative modeling approaches when simulating market competition.

5.3.1. Base case

The first case considered in [93] is the no regulation case ($c^R = 0$), which provides a baseline comparison for results obtained under different regulatory policies. In the absence of fuel economy and emissions regulations the model predicts 10 producers in the small car market. Each producer manufactures a single vehicle with the design variables, product characteristics, and costs shown in Table 3. The resulting vehicle has a spark ignition engine with a fuel economy of 20.2 miles per gallon.

5.3.2. CAFE

Table 3 shows that the CAFE standard succeeds in increasing resulting vehicle design fuel economy to 22.0 mpg with roughly a half-second increase in 0–60 acceleration time and a \$172 increase in vehicle price. The vehicle production cost drops by \$132 per vehicle relative to the baseline case due to the smaller engine size; however, regulation costs are approximately \$304 per vehicle. The CAFE standard is not attained at equilibrium because, unlike the real automobile market, the model does not capture intangible costs to companies who do not meet the CAFE standard. According to the model there is significant risk for a company that would attempt to produce a vehicle at 27.5 mpg, since its market share would be captured by more powerful, less fuel-efficient competitor vehicles. This is a direct consequence of Eq. (4), which shows that consumers receive more utility from improvements in acceleration than they do from improvements in fuel economy. Making matters worse, for a given powertrain technology there is a negative trade-off between acceleration and fuel economy. In fact, for the gasoline engine favored by producers in the modeling results, a regression between z_1 and z_2 through the optimal designs in Table 3 (an estimate of the Pareto surface) yields,

$$z_2 = az_1^2 + bz_1 + c, \tag{10}$$

where $a = 0.0159$, $b = -0.380$, and $c = 8.64$ ($R^2 = 0.99$).

Given that Eq. (10) expresses the relationship between fuel economy and acceleration, one can utilize Eq. (4) to calculate the change in vehicle price p that would be necessary to maintain constant utility to the consumer as the fuel economy z_1 is increased:

$$\frac{\partial p}{\partial z_1} = \frac{1}{\beta_1} \left(\frac{100\beta_2}{z_1^2} + \frac{60\beta_3(2az_1 + b)}{\left(az_1^2 + bz_1 + c\right)^2} \right). \tag{11}$$

Using the baseline engine from the no regulation scenario as the evaluation point for Eq. (11), it is observed that the producer must *lower* the asking price by \$136 per mpg increase in fuel economy to maintain equal utility to the consumer. This result provides a quantified expression of a trend currently observed the US: In many cases higher fuel economy actually brings with it reduced utility to individual consumers, a fact which is consistent with the observation that the average fuel economy of the US fleet is in decline. While a

Table 4
Sample List of desirable automobile features and their relationship to lower fuel efficiency

Desirable Feature for Consumers	Engineering Solution	Impact on Vehicle
Engine Performance		
Quiet engine compartment	Add padding and deadener	Adds weight, material
Strong engine performance	Robust design, larger engine size	Adds weight, material
Passing power	Robust mounts, larger engine	Adds weight, material
Ride, handling, and braking		
Quick, safe braking	Robust brake design	Adds materials
Quiet ride during highway driving	Add padding and deadener	Adds materials
Quiet ride over harsh bumps	Add padding, better shocks	Adds materials
Power steering with minimal effort	Always-on fluid pump	More power, fluids used
Comfort and convenience		
Front leg/foot room	Move engine, lengthen vehicle	Adds material
Headroom	Taller vehicle	Adds material
Side mirror controls	More electronics	Increased current draw
Well lit gauges and instruments	Add materials	Increased current draw
Ability to watch movies; internet	DVD player, WIFI, Bluetooth	Increased current draw
Navigation system	More electronics	Increased current draw
High-quality and powerful stereo	Powerful amplifier and speakers	Increased current draw
Heated/cooled seats	Add electronics and content	Increased current draw
Adjustable with controls	Add electronics and content	Increased current draw

vast number of attributes are not considered in Eq. (11), Table 4 suggests that many desirable attributes, like acceleration, are negatively correlated with fuel economy.

5.3.3. "Strict" CAFE
As shown in Table 3, a fuel economy of 25.5 mpg is achieved by the strict CAFE standard, with a consumer vehicle price $114 less than the baseline case. The 0–60 acceleration time, however, is approximately 1.8 s higher. Perhaps surprisingly, this "strict" CAFE policy *reduces* regulatory costs for each producer. The reduction in regulatory costs follows from the fact that under the previous CAFE model (remembering that unmodeled "reputation and image" costs associated with CAFE violations are not captured), it is profitable for manufacturers to violate CAFE and pay the penalty in order to increase market share by selling powerful vehicles. However, when CAFE penalties are increased substantially, producers are forced to meet the standard in order to stay in business. In this case there is little danger of losing significant market share to a competitor who sells more powerful engines because none of the producers can

afford to sell such engines; therefore all of the producers design smaller, less expensive engines. As such, the strict CAFE standard serves to remove risks associated with producing more fuel efficient vehicles by increasing the penalty for deviation from the CAFE standard 27.5 mpg. The desired eco-efficient result is achieved. Company profits are unaffected, vehicles are less expensive, and fuel economy is increased. However, as Table 4 indicates, consumers and society at large lose out on benefits associated with the engineering design characteristics that reduce fuel economy (e.g. acceleration). Here we see the trade-off between desirable attributes as perceived by individuals acting in the marketplace and individuals acting as members of society, with preferences for resource conservation, lower air pollution, and reduced life cycle carbon dioxide emissions, as expressed by their support for government regulations on fuel economy.

5.3.4. Use-phase CO_2 emissions tax

It is seen in Table 3 that the use-phase CO_2 emissions tax is a considerably less eco-efficient policy than the CAFE standards. As the tax increases, producers do tend to design smaller, more fuel-efficient engines. However, the low valuation penalty (\$2/ton) has little impact on fuel economy relative to the baseline case, with the only notable effect being added regulation costs that are passed on to consumers. The median valuation (\$14/ton) has a larger impact, increasing fuel economy by 1.5 mpg, while the high valuation (\$22/ton) adds only slight additional improvement in fuel economy relative to the median level (0.6 mpg) at a substantial added regulation cost per vehicle (\$540) relative to the median tax level. The results suggest not only that the use-phase CO_2 emissions tax is less eco-efficient than CAFE standards, but that it is also dangerous as a policy approach: In this policy, vehicle prices increase, performance is lower, and fuel economy is increased only marginally. Therefore it would appear that there are no major winners with this policy. Based on these results, the policy might only be expected in practice to lower the demand and sales of vehicles relative to other modes of transportation or market segments not subject to the tax (e.g. in the current CAFE standards, more lax standards exist for light trucks relative to automobiles).

5.3.5. Quota

Although diesel engines have higher fuel economy than gasoline engines for equivalent acceleration performance, the model predicts that they are only manufactured under the quota policy (which bears similarity with the current situation in the US small car market). Additionally, the model predicts that under the quota policy only the minimum number of diesel vehicles is produced to exactly meet the standard. This is due to the higher costs associated with producing diesel engines, and the greater profitability of gasoline engines prompting producers who are forced to sell diesel engines to also produce as many gasoline engines as allowed.

The diesel engines produced in this scenario have a fuel economy 9.5 mpg higher than the baseline (no regulation) scenario. This is achieved at a substantial increase in diesel vehicle price relative to the baseline spark ignition vehicle ($3197). However, only a small reduction in vehicle acceleration relative to the no regulation case is observed (0.38 s). On the other hand, for a 150,000 mile life of the vehicle, the 29.8 mpg diesel vehicle consumes about 2400 fewer gallons of fuel, which means that at fuel prices above $1.33 per gallon, the initial cost of the vehicle is recovered over its life (not accounting for time-based discounting). This suggests that the quota policy is a reasonably eco-efficient approach, albeit one that could not spontaneously self-assemble in today's market place. Consistent with these observations, Sullivan et al. (2004) has recently suggested that the increased adoption of diesel engines into US vehicles is worthy of consideration for its potential to economically reduce CO_2 emissions produced by the vehicle fleet [106].

5.3.6. Analysis of eco-efficiency for selected policies

Eco-efficiency in the context of policy evaluation implies that a re-evaluation of the balance between private and public value is occurring with the aim of simultaneously minimizing environmental impact and the costs to society necessary to achieve the reduction in environmental impact. To capture this trade-off, Fig. 13 attempts to illustrate the performance of the individual policies as a

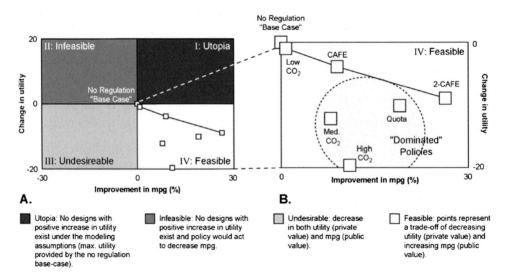

Fig. 13. Change in utility versus percentage change in fuel efficiency relative to baseline no-regulation case. (a) Definition of four quadrants in trade-off analysis. (b) Quadrant IV policies.

function of both their public and their private values. On the vertical axis, the change in utility relative to the no regulation case is given. Utility as calculated by Eq. 4 is utilized because it captures trade-offs between key attributes in a manner related to private value (although it does not express private value directly). On the horizontal axis, the change in fuel economy resulting from each policy relative to the no regulation case is given (expressed as percentage). This relative improvement in fuel economy is the core intent of the policy, although like utility, it is not a direct expression of value. Ideally, the public and private value of the specific policies would be estimated directly, in which case plots such as Fig. 13 could provide a direct measure of eco-efficiency that could be compared across applications as necessary for the systematic development of sustainability targets across industries (see Section 3).

In Fig. 13b, the trade-off between public and private value is expressed by the set of non-dominated policies, which are the policies for which no alternative policies achieve higher fuel economy without sacrificing utility or vise versa. Given the definition of the origin as the reference no regulation case, from which deviations in utility and fuel economy are defined, it is evident that policies in Quadrants II and III, which result in inferior fuel economy, would not be of interest to policymakers. Also, Quadrants I and II contain no feasible designs because the base no regulation (free market) case results in the design with highest feasible (private) utility. In other words, consumers asked to choose their most preferred design while paying only the cost to manufacture that design will choose the same design as that produced by the unregulated market. Thus, viewing the set of non-dominated policies as a (Pareto) trade-off curve between private and public preferences, we see that the unregulated free market yields an extreme point on this trade-off curve such that private preferences (utility) are valued exclusively over public preferences (mpg improvement), while non-dominated policy alternatives allow exploration of the best alternative policies as modeled in the investigation. It can be seen in Fig. 13b that the Pareto set of non-dominated policies include the no-regulation case, the low CO_2 taxation case (which is much the same as no-regulation case), the CAFE standard, and the Strict CAFE standard. Depending on the interpretation of "acceptable utility loss to consumers" (recall Section 3), one of these policies would be best under the modeling assumptions and scope of the model.

5.4. Remarks on policy-driven eco-design and eco-efficiency analysis

This section has described an optimization framework to analyze the impact of fuel economy regulations on the design decisions made by automobile manufacturers from an eco-efficiency perspective. It was observed that government policies are necessary to provide incentives for producers to design alternative fuel vehicles (e.g. diesels) that cost more to produce. Without a regulatory standard, producers cannot afford to make smaller, less expensive, and more fuel-efficient engines. Under the modeling assumptions, it is observed that some policies can result in cost savings for all parties (e.g. CAFE) and do not affect profitability within the market segment. On the other hand, certain regulations can also lead to higher costs, diminishing returns, and little environmental improvement (e.g. CO_2 tax).

Such results indicate that the cost-benefit characteristics of policy alternatives can be modeled in a realistic and quantitative way, and that a holistic integration of costs, performance, consumer preference, and competition can facilitate the selection of effective policies, while helping to determine how policy parameters should be set. Additional investigations that combine engineering, marketing, and policy models with models of changing consumer preferences and driving habits could be used to predict trends regarding the diffusion of alternative fuel vehicles into society, possibly avoiding costly investments in products that are unlikely to achieve wide acceptance and helping to focus resources and incentives toward sustainable design solutions that will make the most impact.

6. Summary and conclusions

The very need for sustainable design research implies that an imbalance exists between private value captured in the marketplace and its consequential impacts on societal and environmental systems. It also implies an imbalance between the incentives and inhibitors to sustainable design outlined in this chapter (Section 2). Specifically, the chapter has discussed six challenges to sustainable design on which academic research, performed in conjunction with industrial partners and governments, could have a major impact:

1 Understanding incentives and inhibitors to sustainable design
2 Establishing targets, metrics, and strategies for sustainable design
3 Understanding variability in product-user interactions
4 Evaluating alternative technologies for sustainability characteristics
5 Estimating the market value of sustainable design attributes
6 Developing market-conscious policies to encourage sustainable design

When sustainable designs do not spontaneously self-assemble in the marketplace, corrective action is generally left as the responsibility of governments, who can either (1) create market conditions that allow sustainable designs to proliferate on their own, (2) restrict the feasible space of options available to designers, or (3) establish tangible sustainable design targets with which individual products must comply. Each of these approaches requires a profound understanding of the relationship between engineering design options, market costs and revenues, available alternative technologies, and societal and environmental impacts. Research needs in these areas have been outlined in Section 3.

Once sustainable design is defined in the context of a specific product, it is necessary to develop tools that facilitate the seamless incorporation of sustainability metrics into the design process. The development of appropriate metrics, and the establishment of trade-offs with other cost and performance aspects of the design were underlying themes of the case studies presented in this chapter. The first case study (Section 4) provided an overview of economic, environmental, and societal factors related to mobile telephone production, use, and remanufacturing. Empirical observations of current activities revealed the complexity associated with sustainability assessment, as well as the critical importance of considering the specific circumstances and drivers surrounding individual products, activities, or services that are being evaluated.

The second case study (Section 5) provided a quantitative framework for the evaluation of sustainable design policies related to automotive fuel economy. This case study demonstrated the possibility of capturing market forces, technology realities, and environmental considerations within a model suitable for policy evaluation. The complexity and data challenges involved with developing viable quantitative models were demonstrated. The necessary simplifications required in the model development made the conclusions most valuable in terms of their trends, and the realistic nature of these trends suggest that the mathematical approach developed here would be helpful to consider during the development of policies intending to encourage sustainable design.

We conclude this chapter by recalling from Fig. 1 the perspective of design as a flow from abstract societal values to products and services with impacts on the sustainability triangle. This flow is, at its root, influenced by the knowledge base of society as a whole. As evidenced by the change in attitude toward safety design over the past century, ethical systems are evolutionary in nature and can impact design processes positively. A similar change of heart and practice is needed in the sustainability realm. However, quantitatively measuring the sustainability of products, processes, and services is a very difficult task and will remain a primary focus of sustainable science and engineering for years to come. As the field matures, it will be necessary for researchers and practitioners alike to provide education to the general public, as well as to engineers, designers, and policymakers, in order to create the conditions necessary for sustainable design to flourish.

Acknowledgements

The research described in Section 4 was performed in collaboration with Professor Günther Seliger, Dipl-Ing. Bahadir Basdere, Dipl.-Ing Marco Zettl, and

Global Production Engineering graduate Aviroot Prasitnarit of the Technical University Berlin (TUB). The research described in Section 5 was performed as part of the Antilium Project at the University of Michigan in collaboration with Professor Panos Y. Papalambros. We are grateful to these colleagues for their support and contributions to the results described here, without which this presentation would not have been possible. The authors are also grateful for the contributions of the following individuals. Section 3: Andres F. Clarens, Steven Lauritzen, David P. Morse, and Julie B. Zimmerman; Section 4: Kuei-Yuan Chan and Aaron Hula; Section 5: Professor Fred Feinberg, Sarah Bahrman, Carl Lenker, Jeron Campbell, and Sneha Madhavan-Reese.

References

[1] European Consultative Forum on Sustainability and the Environment, Proceedings of the Sustainability 21 Conference, Helsinki, 1999, http://www.eeac-network.org/workgroups/docs/proc-sust21_en.pdf, accessed 06.26.2004.

[2] Directive 2002/96/EC of the European Parliament and of the Council of 27 January 2003 on waste electrical and electronic equipment (WEEE), http://europa.eu.int/eur-lex/pri/en/oj/dat/2003/l_037/l_03720030213en00240038.pdf, accessed 06.26.2004.

[3] Directive 2002/95/EC of the European Parliament and of the Council of 27 January 2003 on the restriction of the use of certain hazardous substances in electrical and electronic equipment, http://europa.eu.int/eur-lex/pri/en/oj/dat/2003/l_037/l_03720030213en00190023.pdf, accessed 06.26.2004.

[4] Directive of the European Parliament and of the Council on establishing a framework for the setting of Eco-design requirements for Energy-Using Products and amending Council Directive 92/42/EEC, Brussels, 01.08.2003, 453 final 2003/0172, http://europa.eu.int/eur-lex/en/com/pdf/2003/com2003_0453en01.pdf, accessed 06.26.2004.

[5] C. Handleya, N.P. Brandonb, R. van der Vorsta, Impact of the European Union Vehicle Waste Directive on End-Of-Life options for Polymer Electrolyte Fuel Cells, J. Power Sources 106 (2002) 344–352.

[6] http://www.ford.com/en/goodWorks/environment/cleanerManufacturing/default.htm, accessed 06.26.2004.

[7] S. Pearce, Another Approach to Wind, Mech. Eng. 126 (6) (2004) 28–31.

[8] J. Ottman, Green Marketing: Opportunity for Innovation, J. Ottman Consulting, New York, 1998 http://www.greenmarketing.com/Green_Marketing_Book/Chapter02.html, accessed 06.26.2004.

[9] US Department of Agriculture London Embassy, US Organic Information Consumption, http://www.usembassy.org.uk/fas/us_organic_consu.htm, accessed 06.26.2004.

[10] B.R. Allenby, T.E. Graedel, Industrial Ecology, Prentice-Hall, Engelwood Cliffs, NJ, 1995.

[11] R.T. Lund, The Remanufacturing Industry: Hidden Giant, Boston University, 1996.

[12] Committee on Science, US House of Representatives Hearing Charter, Green Chemistry Research and Development Act of 2004, March 17, 2004, http://www.house.gov/science/hearings/full04/mar17/charter.pdf, accessed 06.26.2004.

[13] N. Rajagopalan, V.M. Boddu, S. Mishra, D. Kraybill, Pollution Prevention in an Aluminum Grinding Facility, Met. Finish. 96 (11) (1998) 18–24.

[14] N. Rajagopalan, T. Lindsey, S.J. Skerlos, Engineering of Ultrafiltration Equipment in Alkaline Cleaner Applications, Plating Surf. Finish. 88 (12) (2001) 56–60.

[15] S.J. Skerlos, F. Zhao, Economic Considerations in the Implementation of Microfiltration for Metalworking Fluid Biological Control, J. Manuf. Systems, 22 (2003) 202–219.

[16] T. Rusk, N. Rajagopalan, T. Lindsey, Investigating Some Factors Affecting First Pass Transfer Efficiency, Powder Coating 11 (4) (2000) 33–42.

[17] US EPA Office of Water, Factor 2 Analysis: Technology Advances and Process Changes, December 30, 2003, http://www.epa.gov/waterscience/guide/304m/factor2.pdf, accessed 06.26.2004.

[18] J.M. DeSimone, Practical Approaches to Green Solvents, Science 297 (2002) 799–803.

[19] J.A. Behles, J.M. DeSimone, Developments in CO_2 Research, Pure Appl. Chem. 73 (8) (2001) 1281–1285.

[20] US Environmental Protection Agency, Office of Policy Planning and Evaluation, Environmental Investments: The Cost of a Clean Environment, EPA-230-11-90-083, November 1990.

[21] US EPA, Final Report to Congress on Benefits and Costs of the Clean Air Act: 1970 to 1990, EPA 410-R-97-002, October 15, 1997, http://www.epa.gov/air/sect812/copy.html, accessed 06.26.2004.

[22] UN Global Compact program, www.unglobalcompact.org, accessed 06.26.2004.

[23] G. Gonzalez, Symposium Makes Business Case for Sustainability, Stanford Report May 14 (2003), http://news-service.stanford.edu/news/2003/may14/sustainable-514.html, accessed 06.26.2004.

[24] E. Eldridge, Ford borrows from Toyota's blueprints for new hybrid Escape, USA Today, Wednesday, March 10, 2004, http://www.detnews.com/2004/autosinsider/0403/11/autos-87820.htm, accessed 06.26.2004.

[25] D.A. Gross, Hummer vs. Prius: the Surprising Winner, National Public Radio, March 1, 2004, http://slate.msn.com/id/2096191, accessed 06.26.2004.

[26] European Union Eco-Label Program, http://europa.eu.int/comm/environment/ecolabel/index_en.htm, accessed 06.26.2004.

[27] The Swedish Environmental Council, Environmental Declarations Program, http://www.environdec.com/, accessed 06.26.2004.

[28] T. Tietenberg, Environmental and Natural Resource Economics, 6th Edition, Addison-Wesley, Boston, 2003, 72 pp.

[29] National Academy of Sciences Committee, Committee on the Effectiveness and Impact of Corporate Average Fuel Economy Standards, Board on Energy and Environmental Systems, Transportation Research Board, National Research Council Effectiveness and Impact of Corporate Average Fuel Economy (CAFE) Standards, National Academies Press, 2002.

[30] G. Hardin, The Tragedy of the Commons, Science 162 (1968) 1243–1248.

[31] S.J. Skerlos, G. Seliger, B. Basdere, W.R. Morrow, A. Hula, A. Prasitnarit, Evaluating the Profit and Environmental Characteristics of Global Mobile Phone Remanufacturing, Proceedings of the Electronics Goes Green 2003 International Congress and Exhibition: Life-Cycle Environmental Stewardship for Electronic Products, Boston, MA, May 19–22, 2003.

[32] F. Cairncross, Costing the Earth, Harvard Business School Press, Boston, 1991.

[33] T.C. Lindsey, Diffusion of P2 Innovations, Pollution Prevention Review 8 (1) (1998) 1–14.

[34] P. Bishop, Pollution Prevention: Fundamentals and Practice, McGraw-Hill, New York, 2000.

[35] P.T. Anastas, J.B. Zimmerman, Design Through the 12 Principles of Green Engineering, Environ. Sci. Tech. 37 (5) (2003) 94A–101A.

[36] W. McDonough, M. Braungart, P.T. Anastas, J.B. Zimmerman, Applying the Principles of Green Engineering to Cradle-to-Cradle Design, Environ. Sci. Tech. 37 (23) (2003) 434A–441A.

[37] J. Rodrigo, F. Castells, Electrical and Electronic Practical Ecodesign Guide, University Rovira I Virgili, Spain, 2002.

[38] M.S. Hundal (Ed.), Mechanical Life Cycle Handbook: Good Environmental Design and Manufacturing, Mercel Dekker, Inc., New York, 2001.

[39] Environmental Protection Agency, Office of Research and Development, Life Cycle Design Guidance Manual, EPA/600/R-92/226, 1993.

[40] Hewlett-Packard, HP General Specifications for the Environment (GSE), http://www.hp.com/hpinfo/globalcitizenship/environment/pdf/gse.pdf, accessed 06.26.2004.

[41] VDI 2243, Konstruieren recyclingerechter technischer, VDI, 1991.

[42] G. Boothroyd, L. Alting, Design for Assembly and Disassembly, Annals of CIRP 41 (2) (1992).

[43] P. Dewhurst, Design for Disassembly, Boothroyd Dewhurst Inc., 1993.

[44] Global Development Research Center, Design for Environment Guidelines on Transportation, http://www.gdrc.org/uem/lca/g-tre.html, accessed 06.26.2004.

[45] USCAR Vehicle Recycling Partnership, Preferred Practices, http://www.uscar.org/consortia&teams/VRP/preferredpractices.pdf, accessed 06.26.2004.

[46] T. Amezquita, R. Hammond, M. Salazar, B.A. Bras, Characterizing the Remanufacturability of Engineering Systems, ASME Advances in Design Automation Conference, Boston, MA, 1995, 271–278, http://www.srl.gatech.edu/education/ME4171/DETC95_Amezquita.pdf, accessed 06.26.2004.

[47] R. Hammond, T. Amezquita, B.A. Bras, Issues in the Automotive Parts Remanufacturing Industry: Discussion of Results from Surveys Performed among Remanufacturers, Int. J. Eng. Des. Automat. Special Issue on Environmentally Conscious Design and Manufacturing, 4 (1998) 27–46,http://www.srl.gatech.edu/education/ME4171/Reman-Survey-21Nov96.pdf, accessed 06.26.2004.

[48] R. Hammond, B.A. Bras, Towards Design for Remanufacturing: Metrics for Assessing Remanufacturing, Proceedings of the 1st International Workshop on Reuse, in: S.D. Flapper, A.J. de Ron (Eds), Eindhoven, November 11–13, 1996, Eindoven, The Netherlands, 5–22, http://www.srl.gatech.edu/education/ME4171/Metrics-paper-19July96.pdf, accessed 06.26.2004.

[49] http://www.motorola.com/EHS/environment/products/, accessed 06.27.2004.

[50] C.A. Rose, A. Stevels, K. Ishii, Method for Formulating Product End-of-Life Strategies for Electronics Industry, J. Electron. Manuf. 11 (2) (2002) 185–196.

[51] Fraunhofer IZM/EE Toolbox, http://www.pb.izm.fhg.de/ee/070_services/75toolbox/010_Einleitung.html, accessed 06.27.2004.

[52] Environmental Protection Agency, Life Cycle Assessment: Inventory Guidelines and Principles, EPA/600/R-92/245, (1992).

[53] D. Hoffman, Design for Environment at Motorola: Integration of Environmental Aspects into Product Design, http://www.epa.gov/performancetrack/events/design.pdf, accessed 06.27.2004.

[54] Environmental Design of Industrial Products, http://www.ipt.dtu.dk, accessed 06.27.2004.

[55] Umberto, http://www.umberto.de/english/, accessed 06.27.2004.

[56] Simpro, http://www.pre.nl/, accessed 06.27.2004.

[57] Tool for Environmental Assessment and Management (TEAM), http://www.ecobilan.com/uk_team.php, accessed 06.27.2004.

[58] GaBi, http://www.gabi-software.com/, accessed 06.27.2004.

[59] J.B. Zimmerman, Formulation and Evaluation of Emulsifier Systems for Petroleum- and Bio-Based Semi-Synthetic Metal Working Fluids, Ph.D. Thesis, The University of Michigan at Ann Arbor, 2003.

[60] S.O. Ryding, Editorial: ISO 14042, Int. J. LCA 4 (6) (1999) 307, http://www.scientific-journals.com/sj/lca/Pdf/aId/1363, accessed 06.27.2004.

[61] US Congress, Office of Technology Assessment, Green Products by Design: Choices for a Cleaner Environment, OTA-E-541. Washington, DC: US Government Printing Office, October 1992.

[62] S.J. Skerlos, K.F. Hayes, W.R. Morrow, J.B. Zimmerman, Diffusion of Sustainable Systems through Interdisciplinary Graduate and Undergraduate Education, Proceedings of the ASME: Manufacturing Science and Engineering Division, Washington, D.C., November, 2003.

[63] E.D. Williams, R. Ayres, M. Heller, The 1.7 Kilogram Microchip: Energy and Material Use in the Production of Semiconductor Devices, Environ. Sci. Tech. 36 (24) (2002) 5504–5510.

[64] J.C. Bare, G.A. Norris, D.W. Pennignton, TRACI: The Tool for the Reduction and Assessment of Other Environmental Impacts, J. Ind. Ecol. 6 (3–4) (2002) 49–78.

[65] V. Peart, The Refrigerator Energy Use Study: Leaders Guide, Florida Cooperative Extension Service, Institute of Food and Agriculture Sciences, University of Florida, June, 1993. http://edis.ifas.ufl.edu/BODY_EH232, accessed 06.27.2004.

[66] Department of Energy, Energy Information Administration, Emissions of Greenhouse Gases in the United States 1995, http://www.eia.doe.gov/oiaf/1605/gg96rpt/chap2.html, 2001 Residential Energy Consumption Survey: Household Energy Consumption and Expenditures Tables, http://www.eia.doe.gov/emeu/recs/recs2001/ce_pdf/appliances/ce5-7e_4popstates2001.pdf, accessed 06.27.2004.

[67] G. Marland, T.A. Boden, R.J. Andres, Global, regional, and national CO_2 emissions. Trends: A Compendium of Data on Global Change, Carbon Dioxide Information Analysis Center, Oak Ridge National Laboratory, US Department of Energy, Oak Ridge, TN, USA, 2003.

[68] J.B. White, EPA Weighs Changing the Way it Tests Gas Mileage, Wall Street Journal, March 26 (2004) A3.

[69] Environmental Protection Agency, Petition to Amend Fuel Economy Testing and Calculation Procedures, Request for Comments, Federal Register 40 CFR Chapter I, OAR-2003-0214, FRL-7640–7643 69 (60) (2004), http://www.epa.gov/fedrgstr/EPA-AIR/2004/March/Day-29/a6827.htm, accessed 06.27.2004.

[70] J.V. Carnahan, D. Thurston, Tradeoff Modeling for Product and Manufacturing Process Design for the Environment, J. Ind. Ecol. 2 (1) (1998) 79–92.

[71] L. van Grinsven, Mobile phone sales boom in 2003, and more is coming, USA Today, February 3, 2004, http://www.usatoday.com/tech/techinvestor/2004-02-03-cell-sales-boom_x.htm, accessed 06.27.2004.

[72] Associated Press, Discarded Mobile phones Pose Health Hazard, USA Today, May 7, 2002, http://www.usatoday.com/tech/news/2002/05/07/cell-phone-pollution.htm, accessed 06.27.2004.

[73] F. Taiarol, P. Fea, C. Papuzza, R. Casalino, E. Galbiati, S. Zappa, Life Cycle Assessment of an Integrated Circuit Product, In Proceedings of IEEE Intl Symposium on Electronics and Environment, IEEE, Piscataway, NJ, USA, 1999.

[74] O. Kharif, Where Recycled Mobile phones Ring True, BusinessWeek Online, July 26, 2002, http://www.businessweek.com/technology/content/jul2002/tc20020725_6433.htm, accessed 07.27.2004.

[75] C.H. Marcussen, Mobile Phones, WAP, and the Internet The European Market and Usage Rates in a Global Perspective 2000–2003, http://www.crt.dk/uk/staff/chm/wap.htm, accessed 06.27.2004.

[76] V. Thomas, Demand and Dematerialization Impacts of Second-Hand Markets, J. Ind. Ecol. 7 (2) (2003) 65–78.

[77] Ericsson, 2001 Sustainability Report, http://www.ericsson.com/sustainability/download/pdf/Ericsson_sustainable_2001.pdf, accessed 06.27.2004.

[78] B. Ram, A. Stevels, H. Griese, A. Middendorf, J. Muller, N.F. Nissen, H. Reichl, Environmental performance of mobile products, In Proceedings of IEEE Intl Symposium on Electronics and Environment, IEEE, Piscataway, NJ, USA, 1999.

[79] A. Prasitnarit, Life Cycle and Social Implications of Mobile phone Remanufacturing and Recycling, Masters Thesis, Technical University Berlin, Germany, 2003.

[80] I. Nicolaescu, W.F. Hoffman, Energy Consumption of Mobile phones, In Proceedings of IEEE Intl Symposium on Electronics and Environment, IEEE, Piscataway, NJ, USA, 2001.

[81] IDEMAT Database, Emissions Associated with Air Travel, Technical University of Delft, The Netherlands, 2001.

[82] C.C. Reyes-Aldasoro, F. Kuhlmann, Telecommunications and Internet in the Future Society: Myths and Realities, Computers and Communications, Proceedings. IEEE International Symposium, 1999.

[83] L. Lynton, Nations in Race to Provide the Best in Essential Services, Global Logistics and Supply Chain Strategies, Keller International Publishing, 1997, http://www.glscs.com/archives/11.97.nationsrace.htm?adcode=75, accessed 06.27.2004.

[84] United Nations, Supporting Afirca's Efforts to Achieve Sustainable Development. United Nations, New York, 2002, http://www.un.org/issues/docs/d-afric.asp, accessed 06.27.2004.

[85] D. Nik, K. Nir, The Global Digital Divide and Mobile Business Models: Identifying Viable Patterns of e-Development, IFIP WG9.4 Conference, Bangalore, India, May 29–31, 2002, 528–540.

[86] A. Bayes, Infrastructure and Rural Development: In-sights from a Grameen Bank village phone initiative in Bangladesh, Agric. Econ. 25 (2001) 261–272.

[87] J. Fiksel, Emergencye of a Sustainable Business Community, Pure Appl. Chem. 73 (8) (2001) 1265–1268.

[88] J. Schwarz, B. Beloff, E. Beaver, Use Sustainability Metrics to Guide Decision-Making, Chem. Eng. Prog. (July 2002) 58–63.

[89] R. Veenhoven, Why Social Policy Needs Subjective Indicators, Soc. Indic. Res. 58 (2002) 33–45.

[90] United Nations Development Program, Human Development Report, 2002, http://www.undp.org/hdr2002/, accessed 06.27.2002.

[91] M. Munasinghe, Is Environmental Degradation an Inevitable Consequence of Economic Growth: Tunneling through the Environmental Kuznets Curve, Ecol. Econ. 29 (1) (1999) 89–109.

[92] G.A. Keoleian, K. Kar, M. Manion, J. Buckley, Industrial Ecology of the Automobile: A Life Cycle Perspective, Warrendale, PA, Society of Automotive Engineers, 1997a.

[93] J. Michalek, P.Y. Papalambros, S.J. Skerlos, A Study of Fuel Efficiency and Emission Policy Impact on Optimal Vehicle Design Decisions, J. of Mech. Des. 126 (2004) 1062–1070.

[94] National Renewable Energy Laboratory, http://www.ctts.nrel.gov/analysis, accessed 06.27.2004.

[95] D. McFadden, Quantal Choice Analysis: A Survey, Ann. Econ. Soc. Meas. 5 (1976) 363–390.

[96] I. Currim, Predictive Testing of Consumer Choice Models that are Not Subject to Independence of Irrelevant Alternatives, J. Market. Res. 19 (1982).

[97] H.J. Wassenaar, W. Chen, An Approach to Decision-Based Design, Proceedings of DETC'01 ASME Design Engineering Technical Conferences, DTM-21683, Pittsburgh, PA, September 2001.

[98] J.H. Boyd, R.E. Mellman, The Effect of Fuel Economy Standards on the US Automotive Market: An Hedonic Demand Analysis, Transport. Res. A, 14 (1980) 367–378.

[99] US Department of Transportation, Federal Highway Administration, Office of Highway Policy Information, Source: Ward's 2001 Automotive Yearbook and the Annual Ward's Motor Vehicle Facts and Figures Publication, http://www.fhwa.dot.gov///////ohim/onh00/line4.htm, accessed 06.27.2004.

[100] K. Train, Qualitative Choice Analysis: Theory, Econometrics, and an Application to Automobile Demand, The MIT Press, Cambridge, MA, 1986.

[101] M. Delucchi, T. Lipman, An Analysis of the Retail and Lifecycle Cost of Battery-Powered Electric Vehicles, Transport. Res. D-Transport and Environment 6 (6) (2001) 371–404.

[102] D. Whitney, Cost Performance of Automobile Engine Plants, International Motor Vehicle Program, Working Papers, Massachusetts Institute of Technology, 8.22.2001, http://web.mit.edu/ctpid/www/Whitney/morepapers/Engine.pdf, accessed 06.27.2004.

[103] S. Matthews, L. Lave, Applications of Environmental Valuation for Determining Externality Costs, Environ. Sci. Technol. 34 (2000) 1390–1395.

[104] L.S. Dixon, S. Garber, M. Vaiana, California's Ozone-Reduction Strategy for Light-Duty Vehicles: An Economic Assessment, Rand, 1996, http://www.rand.org/publications/MR/MR695.1/, accessed 06.27.2004.

[105] J. Tirole, The Theory of Industrial Organization, The MIT Press, Cambridge, MA, 1988.

[106] J.L. Sullivan, R.E. Baker, B.A. Boyer, R.H. Hammerle, T.E. Kenney, L. Muniz, T.J. Wallington, CO_2 Emission Benefit of Diesel (versus Gasoline) Powered Vehicles, Environ. Sci. Technol. 38 (12) (2004) 3217–3233.